Carl Middleton · Vanessa Lamb
Editors

Knowing the Salween River: Resource Politics of a Contested Transboundary River

Editors
Carl Middleton
Center of Excellence for Resource Politics in Social Development,
Center for Social Development Studies,
Faculty of Political Science
Chulalongkorn University
Bangkok, Thailand

Vanessa Lamb
School of Geography
University of Melbourne
Melbourne, VIC, Australia

https://doi.org/10.1007/978-3-319-77440-4

Series Editor: PD Dr. Hans Günter Brauch, AFES-PRESS e.V., Mosbach, Germany
Editors: Dr. Carl Middleton and Dr. Vanessa Lamb

Foreword

A Reflection on the Role of Researchers and Research on the Salween River: Past, Present and Future

What we highlight in this edited volume are the ecological, social, cultural and political facets of the Salween River Basin. I say "we" because, as I detail below, I have been involved in supporting the important work of organizing and sharing knowledge about the Salween, also known as Thanlwin in Myanmar, for many years alongside this excellent group of contributors.

In this and in previous work, we as scholars and civil society actors have considered the various tools that can be used for research and learning about the Salween River. There are multiple approaches to research the Salween River, and multifaceted systems of governance in this River Basin. This is the time to consider the insights of our research, the tools we use, and the systems operating in the context of development of the river basin, and the region more broadly. It is a time when the world is turning to look at Myanmar and the Salween River as one of the major river systems in the region.

At present, there are 20 proposed hydropower projects along the Salween River. These have the potential to transform agrarian livelihoods, fisheries, migration, and to change political and economic relationships of the Salween River Basin. Of these projects, up to seven are proposed in Myanmar or along the Thai–Myanmar border. But, the majority of the electricity produced would not be for domestic consumption, which is urgently required, but would be sold to Thailand and China.

During the 2016 Salween Studies Network Meeting, held in Chiang Mai, Thailand, it became apparent that the Salween River Basin has only recently attracted the attention of international academic researchers. This can be seen, for instance, in the number of academic publications on the Lower Salween River since 2010. This work on the Lower Salween River is related to a greater focus on Myanmar through the economic opening up of country. Yet, while academic research has only begun to emerge, civil society research has been pursued for decades. This includes work by organizations such as KESAN, International

Rivers, Salween Watch, Southeast Asia Rivers Network, Towards Ecological Recovery and Regional Alliances (TERRA), and Mekong Energy and Ecological Network (MEE Net). This research is important to recognize and use alongside academic work to reinforce and improve the quality of research being conducted now.

The increased attention to the Salween River Basin would not have been possible without the work that has been done over the past ten years, mainly by civil society groups, to raise awareness on these issues. Indeed, this book would not be possible without the sustained effort and collaborations across academic and civil society researchers.

The 2016 Salween Studies Network Meeting, or Salween University Network (SUN), I mentioned above is a good example of one such collaborative effort. The meeting, as evidenced in the title, was focused on moving 'Towards a Shared Vision for the Salween-Thanlwin-Nu'. A wide variety of active participants from many sectors in the Salween-Thanlwin-Nu River Basin region gave presentations over two days. The meeting was an opportunity to review the research that came out of the 2014 Salween Studies Conference and to discuss what research can be done in the future, including a book like this edited volume. More than fifty participants attended the meeting, including professors, researchers, experts, community supported organizations (CSO) and non-government organization (NGO) representatives, and journalists from a range of countries – not limited to the Salween River Basin countries of Thailand, Myanmar, and China.

The 2016 SUN meeting, however, is only one in a long line of conferences and meetings. There are too many to list all here, but some of the additional meetings include, for instance, an October 2012 meeting where there were twelve people – six of whom are also part of this book (Pai Deetes, Dr. Lamb, Prof. Saw Win, Prof. Chayan, Dr. Middleton and Dr. Yu Xiaogang) – who met to discuss the future of the Salween River Basin, supported by International Rivers and Chiang Mai University. Concerns revolved around the mega projects planned for the Upper Salween River and the downstream impacts of these projects as well as logging and mining in the river basin.

The year 2014 saw two significant Salween Meetings organized. This was indeed a rare event. The first, held in Myanmar, was the workshop, *Socio-economics and Ecosystem Values of the Salween River: Towards Transboundary Management Framework and Research Collaboration Network* jointly coordinated by REAM, TERRA, and MEE Net at the Mawlamyine University, Mon State on 2–3 September. About twenty-five presentations were presented by both national and international scholars, researchers, and CSOs during the two-day workshop with special emphasis upon topics, such as: Ecosystems and Socio-economics, Inhabitant People and Resource-Based Livelihoods, Hydropower and River Resources of the Salween River Basin, Upstream Salween River and Sino-Asian Affairs, and Community Research Approach. During the workshop, I presented '*A Geographical Evaluation on the Natural and Human Resources of the Salween Drainage Basin*'. On the final day of the workshop, participants had the opportunity of paying a field visit to some river mouth islands and villages in the Salween Estuarine Region near Mawlamyine City.

Later that year, in November 2014, *the First International Conference on Salween/Thanlwin/Nu Studies: State of Knowledge: Environmental Change, Livelihoods and Development* was hosted by the Regional Center for Social Science and Sustainable Development (RCSD), Faculty of Social Science, Chiang Mai University, Thailand, organized by the Salween-Thanlwin-Nu (STN) Studies Group. With support of the Heinrich Böll Foundation and RCSD at the Chiang Mai University, the First International Salween Studies Conference was a significant conference for the Salween River Basin region and brought together over two hundred participants, not limited to local communities, NGOs, CSOs, universities, researchers, academics, and youth. The conference highlighted that 'universities, and others, can work together to conduct academic research and collect information to support the Salween River Basin'. The presentations at the conference raised important questions about who can create the knowledge used for decision-making in the Salween River Basin. The biggest take-away from the conference was that there is a need to link researchers to broaden decision-making processes. To allow for effective collaboration, everyone must work together to inform each other of existing research. About sixty-three research papers were presented to the Conference. Fifteen major topics were discussed during this event. During this International Conference, I personally contributed, at the kind request of the event organizers, three presentations, including the keynote titled '*Earthquake Hazards: A Brief Analysis of Seismo-tectonic Activities in Myanmar*'.

Prior to the 2014 Salween Studies Conference, there were concerns over how little attention was being paid to the Salween River Basin compared with the Ayeyarwady (Irrawaddy) and Mekong River Basins. But, due to successful mobilization efforts and collaborations in the region, the 2014 Salween Studies Conference was a big success. The biggest achievement of the conference was the knowledge sharing between international experts and local people. For civil society members from Myanmar, it was a wakeup call. The Network's success was shown in how it was working to share knowledge instead of restricting knowledge to a small group of experts. The Network is important for the region and has provided new opportunities for Myanmar academics and civil society members to work internationally. One of the biggest strengths of the Network right now is that it is composed primarily of people from the Salween River Basin region. The Salween University Network (SUN) as a network primarily of non-state actors differs then from the inter-governmental Mekong River Commission (MRC). While the MRC produces important research, much is produced internally and/or by government agencies rather than directly together with local universities. Even as the MRC hosts various public consultations, this risks a perception that the MRC is distant from the region's researchers, as well as more broadly the people who live in the basin. Another challenge for the MRC is that the China and Myanmar governments are not full members of the MRC, but rather hold observer status. For SUN, researchers have sought to bridge across countries; for example, RCSD of Chiang Mai University now works with universities in Myanmar to bring people together.

The importance of these relationships should be highlighted by noting the majority of Myanmar scholars are working on natural science, geology, geography, geomorphology, geobiology, and marine sciences. In contrast, scholars generally specialize in social science and highlands research at Chiang Mai University. This is one area where different universities can collaborate and bring together unique expertise to share with policymakers within Myanmar and throughout the Salween River Basin region as a whole.

We must collaborate so that we can save the Salween River for the future. One problem which academics want to tackle in Myanmar is the deep divide between academia and civil society. This should be attributed to not listening to the voices of people in the Salween River Basin. Research needs to think about people's needs and, in the case of energy demands, provide alternative energy options. Listening and paying attention to needs should be accompanied by scientific knowledge. This is another potential implication of the Network: providing the opportunity for scientific knowledge and community-oriented research to come, taking it together to inform national policies. The SUN meeting proved effective in bringing together academics and creating a space for collaboration through various group work sessions. Following a full day of presentations, participants broke out in groups to identify knowledge gaps, current research, collaborations and priorities for future work. Four areas were identified for possible future collaborations: the environment, law and policy, economics, and society and culture. With a diverse collection of participants, it came as no surprise that each group conducted very different sets of analysis. Through collaborative efforts, each group came up with thorough and comprehensive examinations of gaps, opportunities, and ways forward in the research.

Many of the presentations, and much of the work in this book, focused on challenges faced by local communities and methods used to promote community empowerment and traditional knowledge research. The many forms and values of traditional knowledge is being highlighted by some researchers. Their work shows the importance of complementing social sciences with natural sciences research. All these discussions have emphasized how more focus needs to be placed on this type of research in the future. Collaborative studies linking local to scientific knowledge allows for community members to become researchers themselves. Beyond being useful for future data collection, collaborative research can work to build the confidence of local community members to work side-by-side with academic researchers.

On the topic of sharing knowledge, it is also clear that knowledge links to responsibility in decision-making. At these conferences, there were also presentations examining the environmental impacts on the river's diverse ecosystems. Hopefully, the network members who have experience linking research with the communities and knowledge with policy continue to share their ideas. Reflecting on their own experiences, one researcher remarked that "policy gaps are more a result of poor ears than poor policies." This statement resonated with the meeting participants and acted as a lasting takeaway message.

Moreover, in this volume we see Thai Scholars (Hengsuwan, Chap. 11, and Bundidterdsakul, Chap. 9, this volume) speak to the importance of prioritizing research that looks at economic and social dimensions of the Salween due to the limited information available on those living there. Others have noted that when it comes to the Salween, there is a need for investigation, particularly in Myanmar, in relation to domestic energy supplies, *but* there is still a lack of baseline data on "everything"—including water levels.

Looking towards how data is released through the media, scholars from Myanmar have stressed how there is a gap on sharing information between academia and other communities that needs to be filled. For example, the proposed Hatgyi Dam in Kayin State also sits on a fault line, and a reservoir in this area could create a higher earthquake risk, but we do not have the full information.

Overall, *it is important to think outside the box and to pursue new paths* that could lead to more cooperation and collaboration between many different levels—international, regional, national, and local. It is imperative to include the local community perspective in all policy processes. When local communities gain inclusion into policy processes, other gaps existing between the local and the national levels will become obvious.

My brother-like-friend, and colleague Prof. Chayan, once reflected on the first meeting (and in this book's concluding commentary, see Sect. 16.1) on the "Geography of Knowing and Geography of Ignorance." Hopefully, all readers of this volume will agree that this makes an excellent contribution to knowledge, but in "Knowing the Salween" there is still much work to do for its communities, understanding its place, geography and history.

Yangon, Myanmar
February 2019

Maung Maung Aye
Professor and Rector-in-Charge (Ret.),
Yangon University Distance Education
maungaye3246@gmail.com

Acknowledgements

The preparation of this edited volume reflects a long journey to which many have contributed. First, we would like to express our greatest thanks to all the contributors and collaborators working together on "Salween Studies" for the past ten years. The list is a long one and continues to grow. We thank Pai Deetes and International Rivers for catalysing a meeting of Salween experts in 2012 at Chiang Mai University. We also thank the participants of the conferences and workshops organized over the past five years, and all those who helped organize these milestones. They include the Salween Studies Conference in 2014 and the Salween Studies Network meeting in 2016 at the Faculty of Social Sciences, Chiang Mai University, with particular thanks to Kanchana Kulpisithicharoen, Chanida Puranapun and Ajarn Chayan Vaddhanaphuti of the Regional Center for Social Science and Sustainable Development (RCSD). The Heinrich-Böll-Stiftung and International Rivers provided partial support for these events, for which we are grateful. We thank colleagues who helped organized the Salween Studies Research Workshop at Yangon University in 2018, in particular Prof. Htun Ko from the Department of Geography and Tay Zar Myo Win. There have been many people who although not authoring chapters in the book project, have helped shape the groups wider thinking about "Salween Studies" or supported the ongoing work including Witoon Permpongsacharoen, Montree Chantavong, Luntharimar (Tidtee) Longcharoen, Philip Hirsch and Sarah Allen.

We want to acknowledge and sincerely appreciate the teams at our own universities. At the Center for Social Development Studies (CSDS), Faculty of Political Science, Chulalongkorn University, this includes Orapan Pratomlek, Bobby Irven, Claudine Claridad and Anisa Widyasari, and our colleagues in the Faculty of Political Science including Dean Ake Tangsupvattana and Ajarn Naruemon Thabchumpon. Orapan in particular has been central to coordinating our research and fellowship programme on the Salween. Park Rojanachotikul also collaborated with CSDS to produce the excellent short films at www.SalweenStories.org.

At York University, York Center for Asian Research, we appreciate the support of Directors Philip Kelly and Abidin Kusno, and of course, Alicia Filipowich, in particular, for managing so many of the fun aspects of administration. At the University of Melbourne, the School of Geography offered support and the kind and generous cartographer Chandra Jayasuriya created the superb maps for this book.

A large part of the impetus for many of the chapters presented in this volume was the generous support from the Salween Water Governance Project under the CGIAR Research Programme on Water, Land and Ecosystems (WLE) in the Greater Mekong with support from CGIAR Fund Donors and the Australian Department of Foreign Affairs and Trade (DFAT). They supported two research for development projects fondly referred to as "MK21" or Matching Policies, Institutions and Practices of Water Governance in the Salween River Basin: Towards Inclusive, Informed, and Accountable Water Governance, and the "MK31" Professional Development of Water Governance and Regional Development Practitioners in the Salween Basin which ran from 2015–2018. We thank in particular: Kim Geheb, David Clayton, Stew Motta, Rattamanee Laohapensang (Ruby) and Mayvong Sayatham at WLE-Greater Mekong, and John Dore, and Ounheuan Saiyasith at DFAT.

Key partners for the "MK21" research project included the aforementioned York Centre for Asian Research, York University, and the Centre for Social Development Studies (CSDS), Chulalongkorn University. In addition, the project included colleagues and friends at the Karen Environment and Social Action Network (KESAN), in particular Saw John Bright, Jeff Rutherford, Saw Tha Phoe, and Naw Aye Aye Myaing; Nang Shining at Weaving Bonds Across Borders/Mong Pan Youth Association; Nang Aye Tin, Nang Hom Kham, Nang Sam Paung Hom, and Sai Aum Khay at Mong Pan Youth Association; Prof. Saw Win of the Renewable Energy Association Myanmar (REAM); Diana Suhardiman at the International Water Management Institute; and Prof. Chayan Vaddhanaphuti at the Regional Center for Sustainable Development (RCSD), Chiang Mai University, Thailand. We also thank Paung Ku and in particular Sayar Kyaw Thu, Myint Zaw, and Shwe Sin, for their continued support. We also appreciate the work done by independent researcher April Kyu Kyu, for his research and organizing support at crucial moments, and Tay Zar Myo Win, for his ongoing logistics assistance. There are other close collaborators who cannot be named, but who know who they are.

Key partners for the "MK31" Salween Fellowship programme include the mentors, many of whom also doubled up as workshop trainers, namely: Naruemon Thabchumpon, Stew Motta, Diana Suhardiman, Chantana Banpasirichote Wungaeo, Lyu Xing, John Dore, Prof. Saw Win, Jakkrit Sangkamanee, Jeff Rutherford, and Michael Medley. Others include our partner fellowship programmes on the Mekong and Red Rivers, with whom we co-organized trainings and field visits. These include Ajarn Kanokwan Manorom and Chawirakan Nomai (Ploy) from Ubon Ratchathani University, and Nguyen Tung Phong, Ha Hai Duong, and their team at Vietnam Academy of Water Resources.

General project support, for both the academic and civil society networks for the two MK projects above, was provided by Prof. Saw Win, who has been an amazing facilitator, and we have been very happy to know since that 2012 meeting. Thank you, Sayar-gyi!

Finally, with regard to this book, it would not have been possible without the assistance (research and copyediting support) from K. B. Roberts and her 'mobile mentorship', and from Sabrina Gyorvary and her mellifluous prose. Elena Tjandra transcribed talks and offered referencing assistance at very important times. The book's two anonymous peer reviewers also worked tirelessly to provide constructive and critical feedback to the book's contributors. We also thank Michael Cook for the use of his photograph on this book cover, and appreciate his unparalleled photographic ability to capture the banal moments of everyday river life in visual art.

We also very much appreciate the team at Springer for their patient support, and in particular the independent series editor, Dr. Hans Günter Brauch as well as Dr. Johanna Schwarz and Dr. Christian Witschel in Heidelberg (Germany) and the book producers Ms. Aurelia Heumader (Heidelberg) and Mr. Arulmurugan Venkatasalam (Chennai) and their team of typesetters, graphic designers, webmasters and many others that have made it possible that our research results are now available to a global audience.

Bangkok, Thailand
Melbourne, Australia
February 2019

Carl Middleton
Vanessa Lamb

Endorsements for *Knowing the Salween*

"A fascinating and detailed analysis of how development decisions about a river impact on local people's lives and pose important challenges for politics. Strongly recommended. A must-read for anyone interested in transboundary water politics and the urgent environmental challenges now facing Southeast Asia." – Professor Tim Forsyth, Professor of Environment and Development, Department of International Development, London School of Economics, UK.

"This book shares detailed and diverse knowledge and experience about the people of the Salween basin and their river. It reveals what is at stake if plans for large dams go ahead, especially for the communities who would be directly affected by them." – Associate Professor Kanokwan Manorom, Faculty of Liberal Arts, Ubon Ratchathani University, Thailand.

"This important collection of essays on the Salween (also known as Gyalmo Ngulchu – Nu Jiang – Nam Khone – Thanlwin) demonstrates how important it is to understand a watercourse, and its meaning for those who depend on it, well before ill-conceived development plans become etched into the river's future pathways. The knowledge behind such understanding is produced through science, experience and socio-political analysis. In this book, an impressive range of authors articulate their research-based visions for a different kind of future for the Salween and provide grounds for optimism – despite the severe challenges that lie ahead for this transboundary river system." – Emeritus Professor Philip Hirsch, School of Geosciences, University of Sydney, Australia.

Contents

Abbreviations

ADB	Asian Development Bank
AG	Advisory Group
AIRBM	Ayeyarwady Integrated River Basin Management
ASEAN	Association of Southeast Asian Nations
BGFs	Border Guard Forces
CBOs	Community-Based Organizations
CBWG	Community-Based Water Governance
CLFs	Community Learning Facilitators
CODI	Community Organizations Development Institute
CPC	Communist Party of China
CSOs	Civil Society Organizations
CTGC	China Three Gorges Company
DKBA	Democratic Karen Benevolent Army (before 2010, known as the Democratic Karen Buddhist Army)
DNP	Department of National Parks, Wildlife and Plant Conservation
DP	(China) Democratic Party
DWRI	Directorate of Water Resources and Improvement of River Systems
EAOs	Ethnic Armed Organizations
EGAT	Electricity Generating Authority of Thailand
EGATi	EGAT International Company
EIA	Environmental Impact Assessment
FAO	Food and Agricultural Organization
GAD	General Administration Department
GIS	Global Information System
GMS	Greater Mekong Subregion
GW	Green Watershed
IFC	International Finance Corporation
IFIs	International Financial Institutions
IGE	International Group of Entrepreneurs
INGOs	International Non-Governmental Organizations

IWRM	Integrated Water Resources Management
KESAN	Karen Environmental and Social Action Network
KNLA	Karen National Liberation Army
KNLP	Kayan New Land Party
KNU	Karen National Union
MNDAA	Myanmar National Democratic Alliance Army
MoA	Memorandum of Agreement
MOEE	Ministry of Electricity and Energy
MONREC	Ministry of Natural Resources and Environmental Conservation
MRC	Mekong River Commission
MW	Megawatts
MWR	Ministry of Water Resources
NCA	National Ceasefire Agreement
NDAA-ESS	National Democratic Alliance Army of Eastern Shan State
NEA	National Energy Agency
NGOs	Non-Governmental Organizations
NHRCT	National Human Rights Commission of Thailand
NLD	National League for Democracy
NTFP	Non-Timber Forest Products
NWFD	National Water Framework Directive
NWP	National Water Policy
NWRC	National Water Resources Committee
ONWR	Office of National Water Resources
PDP	Power Development Plan
RCSS	Restoration Council of Shan State
RFD	Royal Forest Department
RMB	Renminbi
RS	Remote Sensing
SEIA	Strategic Environmental Impact Assessment
SEPA	State Environmental Protection Administration
SIA	Social Impact Assessment
SLORC	State Law and Order Restoration Council
SMEC	Snowy Mountain Engineering Corporation
SOEs	State-Owned Enterprises
SPDC	State Peace and Development Council
SPP	Salween Peace Park
SSPP	Shan State Progressive Party
TNLA	Ta'ang National Liberation Army
UNESCO	United Nations Educational, Scientific and Cultural Organization
USDP	Union Solidarity and Development Party
UWSA	United Wa State Army
WEPT	West-East Power Transfer
WHC	World Heritage Committee

Chapter 1
Introduction: Resources Politics and Knowing the Salween River

Vanessa Lamb, Carl Middleton and Saw Win

1.1 Introduction

The Salween is a transboundary river connecting the people, ecosystems, and nation-states of China, Myanmar (Burma), and Thailand. Over its 2,820 kilometer course, it flows from the Tibetan Plateau to Yunnan Province in China, then connecting to Myanmar via Shan State. The river continues, through Karenni and Karen States, also forming the border with Thailand, before joining Mon State and coming together with the Andaman Sea.[1] Over 10 million people live throughout the basin, comprising at least 16 ethnic groups, many of whom depend on river resources for livelihood and food (Johnston et al. 2017). These livelihoods are diverse; they range from fishing-based livelihoods practiced by communities in the Salween estuary in Myanmar, to farmers who practice swidden agriculture and paddy rice cultivation in Myanmar and Thailand, to herders who raise livestock and manage the rangelands at the very start of the river on the Tibetan Plateau. There is also remarkable biodiversity within the basin. It is these livelihoods, peoples, and the river itself that are the focus of this edited volume.

[1]The Salween Basin covers 283,500 km^2, of which 48% is in China, 7% is in Thailand and 44% is in Myanmar (Johnston et al. 2017).

Dr. Vanessa Lamb, Lecturer, School of Geography, University of Melbourne, Melbourne, Australia; Email: vanessa.lamb@unimelb.edu.au.

Dr. Carl Middleton, Director, Center of Excellence for Resource Politics in Social Development, Center for Social Development Studies, Faculty of Political Science, Chulalongkorn University, Bangkok, Thailand; Email: carl.chulalongkorn@gmail.com.

Dr. Saw Win, Senior Research Associate, Center for Social Development Studies, Faculty of Political Science, Chulalongkorn University, Bangkok, Thailand; Email: prfswwin@gmail.com.

C. Middleton and V. Lamb (eds.), *Knowing the Salween River: Resource Politics of a Contested Transboundary River*, The Anthropocene: Politik—Economics—Society—Science 27, https://doi.org/10.1007/978-3-319-77440-4_1

While some areas of the basin are undergoing dramatic transformations, other areas appear to be almost at an impasse. This is due to a combination of factors, including the transformations within the basin, and in the context of the broader political, economic and social changes in the region (e.g., Simpson 2016; Egreteau 2016; Farrelly/Gabusi 2015). There is also a long-standing call to "know more" about this understudied basin, often associated with a whole range of development plans and visions for enhancing economic integration, nationally, regionally, and globally (Johnston et al. 2017).

In this context, we see this edited volume as the first to consider and link these concerns across the basin. With a focus on the contested politics of water and associated resources in the Salween basin, this book offers a collection of empirical case studies highlighting local, regional, and international knowledge and perspectives. Given the paucity of grounded social science studies in this contested basin, this book provides conceptual insights at the intersection of resource governance, development, and politics relevant to researchers, policy-makers and practitioners at a time when rapid change is underway. It also offers something more: a call to study, collaborate, and appreciate the range of efforts and actors necessary to do this. As such, we also present a proposal to study the Salween as an 'area' for continued critical engagement.

We recognize that the present state of the Salween Basin is informed by both contemporary transformations and historical dynamics. It has been shaped by—although, we note, not *determined* by—histories of colonial resource extraction. For instance, particularly in present-day Myanmar (formerly, British Burma), an emphasis on timber production, not for the improvement of the Salween Basin, but for support of the colonial center is evident (Bryant 1997). This constituted, as well, the associated establishment of the Salween as a 'periphery' or 'frontier' (Leach 1960; Scott 2009).

More contemporary policies and processes of regional economic integration also play a role in delimiting and creating the Salween as part of, if not constitutive of, an area or region. For example, as discussed by Middleton, Scott, and Lamb in Chap. 3, this volume, the 1990s saw the Asian Development Bank work with regional governments to shape and link the Greater Mekong Subregion (ADB 2012). In addition, since the mid-2010s, it now seems that China's Belt and Road initiative is building momentum, and building a vision for a global 'belt' of interconnection. These, and multiple other regionalization initiatives, represent new planned connections for the Salween, connections to and from distant sites, territories, and markets, via new infrastructures, both hard and soft. Yet, as these plans are promoted and pursued by their proponents, there is a limited understanding of what precisely these new connections and developments would mean for the people, ecologies, and localities that they 'intersect' with in the Salween basin. There is also, it seems at least in some senses, a lack of appreciation that this would be something that is valuable to know.

We see these as pressing concerns in the current context of a basin and a region that is experiencing often precipitous political, economic, and ecological change. How can this accumulated work to construct the Salween, often framed to be at the 'margins' (Scott 2009), be transformed or even overcome—so that scholars, and the broader public might consider more diverse collaborations and ways of knowing the Salween, which also take into account the change and dynamism of people, states, and ecologies?

van Schendel and others (e.g., Scott 2009; Michaud 2010; Turner 2010) have attempted to overcome what they refer to as "geographies of ignorance" produced as a result of studies focused on centers, heartlands, and academic narrowness of field, by developing a new concept of region and a novel way to approach the study of an 'area': Zomia.[2] As Michaud explains, Zomia is

> a neglected—an invisible—transnational area, which overlapped segments of all four sub-regions without truly belonging to any of them. It is an area marked by a sparse population, historical isolation, political domination by powerful surrounding states, marginality of all kinds, and huge linguistic and religious diversity. (2010: 187–188)

van Schendel underlines that, "In order to overcome the resulting geographies of ignorance, we need to study spatial configurations from other perspectives as well" (2002: 664). He proposes three alternatives, in what was seen as a moment of opportunity post-WWII, as the globe and its territories were being transformed. We also re-consider these three propositions here for our own purposes, in the contemporary moment, for the Salween.

First, van Schendel considers suggestions to construct "regions crosscutting the conventional ones" but notes that this is unlikely to enact lasting change, and likely to reinforce new margins and centers (2002: 665). Appreciating this point of critique, he (and we) more seriously consider, the second alternative, studying regions in a way that highlights new spatial arrangements or networks. This invokes not discrete, bounded territories in the traditional sense, but instead, continuous, connected spaces with no particular center. The third alternative proposed "goes further" to develop the study of region via flows and their resulting architectures and infrastructures, which cross (and shape) conventional territories and borders, and which van Schendel characterizes as "more ephemeral" and emergent (2002: 665).

This is an intentional shift, then, by scholars to move beyond areas or territories as static spatial configurations, and beyond regions as depicted in, any of the colonial maps of Southeast Asia, or the development planning devices of regional and world development agencies. It is a shift to study the processes, flows, and connections as way to "develop new concepts of regional space" (van Schendel 2002: 665) as dynamic and changing.

This book builds upon the opportunity presented by van Schendel and related scholars' insights. As scholars and individuals who have also spent large parts of

---

[2]Michaud (2010: 187–188) helpfully identifies that Zomia includes "the highlands of Asia, from the western Himalayan Range through the Tibetan Plateau and all the way to the lower end of the peninsular Southeast Asian highlands, as a political and historical entity significantly distinct from the usual area divisions of Asia: Central (Inner), South, East, and Southeast." But also, that this has also undergone a shift in 2007 to include areas further west and north, "including southern Qinghai and Xinjiang within China, as well as a fair portion of Central Asia, encompassing the highlands of Pakistan, Afghanistan, Tajikistan, and Kyrgyzstan."

our careers outside academia working with civil society organizations, we (the co-authors of this introduction to the book and editors of this book) have been working over the past 10 years under a broader and shifting set of ideas, and among a cohort of concerned individuals, towards an understanding of what "Salween Studies" (or Nu-Thanlwin-Salween Studies) could be. This is part conscious and conceptual, but also, partly practical. Indeed, practically, there is a need for some sort of organizing principle as even the various names of the river and its cartographies have been disconnected and fragmented through time (see Lamb, Chap. 2, this volume), with the academic literature, for instance, on China's Nu Jiang almost entirely distinct from research on Myanmar's 'Thanlwin' River.

Of course, we recognize that an area of study is not made by maps, names, or words alone.[3] Conceptually, we have been thinking about collaborations, while simultaneously practicing the work of collaboration, to consider what a "Salween Studies" might include, follow, connect, and entail. Areas of study or interest, we argue, should be understood as inherently transdisciplinary, and require collaborators including civil society groups and researchers, academic experts (and generalists), interdisciplinary experts, language interpreters and knowledge interlocutors, as well as interested friends, coalitions, and at times, governments and their representatives, who see enough value in understanding an 'area' (as a set of processes, not necessarily geographically fixed) that they can come together and emerge with new understandings of the world and our place within it.

The collaborators and community familiar with and mobilizing around the Salween and Salween Studies is growing. This is evidenced, for example, in the work to establish international networks – Save the Salween, Salween Watch, Salween University Network – and also in the continued efforts of a large range of collaborators who have organized various Salween gatherings. These have ranged from conferences, such as the 2014 *First International Conference on Salween Studies*, to smaller workshops, road trips, meetings, panels, protests, and ceremonies along the river over the past ten years or more. These efforts aim to understand and position Salween, its peoples, politics, and ecologies, through multiple epistemological and ontological approaches, as an 'area' to know and in its connections with people and places outside the physical, or 'natural', area of the basin.

We position this edited volume, then, as the first to seriously center on the Salween as a site for critical consideration and interrogation. It provides a wide range of studies presenting rich empirical work and conceptual facets that reflect the varied ways of knowing the Salween (see Fig. 1.1: Map of Salween River Basin indicating the range of "Salween Studies" in this volume). This book provides these at a key moment in the history of the Salween, and which we anticipate will therefore be of interest to a widening audience of academics, researchers, policy-makers and practitioners. It is also written by a wide range of contributors, many of whom are based in

[3]Even if it has been successfully argued that a nation can be made by maps alone (Winichakul 1994).

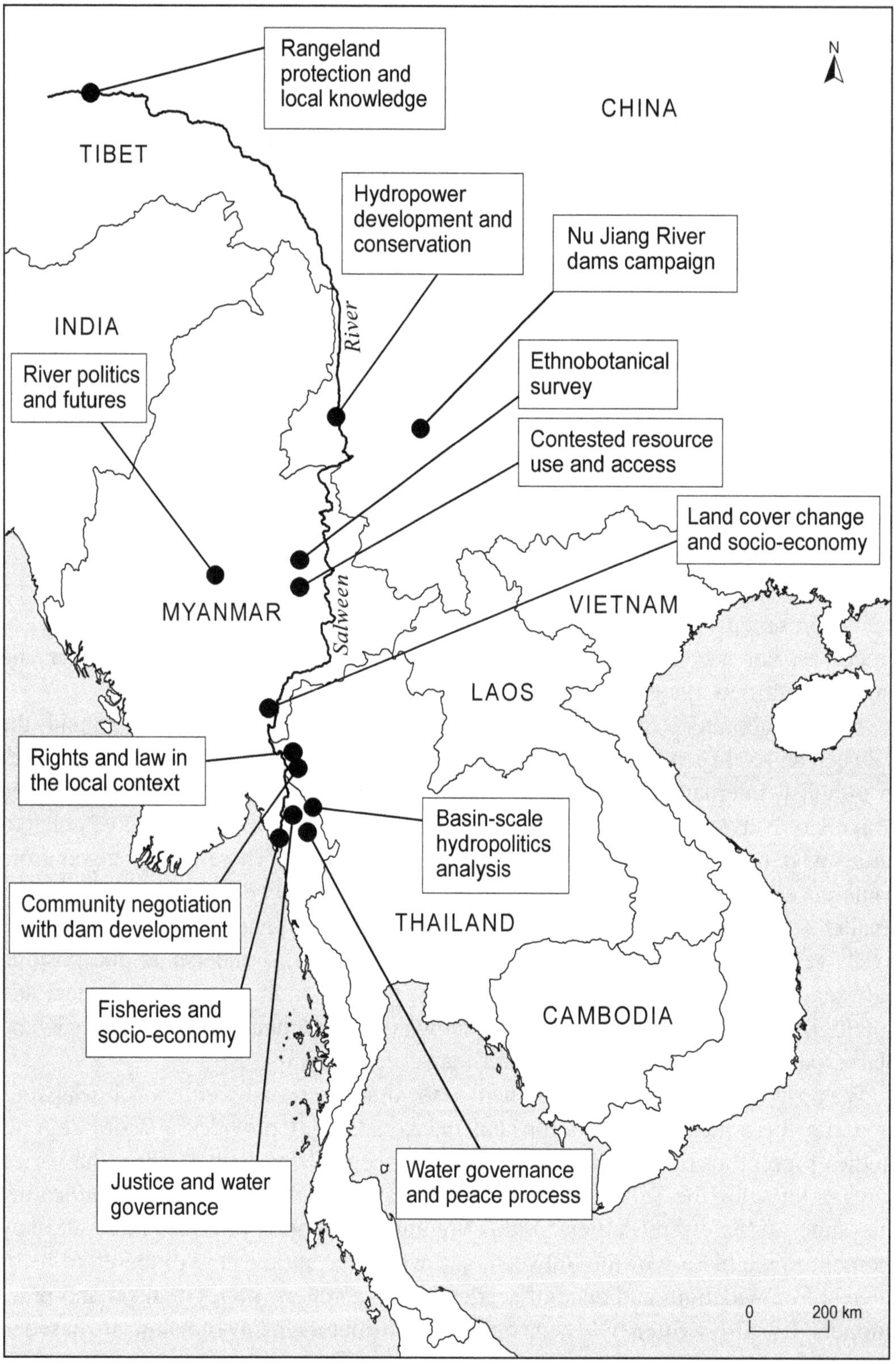

Fig. 1.1 Map of Salween River Basin indicating the range of "Salween Studies" in this volume. *Source* Cartography by Chandra Jayasuriya, University of Melbourne, with permission

the basin. In the following sections, then, we introduce this range of collaborators and topics of study via three key themes, namely: resource politics (theme 1); politics of making knowledge (theme 2); and reconciling knowledge across divides (theme 3).

1.2 Theme 1: Resource Politics

The dynamics of access to, control over, and use of resources are at the heart of resource politics in the Salween basin (Magee/Kelley 2009; Leach et al. 2010; BEWG 2017). The basin is witnessing intensified dynamics of resource extraction, alongside large dam construction, conservation, and development interventions. These are unfolding within a complex terrain of local, national, and transnational governance processes. This intensification raises questions about the contested future visions for the basin, how inclusive they are, and who has the authority to make decisions and on whose behalf? For the Salween, addressing these questions is not straightforward.

Plans for dams in Myanmar have existed since at least the late 1970s (Paoletto/Uitto 1996), and while there is a much longer history of development in the basin, it can be argued that essentially the Salween dam projects were first seriously considered in the late 1980s under Thailand's Prime Minister Chatichai Choonhaven, and have been continually linked to narratives which aimed to transform "battlefields to marketplaces" (Magee/Kelley 2009). These narratives and discourses around 'battlefields' and conflict, as opposed to marketplaces and development for peace, have characterized the Salween since at least this time. The associated frames have also impacted our ability to understand and to 'know' the multitude of ways that the river and the basin matter, and how resource and peoples are governed in multifaceted ways, which as noted above, are not always evident when examination of the 'center' is prioritized. Many of the authors in this volume take up these points, revealing further particulars of the proposed developments, their politics, contests, and shifts over time.

In Chap. 3, titled "Hydropower politics and conflict on the Salween River," Carl Middleton, Alec Scott, and Vanessa Lamb draw on the lens of 'hydropolitics' to analyze the contested nature of dams planned for the Salween River (Sneddon/Fox 2006; Rogers/Crow-Miller 2017). A hydropolitics approach is sensitized to how water and energy are intimately constitutive of politics at multiple scales, and also highlights how these politics foreground certain facts, actors, and agendas, while others are rendered forgotten. Indeed, analysis of foregrounding and forgetting can reveal the power relations at play.

The particular focus of the chapter is the projects proposed in Myanmar and their connections with neighboring China and Thailand via electricity trade, investment, and regional geopolitics. The authors detail the recent political history of conflict in the Salween basin in Myanmar, and relate the contesting claims for political authority and territorial control to plans for hydropower dams. In Myanmar, at present there are five dams at various stages of planning on the Salween River mainstream, and a further two that have been suspended, if not cancelled. On the tributaries, there are a further four medium or large-dams already complete, two under construction, and seven at an advanced stage of planning. As each of the dam projects are located in places where there is either open or recently ended conflict, that the central government has limited ability to fully project its authority is profoundly significant for the decision-making process for each project. Overall, the authors argue that Myanmar's peace negotiations need to be concluded before large dams on the Salween are discussed as a part of a broader discussion on resource governance in the context of federalism.

Moving the focus upstream, Chap. 4 titled "From hydropower construction to national park creation: Changing pathways of the Nu River" by Yu Xiaogang, Chen Xiangxue and Carl Middleton, explores the politics of the evolving visions, narratives, and decision-making processes for the Nu River and the extent to which they have materialized through five 'pathways'. This includes: a hydropower construction pathway that led to plans for 13 large dams on the mainstream, of which two to date have been built in the most upper stretch; a civil society protection pathway that emphasizes their role both in resistance to the hydropower construction pathway, and in support of environmental protection via national parks and energy reform; an energy reform pathway that highlights the current oversupply in China of electricity, and other (better) future options; a national park pathway that foresees the Nu River giving priority to ecological protection; and a water conservancy pathway recently proposed by the government to build dams as multi-purpose projects that include irrigation and flood/drought management.

The authors utilize this approach to reveal the contested character of the Nu River, and to render visible how there are multiple potential futures. Yet, the authors also highlight that over time not all pathways have been given equal consideration, and for at least a decade until the mid-2000s large hydropower plans were dominant. While now suspended, and the national park pathway appears currently favored by the government, the authors still argue that access to information, transparency and accountability are of the utmost importance in any decision-making towards the river that is fully inclusive of the ethnic communities living along the river.

Work on the Salween Peace Park is also instructive in regard to resource politics. Saw John Bright, in Chap. 5, "Rites, rights, and water justice in Karen State: A case study of community-based water governance and the Hatgyi Dam", juxtaposes two very different plans for the Salween River. Bright details the case of a Community-Based Water Governance project supported by the NGO Karen Environment and Social Action Network (KESAN) around the Daw La Lake that connects to the Salween River, and the Kaw Ku seasonal island located in the Salween itself, to draw out how the communities relate to these resources in terms of human rights and cultural practices, the latter of which he terms 'rites' (Badenoch/Leepreecha 2011). He shows how the communities have organized strategically seeking to gain recognition from the Karen State government on the right to govern their resources, and to protect them from enclosure by outside private actors.

Bright then contrasts this with the planned Hatgyi Dam, around which armed conflict continues to erupt creating severe insecurity for communities nearby. He emphasizes that the project's centralized and opaque formal decision-making process has excluded communities and civil society groups to date, and challenges the project's legitimacy including due to the ongoing conflict. He concludes by calling for more emphasis to be placed on principles of justice in water governance and proposes that both human rights and community rites should be recognized, thus pointing towards the multiple normative values that could or should be in play within resource politics.

Johanna M. Götz, in Chap. 6 titled "Contested water governance in Myanmar/Burma: Politics, the peace negotiations and the production of scale", shows how highly contested power-relations shape decision-making around various scales. Throughout the chapter, Götz draws on a hydrosocial approach (Linton/Budds 2014), which understands water not merely as its chemical form of H_2O, but as always socially embedded and the product of historical processes. Götz utilizes this sophisticated conceptual approach to resource politics to analyze two groups of actors with divergent visions, namely: Myanmar's Union government's National Water Resources Committee and its Advisory Group and its work to create a National Water Policy; and a network including the NGO KESAN, the Karen National Union (KNU) and community groups to create the Salween Peace Park (SPP) within KNU-controlled Mutraw District in Karen State. Overall, Götz shows how scale in water governance cannot be taken as given *a priori*, but that scale's production is in fact a key contestation within the politics of water governance. In the context of Myanmar, Götz concludes that how these scales become consolidated within institutions, laws, and decision-making processes will also hold implications for the outcome of the ongoing peace negotiations, and will inevitably shape the form of federalism that is intended to emerge.

These are significant discourses, narratives, and stories of resource politics that do not generally make it into the news cycle on the Salween, which highlights instead neat, linear narratives of conflict. To complicate these contestation narratives, as our contributors do, does not mean in any way that conflict does not exist. But, in a setting like the Salween River Basin, there are also enormous efforts by the basin's residents to create alternative visions. This includes the SPP and community-based water governance in Karen State, as well as new networks of women and youth as emphasized by Nang Shining in her concluding commentary on Shan State (Sect. 16.3), or the rituals that affirm community relationships with the river in Thailand as discussed by Pianporn Deetes (Sect. 16.4). Thus, to focus only conflict overshadows the immense progressive work and diversity of the Salween.

1.3 Theme 2: Politics of Making Knowledge

In this introduction, we highlight that knowledge making is an active endeavor, a practice, and that making knowledge is about more than addressing a 'gap' or a lack of data, but also about how knowledge boundaries are defined. It has a geography, a politics, and a history. What these chapters present, then, when assembled alongside one another is a provocation about not just what we know, but about who knows and who decides on those boundaries.

In emphasizing knowledge as practice (Jasanoff 2004), we also demonstrate that knowledge about the Salween is very much an ongoing conversation. Chapter 7 is illustrative of what we mean. Titled "A State of Knowledge of the Salween River: An overview of civil society research" this contribution emerged from ongoing conversations between a group of researchers about the "state of the basin". State of the basin reports are generally large tomes, written by leading scientific researchers in the field. When applied to water governance, such studies are intended to serve as key reference points. This chapter discusses what would such a study would look like if civil society and activist research was considered as one (or more) of those leading fields? The chapter presents, partially in response to that question, an argument that much of what we now know about the Salween River Basin, its people, ecosystems, values, and threats, has been documented through civil society research. The chapter is of specific importance in reviewing this research and at the same time, recognizes the need for additional information such as biophysical baseline studies. The chapter, thus, provides an overview of the existing knowledge of the basin and begins to identify the key knowledge gaps in support of more informed and inclusive water governance in the basin.

Hannah El-Silimy's contribution, Chap. 8 titled ""We need one natural river for the next generation": Intersectional feminism and the Nu Jiang Dams campaign in China" focuses on the politics of making knowledge with a particular purpose in advocacy. In taking 'activism' seriously and as worthy of sustained research, El-Silimy interrogates 'who' makes knowledge and does activism, and who can and is involved in advocating for conservation of the river. She focuses on an analysis of the Nu Jiang dams campaign in China via an intersectional feminist lens, writing an eloquent contribution that speaks to debates on the politics of making knowledge. These debates consider that not only is politics inherent in influencing decision-making or ecological management, but that politics are also about who participates and with what implications for the making of subjects and identities (Lamb 2018; Agrawal 2005; Tsing 1999; Li 2000). El-Silimy argues that gender and class matter in the politics of participation; particularly when civil society groups argue for this in the representation of affected groups, it also matters who can participate in campaigns and as part of civil society.

The works in this section recognize a plurality of knowledge making activities. The interpretation, practice, and production of law is another societal process through with knowledge is produced and acted upon by a range of actors including lawyers, judges, government officials and activists. Chapter 9, "Local context, national law: The rights of Karen people on the Salween River in Thailand", by Thai lawyer Laofang Bundidterdsakul, examines how the rules and norms reflected in Thailand's current national law runs up against existing Karen practices in three forest communities on the Salween. He details the existing livelihoods of the Karen communities in relation to land, forests, and river resources, and shows how the creation of the Salween National Park around them and its associated laws have placed constraints on the Karen communities' traditional practices. On the ground, the existing law mandates local government officials to act to constrain the Karen's livelihood practices, creating a sense of insecurity among the communities. Underlying these insecurities, however, are the histories of the communities themselves, within which many members still have been unable to receive full Thai citizenship despite existing for many decades. Overall, Bundidterdsakul argues that the existing national law has not acknowledged the rights to these resources for the ethnic minorities. Yet, rather than take the law as absolute and immutable, Bundidterdsakul sees it as an arena of ongoing contestation in which ongoing court cases test, interpret and on occasion rewrite it. Within these court cases, whose knowledge counts in the eyes of the law is revealed as one of the considerations in their outcome. The stakes are high, as Bundidterdsakul argues that the tension between local practice and national law also reflect a more fundamental question regarding the definition of Karen identity in Thailand.

Also critical in any approach to studying the politics of making knowledge are the long-standing debates in political ecology, anthropology, and geography that have pushed our understanding of what constitutes local, indigenous, and scientific knowledge (e.g., Agrawal 1995, 2002; Forsyth 1996; Tsing 1999, 2005; Li 2000; Santasombat 2003; Vaddhanaphuti/Lowe 2004; Lowe 2013). While early work in anthropology characterized many local or indigenous knowledge systems in Southeast Asia and elsewhere as practice-based or even "fixed in time and space" (Agrawal, 1995), more engaged scholarship has shown that local knowledge systems are *simultaneously* embedded within practice, empiricism, and theory, suggesting that the boundaries between 'local' and 'scientific' knowledge are porous and not all that distinct.

Two chapters that take on the politics of the "local-scientific" knowledge divide, but in very different ways, are the contributions by ethnobotanists Mar Mar Aye and Swe Swe Win in their survey and analysis of plants and health in Myanmar's Shan State (Chap. 10) and by Paiboon Hengsuwan in his study of villagers' research and their perspectives on development along the Thai stretch of the Salween (Chap. 11).

Paiboon's contribution, titled "Not only anti-dam: Simplistic rendering of complex Salween communities in their negotiation for development in Thailand," details his research in Thailand into what it means for local people along the Salween River-border to be "anti-dam" but not necessarily anti-development. His nuanced analysis also shines light on a rather under-considered but significant points in this debate: that individuals and communities are making strategic choices in development and activism.

Following work by Keyes (2014), Hengsuwan explores how villagers have reworked the discourses of development to be meaningful as related to their own experiences. He writes that, "Local communities see their participation in the anti-dam movement as a strategy to improve local development. Participation in the anti-Salween dam movement is one of the villagers' many strategies to articulate their own meaning of local development." He suggests, then, that it is imperative to understand the broader networks of movement and resistance around local development, as a way to produce richer understandings of complex communities.

In Chap. 10, "An ethnobotanical survey in Shan State, Myanmar: Where Thanlwin biodiversity, health, and deforestation meet", the authors perform a mix of methods for their study, from plant species surveys and herbarium stand methods, to interviews with traditional healers. They take these together to understand (and to document and categorize) plant species from the perspective of botanical records, healers, and residents who rely on medicinal plants because of complex histories and geographies. Some interviewees report turning to traditional plant remedies when they lack other options, like hospitals or clinics. Others report a decline in plants themselves due to collection for export and intensification of cash crops nearby the wild areas these plants are collected. The authors then link these pressing issues of health, plant availability, and production of herbal medicines to broader changes to forested land via agriculture intensification and the ability to make a living in Myanmar's changing economy.

1.4 Theme 3: Reconciling Knowledge Across Divides

In this third theme, reconciling knowledge across divides, we consider the divides, spatial and otherwise, that are both imposed on and constituted through research into the Salween. In particular, this work highlights the struggles to consider, appreciate, appraise, and reconcile knowledge or ways of making knowledge that may initially seem irreconcilable or at odds. As such this section begins with an important study of access to Salween River resources, positioned as part of a contested history of authority and control, still manifest in contemporary development debates, particularly in Myanmar. As K. B. Roberts underlines in this study in Chap. 12, "Powers of access: Impacts on resource users and researchers in Myanmar's Shan State", these contests over conflict and development (theme 1) and knowledge making (theme 2) are linked. Roberts shows via field research in Myanmar's restricted stretch of the Salween River in Shan State, that conflict has not only restricted local residents' access to natural resources, but the ongoing conflict also restricts research and researchers, both foreign and domestic. This closing off impacts researchers' ability to understand and operate, necessitating new methodologies and tools, as Roberts details in their innovative study of resource access via collaborative fieldwork and analysis with local Shan youth.

Chapter 14, presenting new research on the connections between forest cover and histories of conflict in Myanmar by Khin Sandar Aye and Khin Khin Htay is one these kinds of knowledge that Roberts refers to. Titled, "The impact of land cover changes on socio-economic conditions in Bawlakhe District, Kayah State," it draws on remote sensing as well as quantitative and qualitative surveys on-the-ground to document and understand the changing land cover and land use in Kayah State, with particular focus on the implications for people's livelihoods. The authors identify three periods during which land cover, livelihoods, and governance changed significantly. These include: first, a period prior to 2010, when armed conflict limited timber companies' access to the area and limited forest loss; second, from 2010 to 2015, since a peace agreement was signed, during which timber companies could undertake logging, and the area's forests rapidly depleted; and third, the period since 2016 when a logging ban was implemented, although much remains to be done to recover and protect forests, and ensure sustainable livelihoods for communities in the area. While not a substitute for long-term ethnographic field study, the methods applied enabled the researchers to generate and triangulate data and insights over a three decade period, covering a period when access to undertake field research would have been challenging and risky, and to integrate natural and social science methods.

Chapter 13 by Professor of Marine Science, Dr. Cherry Aung, also highlights what linking changes that might otherwise be divided as distinct economic, environmental, and social change can reveal. Titled "Fisheries and socio-economic change in the Thanlwin River Estuary in Mon and Kayin State, Myanmar," the chapter begins with an understanding that while the estuary is a well-established fishery, the fishery is in decline for a range of reasons linked to economic, social, and environmental change. As a result, the communities who depend upon the fishery are also in a kind of decline; people of working age have migrated out to Thailand, Malaysia, and other parts of Myanmar, to make a living.

In turn, the fishery has also transformed: a fishery that was in the past a more independent endeavor, is now characterized by larger operations, who hire domestic migrants from across Myanmar. The key linkages made between environment change, labor change, and people illustrates a 'community' in transition in Myanmar. They also point to the broader links between livelihoods, ecosystems, and economic development. Dr. Cherry Aung's research was conducted by interviews and focus groups across four villages located in the estuary in Mon and Kayin States. She is in a unique position to carry out research in this area, which might otherwise be restricted, as an individual who hails from this area and an academic who can speak across natural scientific and social science divides, which have been a long-standing challenge in Myanmar.

This brings us to the final case study in this edited volume, Chap. 15, sited at the highest point of the Salween in the Tibetan plateau where the author, doctoral candidate Ka Ji Jia, is also in a unique position to carry out and present this work. Ka Ji Jia's work reveals key insights into the sacred knowledge and local practices of Tibetan herders. Titled, "Local knowledge and rangeland protection on the Tibetan Plateau: Lessons for conservation and co-management of the upper Nu-Salween and Yellow River watersheds" her chapter argues for greater consideration of the work and knowledge of local residents in the protection and conservation of the rangelands, as opposed to the more conventional "fixed fences" approach preferred at the moment by the Chinese government.

Particularly salient is Ka Ji Jia's presentation of research in relation to the less visible aspects of sacred knowledge linked to environmental protection. While she is optimistic that these forms of local knowledge can be harnessed, even in the context of a fences and fines approach to conservation, she also acknowledges that this would require a divide be crossed and reconciled. It is a divide between what is (and has been) considered 'evident' and visible, and what is, according to outsiders, *less visible* in the nomadic practices of herders: their efforts at environmental conservation. Writing on knowledge and visibility, anthropologist Celia Lowe also recognizes this divide. She explains that in her research in Indonesia on the Sama people, known by some as the "sea nomads", she also saw the ingrained assumptions about how people who move would not, and could not, care about place, separating nomads from sedentary groups:

> I once spoke with a biologist visiting Susunang village who explained to me that since Sama people are "sea nomads" and are always moving from place to place, they can't possibly care about the particular location they happen to find themselves in at a given moment. His implication was that the people of Susunang would not protect the Togean [Indonesian island] environment because they were just going to move one. (Lowe, 2013, 86)

Lowe goes on to point out that visibility and its assumptions "tells us nothing" of the less visible, magic or sacred ways of knowing and being, which may be, in practice, more powerful. Not wholly dissimilar to the divides that Roberts and Cherry Aung connect—spatial/political, economic/social—Ka Ji Jia makes a concerted effort to move forward in an attempt to reconcile plural knowledges and perceptions, which might otherwise not be 'struggled with' and dismissed.

The final chapter of this edited volume offers five concluding commentaries. Each writer holds a long experience with the Salween River, including two academics (Chayan Vaddhanaphuti (Thailand) and R. Edward Grumbine (US)), two civil society leaders (Nang Shining (Myanmar) and Pianporn Deetes (Thailand)), and one government advisor (Khin Maung Lwin). Each was invited to provide their personal reflection on both the current volume, and their experience, analysis of and future hope for the Salween River.

Thus, in what follows in this edited volume is a kind of reconciling of plural knowledges as well as a curation for one possible 'area' of Salween Studies for further consideration, discussion, and collaboration. As further evidenced by the immediately following chapter (Chap. 2), a short introduction to the many names of river we have referred to as 'Salween', we recognize that names, knowledges, and conventions of scholarship privilege certain actors and histories and ways of knowing.

References

ADB (Asian Development Bank). (2012). *The Greater Mekong Subregion at 20: Progress and prospects*. Mandaluyong City, Philippines: Asian Development Bank. Retrieved from: https://www.adb.org/sites/default/files/publication/30064/gms-20-yrs-progress-prospects.pdf.

Agrawal, A. (1995). Dismantling the divide between Indigenous and Scientific Knowledge. *Development and Change* 26, 413–439.

Agrawal, A. (2002). Indigenous knowledge and the politics of classification. *International Social Science Journal* 54, 287–297.

Agrawal, A. (2005). *Environmentality: Technologies of Government and the Making of Subjects.* Durham: Duke University Press.

Badenoch, N., & Leepreecha, P. (2011). Rights and rites: Local strategies to manage competition for water resources in Northern Thailand. In K. Lazarus, N. Badenoch, N. Dao & B. Resurreccion (Eds.), *Water rights and social justice in the Mekong Region*. London, United Kingdom: Earthscan.

BEWG (Burma Environment Working Group). (2017). *Resource federalism: A roadmap for decentralized governance of Burma's natural heritage*.

Bryant, R. (1997). *The political ecology of forestry in Burma: 1824-1994*. Hawaii: University of Hawaii Press.

Egreteau, R. (2016). *Caretaking democratization: The military and political change in Myanmar*. Oxford, UK: Oxford University Press.

Farrelly, N., & Gabusi, G. (2015). Introduction: Explaining Myanmar's Tentative Renaissance. *European Journal of East Asian Studies* 14(1), 7–14.

Forsyth, T. 1996. Science, Myth, and Knowledge: Testing Himalayan environmental degradation in Thailand. *Geoforum* 27(3):375–392.

Johnston, R., McCartney, M., Liu, S., Ketelsen, T., Taylker, L., Vinh, M.K., Ko Ko Gyi, M., Aung Khin, T., & Ma Ma Gyi, K. (2017). *State of knowledge: River health in the Salween*. Vientiane, Lao PDR: CGIAR Research Program on Water, Land and Ecosystems. Retrieved from: http://hdl.handle.net/10568/82969.

Jones, L. (2014). Explaining Myanmar's regime transition: the periphery is central. *Democratization, 21*(5), 780–802.

Keyes, C. (2014). *Finding their voice: Northeastern villagers and the Thai state*. Chiang Mai, Thailand: Silkworm Books.

Lamb, V. (2014). "Where is the border?" Villagers, environmental consultants and the 'work' of the Thai–Burma border. *Political Geography*, 40, 1–12.

Lamb, V. (2018). Who knows the river? Gender, expertise, and the politics of local ecological knowledge production of the Salween River, Thai-Myanmar border. *Gender, Place & Culture*, 1–16.

Leach, E.R. (1960). The frontiers of "Burma". *Comparative Studies in Society and History,* 3(1), 49–68.

Leach, M., Scoones. I., & Stirling, A. (2010). *Dynamic sustainabilities: Technology, environment, social justice*. Abingdon, United Kingdom: Earthscan.

Li, T. (2000). Articulating Indigenous Identity in Indonesia: Resource Politics and the Tribal Slot. *Society for Comparative Study of Society and History*, 149–179.

Linton, J., & Budds, J. (2014). The hydrosocial cycle: Defining and mobilizing a relational-dialectical approach to water. *Geoforum, 57*, 170–180.

Lowe, C. (2013). *Wild profusion: Biodiversity conservation in an Indonesian archipelago*. Princeton and Oxford: Princeton University Press.

Magee, D., & Kelley, S. (2009) Damming the Salween River. In F. Molle, T. Foran & M. Käkönen (Eds.), *Contested waterscapes in the Mekong Region: Hydropower, livelihoods and governance* (pp. 115–140). London, United Kingdom: Earthscan and USER (Chiang Mai University, Thailand).

Michaud, J. (2010). Editorial Zomia and beyond. *Journal of Global History,* 5, 187–214.

Paoletto, G., & Uitto, J.I., 1996. The Salween River: is international development possible?. *Asia Pacific Viewpoint*, 37, 269–282.

Rogers, S., & Crow-Miller, B. (2017). The politics of water: a review of hydropolitical frameworks and their application in China. *Wiley Interdisciplinary Reviews: Water* 4(6): 1239.

Santasombat, Y. (2003). *Biodiversity: Local knowledge and sustainable development*. Chiang Mai, Thailand: RCSD Chiang Mai University.

Scott, J.C. (2009). *The art of not being governed: An anarchist history of upland Southeast Asia*. New Haven, CT: Yale University Press.

Simpson, A. (2016). *Energy, governance and security in Thailand and Myanmar (Burma): a critical approach to environmental politics in the South*. London: Routledge.

Sneddon, C., & Fox, C. (2006). Rethinking transboundary waters: A critical hydropolitics of the Mekong basin. *Political Geography*, *25*(2), 181–202.

Tsing, A. (1999). Becoming a tribal elder, and other green development fantasies. In Tania Li (ed.) *Transforming the Indonesian Uplands: Marginality, Power and Production*. Amsterdam: Harwood, 157–200.

Tsing, A. (2005). *Friction: An Ethnography of Global Connection*. Princeton: Princeton University Press.

Turner, S. (2010). Borderlands and border narratives: a longitudinal study of challenges and opportunities for local traders shaped by the Sino-Vietnamese border. *Journal of Global History,* 5(2), 265–287.

Vaddhanaphuti, C., & Lowe, C. (2004). The Potential of People: An Interview with Chayan Vaddhanaphuti. *Positions: East Asia Cultures Critique* 12(1), 71–91. Duke University Press.

Winichakul, T. (1994). *Siam Mapped.* Honolulu, Hawaii: University of Hawaii Press.

van Schendel, W. (2002). Geographies of knowing, geographies of ignorance: jumping scale in Southeast Asia. *Environment and Planning D*, 20(6), 647–668.

Chapter 2
Salween: What's in a Name?

Vanessa Lamb

2.1 Introduction: Decolonising Development and Renaming the Salween?

In this short chapter, I walk the reader through the ways the river conventionally referred to as 'Salween' is called upon differently in distinct places by many people.[1] I consider the significance of the act of naming a place, a river in particular, and the ways through which this naming can change or influence perceptions of the place being named. I invoke the many names of this river as represented in a series of geographical sites, films, and local knowledge projects, which in many times emerge alongside colonial studies and developmentalist names for the river, internationally known as a single entity by one name: the Salween. I present these euphonious invocations of the river as a way to highlight the multiple meanings of/in place and to consider who or what they privilege, and what they displace.

The debate about the power of place naming is long-running and contentious, engaging residents, states, cartographers, activists, and academics (including Geographers, like myself). The name of the region "Southeast Asia", as Emmerson (1984) argued, for instance, privileges the perceptions of European powers rather than those of the people in the place so-named. 'South' of China, 'East' of India, Southeast Asia was primarily named by outsiders and imagined as a region through acts of war and nation building, which also influenced subsequent studies and invocations of the term. Naming places not only invokes a history, it also has implications, potentially displacing other histories, or in many instances homoge-

[1]The original concept for this chapter came from a shorter commentary written in 2014 on Mekong Commons (http://www.mekongcommons.org/politics-of-place-naming-salween-river/). That piece and this chapter benefitted from editorial comments from Mekong Commons editors.

Dr. Vanessa Lamb, Lecturer, School of Geography, University of Melbourne, Melbourne, Australia; Email: vanessa.lamb@unimelb.edu.au.

C. Middleton and V. Lamb (eds.), *Knowing the Salween River: Resource Politics of a Contested Transboundary River*, The Anthropocene: Politik—Economics—Society—Science 27, https://doi.org/10.1007/978-3-319-77440-4_2

nizing diversity for more legible cartesian lines and states with only one name. Geographer Tuan (1991: 698) contends, however, that "The most striking evidence of naming to create a seemingly coherent reality out of a congerie of disparate parts is the existence of Asia." Named at the end of the 17th century, Asia "was defined negatively as all that was not Europe. Asia's reason for existence was to serve as the backward, yet glamourous ... Other". Foundational work by Said (1979, 1985) took this argument further to consider how on the construction of another geographical region, the "Middle East" (how the West sees the East), was explicitly named as linked to asserting imperial order and defining who could speak on behalf of place. As a result, those who imagined and became experts on the "Middle East" did so by understanding this place as a British colony, and the accompanied configuration of knowledge and power positioned those who resided in that place as less knowledgeable, as less in a position of expertise (Mitchell 2002).

In this short piece, I draw on these approaches to consider what is at stake in naming and developing the 'Salween' River. Seemingly "left off the map" in the development sense for a long time (Paoletto/Uitto 1996), the name and term Salween is now being more visibly contested, mapped, and stabilized. I argue that it is important to pay attention to the names we use, as well as those names we discard, questioning who or what they privilege. In particular, I consider how local residents across the basin differently name and produce knowledge about this river to show the multiple meanings and invocations of 'Salween'. This is increasingly important in the context of increased development and international attention, where there is both risk for homogenizing the multifaceted histories of the river basin, and an opportunity for multiple invocations of the river to be considered and circulated.

2.2 Names of Control and Controversy

To be clear, these controversies over naming and the work they accomplish are not limited to geographical regions like 'Asia' or to just the river Salween. In North America, a river known by hundreds of names by indigenous residents became one river system after colonial contact, known as the Mississippi. This was accomplished as "French explorers in the seventeenth century carried the word 'Mississippi' (of Algonquian origin)[2] all the way from the source of the river in Minnesota to its mouth on the Gulf. In time, 'Mississippi' had displaced all other names (both Indian and Spanish) that applied to only limited stretches of the river" (Tuan 1991: 688). Tuan explains that as a result, the Mississippi going forward "evoked an image of a vast hydrological system" (1991: 688). I raise this example not to contend that the name 'created' the entire river system (for that would mean that it did not exist to indigenous groups prior, or only existed due to the 'material transformation of nature'), but that this naming did the work of bringing together

[2]Baca explains that specifically, Mississippi is Algonquian (Ojibwa/Chippewa) for "big river" – *misi (mshi)*, 'big' and *sipi (ziibi)*, 'river' (Baca 2007: 58).

what might have been seen as disparate places, with different names and histories, under one system, to be acted upon and managed – *the* Mississippi.

At the mouth of 'the' Salween River, where it emerges in the Tibetan uplands, it is and has been better known as *Gyalmo Ngulchu*. In 1938, a British geographer named Kaulback came with an assistant to "explore as much as we could" and to trace and map the river courses. This was written and documented in the *Geographical Journal* published by Great Britain's Royal Geographical Society (Kaulback 1938: 7). Kaulback explained that venturing on previously undocumented routes "enabled us wholly to change the general shape of this piece of country as shown in previously existing maps" (Kaulback 1938: 99). The maps and knowledge produced at the time show *his* journey across the countryside, highlighting 'pieces of country' that were novel to British geographies. As one of the earlier instances I can locate in modern geography which names this stretch of the river 'Salween', it certainly represents one step towards a cartography of a single river system.

In other parts of the upper stretches of the river, however, it is still not known as 'Salween'. In China, for instance, it is known as *Nu Jiang*, and as it flows down through Thailand and Myanmar, the river is known by many names, including *Thanlwin*, *Nam Khone* (Shan State, Myanmar) and *Salawin* (Mae Hong Son, Thailand).

In Myanmar, stories about the way that the name 'Salween' emerged are noteworthy. It is traced back to 1800s British cartographies as a British corruption of the Burmese word *Thanlwin*.[3] Before Kaulback's 1930s visit to the uplands where he mapped the river there as 'Salween', the name 'Salween' was enshrined in colonial documents and maps of Southeast Asia in the 1800s. It was then circulated and propagated across Europe and beyond, reaching the work of Kaulback and others. In many ways, this naming invokes the critiques by Said and Emmerson of constructing regions: it honors outside colonizers, not the peoples and places of the region. Furthermore, as Said argues in *Culture and Imperialism* (1993), the privileging of Western perspectives on the 'East' was part of the accomplishment, alongside or as constitutive of the map: now, the experts who are seen as best positioned to construct knowledge and name places are positioned outside, as objective observers, in service of a distant center.

2.3 Names and Their Implications

Presently, however, the Salween is being defined more in terms of its potential to contribute to economic and energy development for the region (see, for instance, Middleton et al., Chap. 3, this volume), rather than in relation to the British control,

[3]I have heard the words "Than Lwin" linked to the word 'olive' and linked back further to possibly Sanskrit.

or in relation to the river's local values – it is a different set of relations invoked. As the region's nation-states pursue economic growth via resource extraction, the Salween is being invoked as a 'natural resource' and energy development planners are increasingly paying attention to the basin, with at least 20 large dams now proposed by investors from Thailand, China, and Myanmar for this as yet still un-dammed mainstem of the river (Middleton et al., Chap. 3, and Yu et al., Chap. 4, this volume).

This work to develop "the Salween", however, cannot be accomplished without the naming, understanding, and decision-making of the river as a particular kind of system, whether it is a single Salween system, or separate systems divided by political borders, such as *Thanlwin* and *Nu Jiang* systems. The feasibility and efficacy of energy developments on the river for electricity generation require the water flows of the upstream to be managed as a system. Yet, none, of this, can proceed "in the absence of words" (Tuan 1991: 692). And, as noted above, not all names or stories necessarily accomplish these same ends. Across the basin, the river is and has been a significant source of food and livelihoods for residents and this is articulated by residents themselves, in the different ways the rivers is used and invoked (as seen in many of the contributing chapters to this book, for instance, the important study of ethnobotany of indigenous ethnic groups in Chap. 10 by Drs. Mar Mar Aye and Swe Swe Win; Saw John Bright's study of "rights and rites" in Karen State in Chap. 5, or Ka Ji Jia's study of naming and cultural practices in Tibetan uplands in Chap. 15, this volume). I also consider these names and the ways they are used in a range of local knowledge projects which I focus on below.

2.4 Euphonious River Labels

Thanlwin. Mother. Eater of people. Natural resource. Border. Gyalmo Ngulchu. Angry. Energy rich flow resource. China's Grand Canyon. Nam Khone. Salawin.

These are just some of ways I have heard this river invoked in my studies and visits across the basin (see Fig. 2.1: Names of the Salween). These multiple names reflect the river's cross-border position and the region's multifaceted histories and residents' connections to place. What else do these multiple names and phrases tell us about the river? How do these names index different ways of coming into relation with the river? Do the different names of the Salween not just indicate multiple meanings, but also, different ways of relating to place?

In the mid-2000s in Karen State, there was an important naming project conducted near a bend in the river called *Khoe Kay*, where the river forms the political border between Myanmar and Thailand. This collaborative research endeavor by indigenous ethnic Karen residents, activists, and academics focused on biodiversity, emphasizing links to a particular place, Khoe Kay. The team was the first group of researchers to be permitted by the local community to write down and systematize

Fig. 2.1 Names of the Salween. *Source* Cartography by Chandra Jayasuriya, University of Melbourne, with permission

the knowledge and expertise of local residents. Their research was published by the Karen Environment and Social Action Network (KESAN 2008).

One of the most compelling features of this collaborative research project was the way that species identification—a form of naming—was conducted. Local researchers helped conduct research based on a local set of parameters that focused on the researchers' five senses: sight, hearing, taste, smell, and touch. For instance, whether or not a plant species was hard or soft, hairy or itchy, or what sounds tree leaves made in the blowing wind, were relied up on as identification markers. While this is not unlike more conventional scientific assessment that look to properties or particular markers in species identification, it does represent a certain shift in terms of who the experts, or 'namers', are. Even though academics specializing in plant species identification might be better positioned to provide or identify Latin names for local plants, local residents identified species names in local languages and provided identification related to, for instance, the sounds that leaves make.

This link between identification, expertise and "who can name" are seen not only in research, but also in more popular presentations of the river basin. Names can maintain and enhance meaning, as in literature and film. For instance, the names and experts portrayed differ greatly across and within films about or featuring the Salween. This includes a full range, from the Hollywood cinematic production of the Salween River as illustrated in *Rambo IV* (2008) to the earlier Thai film *Salween* (1995), to more recent local documentaries, including *River of Ethnic Minorities* (2007).

One local film of note focused on perceptions of the river-border emerged from a local participatory project called, *Ngan Wijay Thai Baan* or Villager Research (Committee of Researchers of the Salween Sgaw Karen 2005). Not all that different in its approach from the *Khoe Kay* study, Villager Research brought local residents to collaborate with NGO staff and academics to conduct research on issues of ecological significance as defined collaboratively early on. The 'villagers' undertook the research through photography and keeping record of their daily activities, and NGO staff acted as research assistants, assisting in identification, facilitation, and systematization of the villagers' work. In this film, local residents who led in the research invoke the river by multiple names; as, for instance, 'Salawin', 'Khone', and 'Salween', as well as in multiple ways, as the "river of life" and the river as "my home"... "gives us everything we need"... "is not only for us"... "peoples of the Salween basin have been protecting natural resources." In these films, the focus is on the everyday activities of residents and their significance. Multiple names are remarked and the Salween as a basin is invoked as a mobilizing force for local resource protection.

In quite stark contrast, consider Hollywood's rather aggressive portrayal of the Salween in *Rambo IV.* This, the most recent of the *Rambo* films, takes place along the river where it forms the political border between Thailand and Burma, introducing us to Rambo's "temporary home" at the Salween where he has lived since his last exploit. The images and narrative here conjure and emphasize histories of violence and oppression, with the area as largely inhospitable and full of danger, in contrast to the river as 'home' and providing for local residents.

In Rambo's telling, outsiders are both the main characters and, in typical Hollywood fashion, portrayed as the "last hope" for the region. As the narrator notes, "Burma's a war zone"... "You know his name"... "A warrior will come." Residents, their livelihoods, and the river are part of the backdrop for the action to unfold.

Rambo's portrayals of the Salween as a place are distinct from the depictions in the local research film in ways that are both expected and important. When you realize that Rambo is portrayed as living in the same area that the abovementioned film documents, the distinctions are all the more striking. Local residents, when involved in producing films about their own lives and the river, are not shown fleeing violence and are not indistinguishable parts of the landscape. Instead, villagers are positioned as experts who are connected to an important place and who represent themselves as agents of river management. In the vein of the *Khoe Kay* research, local residents in the Villager Research film are themselves agents in naming and place-making.

This relationship between naming, place-making, and expertise—as seen in different names and portrayals of this river across the basin—emphasizes that the distinctions matter. A name does not simply exist; it is constructed and used with meaning, sets of relations, and with implications for who can 'know' about it. These films invoke different histories, different namers, and I would argue, alongside different versions of expertise are different kinds of interventions.

2.5 Decolonising the River's Name, but for What Kind of Change?

While the name Salween has been relied upon in many development agreements, in 2011 the government of Myanmar officially changed the river's name to Thanlwin. This move is related to the way that the name 'Salween' emerged, as described above, which we can also trace back to British *mis*understandings of local invocations of place. It is worth asking what did it take for the river to be re-renamed? For, as Tuan notes, "Normally, only a sociopolitical revolution would bring about a change in the name in a city or a nation" (Tuan 1991: 688). At the time of this 'return' of the name to Thanlwin, a social-political transformation was, in fact, taking place in Myanmar and continues to present. This was the same year that Myanmar moved towards democratic rule (and away from decades of authoritarian regimes) with the first elections held in at least three decades. Even if the opposition party, Aung San Su Kyi's National League for Democracy, could not participate in the 2011 elections, this experience laid the groundwork for their eventual participation and win in 2015. This, of course, does not indicate that the new name, even in its move away from a colonial power's misnomer, is uncomplicated and uncontentious. In fact, it in some ways represents something of an imposition on local indigenous groups, whose own names for the river are obscured for the rise of a common, "Myanmar" name and associated national rules and policies related to one river system: Thanlwin.

2.6 Concluding Points

There is much at stake at the moment for the river. Many of the large projects discussed in this volume would irrevocably change the river flow and function, and are being planned in the basin without adequate studies or knowledge of local ecologies and their existing management regimes, and without the participation and knowledges of local residents. Moreover, there are questions about how the river will be acted upon – will it be seen and studied as a "single river system", the Salween, as a site for large-scale hydropower development? Will the river be mapped as dual systems, first, the *Thanlwin,* with federal or domestic Myanmar policies and processes responsible for its future, and then, the *Nu Jiang,* with separate Chinese policies for hydropower and or conservation? Or, could the river be imagined as 'home' to many, with distinct values, uses, and multiple names, but as part of a broader set of relationships even, 'friendships'?

Moving forward then, as the river is further studied, mapped, and named, in addition to naming, there is at stake the corollary sets of relations and people who will provide expertise and speak on behalf of the river going forward. As Mitchell explains in *Rule of Experts,* the tasks of producing expert knowledge and development outcomes are linked, "expert knowledge works to format social relations, never simply to report or picture them" (2002: 118). In other words, who are included or made as the 'experts' on the river going forward will not only configure the possibilities for development (Paoletto/Uitto 1996) but will also influence who are involved and how. Will the basin's residents be experts and researchers, as residents with strong connections to place, or as victims and part of the backdrop for which national or international 'development' unfolds? To attend to the names we use and consider how this has the potential to honor multiple histories, but also to better acknowledge the transformations taking shape—that rivers, naming, and development can be done differently, is possibly a small contribution to also better understanding the relationships between peoples and ecologies of this multifaceted region.

References

Baca, K. (2007). *Native American place names in Mississippi.* Jackson, MS: University Press of Mississippi.

Committee of Researchers of the Salween Sgaw Karen. (2005). *Thai Baan Research in the Salween River: Villager's research by the Thai-Karen Communities* (in Thai with English summary). [Documentary film]. Published and produced by SEARIN, Chiang Mai, Thailand. Retrieved from: http://www.livingriversiam.org/4river-tran/4sw/swd_vdo2-tb-research.html.

Emmerson, D.K. (1984). "Southeast Asia": What's in a Name? *Journal of Southeast Asian Studies, 15*(1), 1–21.

Kaulback, R. (1938). A Journey in the Salween and Tsangpo Basins, South-Eastern Tibet. *The Geographical Journal, 91*(2), 97–121.

KESAN. (2008). *Khoe Kay: Biodiversity in Peril.* Retrieved from: www.kesan.asia/index.php/resources.

Mitchell, T. (2002). *Rule of experts: Egypt, techno-politics, modernity.* Berkeley, CA: University of California Press.

Paoletto, G., & Uitto, J.I. (1996). The Salween River: is international development possible? *Asia Pacific Viewpoint, 37*(3), 269–282.

Said, E. (1979). *Orientalism.* New York, NY: Vintage Books.

Said, E. (1993). *Culture and Imperialism.* New York, NY: Vintage Books.

Said, E. (1985). Orientalism reconsidered. *Race & Class, 27*(2), 1–15.

SEARIN (Producer and Director). (2007). *Salween: River of Ethnic Minorities.* [Documentary Film]. Retrieved from: http://www.livingriversiam.org/4river-tran/4sw/swd_vdo1.html.

Stallone, S. (Director), & Lerner, A., Templeton, K.K., & Thompson, J. (Producers). (2008). *Rambo IV* [Motion picture]. United States of America: Millennium Films.

Tuan, Y.F. (1991). Language and the making of place: A narrative-descriptive approach. *Annals of the Association of American Geographers*, *81*(4), 684–696.

Yukol, C. (Director), & Yukol, K., & Sethi, K. (Producers). (1993). *Salween (Gunman 2)* [Motion picture]. Thailand: Mangpong.

Chapter 3
Hydropower Politics and Conflict on the Salween River

Carl Middleton, Alec Scott and Vanessa Lamb

3.1 Introduction

This chapter examines the hydropower politics of the Salween River, with a focus on the projects proposed in Myanmar and their connections with neighboring China and Thailand via electricity trade, investment, and regional geopolitics. We do so through a lens of hydropolitics to contextualize and better reveal the contested nature of these projects and their connections. We attend in particular to the Salween River in Myanmar, where there is a complex history of conflict and multiple associated claims for territory, political authority, and legitimacy (Stokke et al. 2018). Underscoring these links, even the 2017 strategic environmental impact assessment (SEIA) baseline study of Myanmar's hydropower electricity sector commissioned by the International Finance Corporation (IFC) states that "natural resource exploitation is linked to armed violence, including hydropower development" (ICEM 2017: 17). In this context, the terrain and assumptions of conventional water governance analysis are unsettled, which in turn supports the value of a critical hydropolitics analysis as undertaken in this chapter.

Plans for dams in the Salween Basin in Myanmar and Thailand have existed since the late 1970s, but gained momentum post-1988 as Myanmar's military government began deepening trade and investment relations with China and Thailand (Paoletto/Uitto 1996; Magee/Kelley 2009). Around 18 large hydropower projects of varying scales have been proposed for the Nu Jiang-Salween basin mainstream (Fig. 3.1: Planned dams on the mainstream Nu Jiang – Salween River). An important context

Dr. Carl Middleton, Director, Center of Excellence for Resource Politics in Social Development, Center for Social Development Studies, Faculty of Political Science, Chulalongkorn University; Email: Carl.Chulalongkorn@gmail.com.

Alec Scott, Regional Water Governance Technical Officer, Karen Environmental and Social Action Network (KESAN): Email: scottsangpo@gmail.com.

Dr. Vanessa Lamb, Lecturer, School of Geography, University of Melbourne, Melbourne, Australia; Email: vanessa.lamb@unimelb.edu.au.

C. Middleton and V. Lamb (eds.), *Knowing the Salween River: Resource Politics of a Contested Transboundary River*, The Anthropocene: Politik—Economics—Society—Science 27, https://doi.org/10.1007/978-3-319-77440-4_3

Fig. 3.1 Planned dams on the mainstream Nu Jiang – Salween River. *Source* Cartography by Chandra Jayasuriya, University of Melbourne, with permission

to these projects is the seven-decade-long history of modern conflict in Myanmar.[1] Indeed, the link between hydropower infrastructure and conflict in the Salween basin is significant (BEWG 2017; Burke et al. 2017). The plans for large hydropower are not a 'new' endeavor, but one that has been pursued (and contested) incrementally over decades, with engagements from multiple actors acting across multiple scales. These actors include the Union government, domestic conglomerates, transnational corporations and consultants, political parties, Ethnic Armed Organizations (EAOs), Myanmar's armed forces (the Tatmadaw) and civil society organizations (CSOs).

To expand on and explore the Salween's multi-faceted hydropolitics, the immediately following section outlines some of the key aspects of the relevant hydropolitics literature and approach. It also identifies what a hydropolitics analysis of the Salween River, with a focus on the Myanmar context, can bring to our understanding of the basin's complex history and range of engagements. The following section then provides a brief history of the plans for large dams on the Salween River mainstream and major tributaries, and the current known project status, before presenting an analysis of how Salween hydropower projects have been incorporated into transboundary and national plans for electricity trade. We then present an analysis of the relationship between Myanmar's peace negotiations, conflict and the hydropower dams, and in the following section the work that has been undertaken by civil society across scales including in the context of this conflict. We also consider how water governance has been structured across scales. Finally, we bring these themes and their associated actors and their engagements over time together in discussion to consider the ways these histories and developments, and their complex relationships with conflict in the area, shape Salween hydropolitics.

3.2 Hydropolitics: Linking Subnational and Cross-Border Governance to Contests and Conflicts over Water

In introducing "a critical hydropolitics", Sneddon/Fox (2006: 183) argue for an approach that "stands in contrast to mainstream studies of water resource development, transboundary waters, and international rivers." Their work is positioned in contrast to, for instance, the "water wars" thesis which argues that there is a looming water crisis at the state-to-state (geopolitical) level linked to water scarcity. Instead, Sneddon/Fox (2006: 182) explain that "These and similar works understand conflicts over water as limited almost exclusively to inter-state conflicts, and thus have very little to say about the multi-scalar, multi-actor character of water

[1] We use the term conflict here to refer to the various conflicts between the Myanmar armed forces (the Tatmadaw), operating under civilian (1948–1962), military-dominated one-party (1962–1988), military junta (1988–2010) and quasi-civilian (since 2010) governments, and various armed opposition movements that in some areas continue to this day (BEWG 2017; ICEM 2017; Bauer et al. 2018).

politics." Instead, hydropolitics understands that the "politics of water" are intimately connected to constructions of territory-making and scale. As Rogers/Crow-Miller (2017: 2) put forward in their review of hydropolitics literature:

> the key contributions of this collection of work [hydropolitics] have been to break down the nature-society dichotomy evident in hydraulic societies, to highlight the politics of access and exclusion, and to introduce into hydropolitical scholarship a more nuanced understanding of power relations by examining various forms of state- and market-driven water management.

Thus, hydropolitics is a useful lens for understanding the debates and contests around hydropower construction and water governance in the Salween basin. As we detail further in the following sections, in Myanmar's Salween River, subnational actors including EAOs, militias, and the Tatmadaw play a pivotal role in resource governance, acting alongside more conventionally-recognized hydropower actors such as government agencies responsible for electricity and water, and construction companies. The role of these armed actors also highlights ongoing (and evolving) concerns around conflict and peace as linked to control, ownership and management of natural resources in the basin. We also call attention to the role of CSOs (similar to Sneddon/Fox 2006), which have increasingly played influential roles in decision-making and shifting the scales of debate and inquiry.

The multi-scalar approach of a hydropolitics lens, which is less concerned with a local-state or local-global dichotomy, rather is interested in the ways that the politics of water are contested across and through multiple scales. Other studies in Southeast Asia have already revealed how civil society activists can contest uneven socio-economic and political processes and garner attention through contests over scale. For instance, Sneddon (2002), Lebel et al. (2005) and Molle (2007) show how civil society actors in the region have been integral not only to contesting scale, but to the production of scale, including through up-scaling efforts from a local scale to the scale of the basin and beyond in order to garner attention for their concerns. In the Salween Basin, 'scale-jumping' and the international-domestic links are useful to consider, but this does not mean that actors simply seek a 'natural' scale to jump to or to be 'chosen' as an existing frame or platform for analysis or political action. Instead, as Lamb (2014: 387) notes "scale must be actively defined, delimited, and populated with goals and concerns" (see also Götz, Chap. 6, this volume). In these ways, scales of analysis and decision-making are both considered as part of the process of understanding governance and politics of the Salween, but also, that these scales (of the nation, sub-nation, and local) are also produced and become subjects of action through these same processes.

In thinking about hydropower's role in particular, looking to the Mekong River, in a now-classic piece on that basin's hydropolitics, Bakker (1999) explains how an ingrained emphasis on hydropower development as 'the norm' in the region, rather than an understanding of hydropower as a contested project, is generative of both a politics of scale and justice. She further explains that the discourse around hydropower,

> operates as a 'politics of forgetting'; in presupposing rational allocation, it ignores the politicisation of water, thereby obscuring questions of equity of access, of control over prioritisation of projects, and of spatial and temporal scales of development. (Bakker 1999: 221)

Such an approach reveals how and why water becomes politicized, as well as who or what is included in these processes. As such, an approach towards the analysis of water that does not interrogate the proposition of hydropower development critically, or ask who is involved (or who is not), makes problematic assumptions about the terrain of debate, including about what constitutes 'cooperation' around water. As Sneddon/Fox (2006) further explain, an apolitical approach,

> obscures the ways in which states, non-state actors and river basins themselves interact to construct 'transnational' basins through institutional and material processes. Cooperation in and of itself is not the desired end for third-world riparian governments who create transboundary governance institutions; rather, cooperation is perceived as the basis for proceeding with the *development* of water resources encompassed by basins. (emphasis in original, Sneddon/Fox 2006: 182)

In other words, an apolitical approach, not considerate of the politics of water, ignores much of the work and efforts by the multiple actors, across various scales, who contest hydropower development and thus may also rationalize overlooking these same actors when defining the terms for 'collaboration' moving forward. What a hydropolitical approach offers, then, is "reflective of critical theoretical and political concerns that go beyond mere calls for 'cooperation'" (Sneddon/Fox 2006: 199). It is a call to critically understand and assess the politics of water, with attention to what it accomplishes for particular actors and across certain scales, as well as what it 'forgets.'

In the following sections, we detail how various schemes, plans and visions towards the Salween River have been formulated, with attention to the multiple actors and scales of analysis. To begin, we consider how these 'hydropolitics' are situated within the broader political economy of the region, particularly as linked to economic integration and (hydro)electricity cross-border trade (see also, Middleton/Allouche 2016).

3.3 Emergence and Current Status of Large Dam Plans on the Salween River

The Salween Basin is located in a region undergoing significant change, including processes of regional economic integration and a growing demand for natural resources, food, electricity and commodities. In the 1990s, economic integration was promoted through the Asian Development Bank's (ADB's) Greater Mekong Subregion (GMS) program (ADB, 2012a) and since the mid-2000s, as a vision of the Association of South East Asian Nations' (ASEAN) economic integration leading up to the ASEAN Economic Community in 2015. Since the mid-2010s, the China-led Belt and Road Initiative and the associated Lancang-Mekong

Cooperation Framework have further framed the region's economic integration, although its implications for the Salween are little known compared to more public plans for the Mekong River (Middleton 2018). These occur alongside other regional initiatives involving Japan, Korea, and the US among others.

As one of the main intended electricity markets, the role of Thailand's government and the fate of the country's economy has been central to plans for large dams on the Salween River mainstream in Myanmar. Preliminary studies for large dams on the Salween were initiated in the late 1970s, and addressed hydropower, irrigation and water transfer potential (Vatcharasinthu/Babel 1999). The Salween dam projects were first seriously considered in the late 1980s under Thai Prime Minister Chatichai Choonhaven (1988–1991), as the policy concept of turning "battlefields to marketplaces" gained traction (Magee/Kelley 2009). Building on this political momentum, in 1989, a Joint Working Group for the Development of Hydroelectric Projects on Border Rivers was established between the governments of Myanmar and Thailand tasked to prepare a feasibility study for projects along the Thai-Burma border (TERRA 2006).[2]

Correspondingly, the momentum for these projects have coincided at least partly with the rise and fall of Thailand's economy. The Asian Financial Crisis of 1997 in particular collapsed electricity demand in Thailand. As a result, Thailand's plans for power import projects from both Myanmar and Laos were put on hold (IRN 1999). At the same time, other bilateral issues were also creating tensions between the Myanmar and Thai governments, including boundary disputes, the production of and distribution of methamphetamines and heroin that were entering Thailand from Myanmar, and the conflict in Myanmar that resulted in refugees entering Thailand (Magee/Kelley 2009: 122). Plans for dams on the Salween mainstream reemerged in 2002 under the Thai Rak Thai government of Prime Minister Thaksin Shinawatra (2001–2006), when EGAT's Governor visited Myanmar to discuss the construction of the Weigyi and Dagwin dams on the border. Prime Minister Thaksin Shinawatra sought to work closely with Myanmar's military government to enable Thai investment, and Thailand affirmed its commitment to the projects on numerous occasions. For example, EGAT incorporated the Weigyi and Dagwin dams into its 2004–2016 Power Development Plan (PDP) as potential projects and in its 2007–2021 PDP included the Hatgyi and Tasang (now Mong Ton) dams as planned projects. Thus, between 1988 and 2010, under Myanmar's State Law and Order Restoration Council (SLORC) and subsequently the State Peace and Development Council (SPDC), plans to construct large hydropower dams on the Salween River accelerated.

Table 3.1: Proposed dams on the Salween River mainstream in Myanmar summarizes the current status of these mainstream hydropower projects on the

[2]Japan's Electric Power Development Company was commissioned by the Joint Working Group to prepare initial studies in the early 1990s and identified around ten dam sites with a total capacity of 6,400 MW at a cost of about USD 5.12 billion (Magee/Kelley 2009). The Joint Working Group was composed of representatives from the Myanmar Electric Power Enterprise, Thailand's National Energy and Policy Office and EGAT.

Table 3.1 Planned dams on the Salween River mainstream in Myanmar

Project	Location	Reported capacity (MW)	Developers	Power market	Status
Hatgyi	Karen State	1,365	Sinohydro (PowerChina), EGATi (Thai), MOEE (Myanmar), International Group of Entrepreneurs (IGE) (Myanmar)	Thailand	Joint Venture Agreement, Memorandum of Agreement (24 April 2010)
Dagwin	Karen State/ Mae Hong Son Province	729	EGATi	Thailand	Cancelled
Weigyi	Karen State/ Mae Hong Son Province	4,540	EGATi	Thailand	Cancelled
Ywathit	Karenni State (approx. 45 km upstream of Thai border)	4,000	China Datang Overseas Investment Co., Ltd., PowerChina, MOEE, Shwe Taung Group	Reportedly China	Memorandum of Agreement (18 January 2011)
Mong Ton (previously Tasang)	Shan State	7,110	CTGC, Sinohydro, China Southern Power Grid, EGATi, MOEE, IGE	Thailand	Memorandum of Understanding (10 November 2010)
Nongpha	Shan State	1,200	HydroChina (PowerChina) MOEE, IGE	China	Memorandum of Agreement (22 May 2014)
Kunlong	Shan State	1,400	Hanergy Holding Group, PowerChina, MOEE, Gold Water Resources (Asia World)	China	Memorandum of Agreement (21 May 2014)

Sources Salween Watch Coalition (2016); ICEM (2017); and IFC (2018)
Acronyms *CTGC* China Three Gorges Company; *EGAT* Electricity Generating Authority of Thailand; *EGATi* EGAT International Company; *IGE* International Group of Entrepreneurs; *MOEE* Ministry of Electricity and Energy (previously Ministry of Electric Power)

Salween River. While Thai and Myanmar government policy has a track record of broadly supporting plans for Salween dams, recent government statements waver between support and reservation. For example, in Myanmar in August 2016 at a press conference marking the National League for Democracy's 100 days in power, the Permanent Secretary of its Ministry of Electric Power said the Salween dams should move forward to address energy shortages in Myanmar (Myanmar Eleven 2016). Subsequently, in September 2016, Thailand's Permanent Secretary for energy announced that the Ministry of Energy planned further discussion with its Myanmar counterparts on Salween River hydropower dams (Deetes 2016). Yet, in 2017 Thailand's Ministry of Energy announced that it would not consider the Hatgyi Dam until there was an end to conflict in the area (Pollard 2018).

At present, four large hydropower projects have already been completed on two tributaries of the Salween River in Myanmar.[3] Three are on the Baluchaung River (Baluchaung 1, 2 and 3) in Karenni State, which is a tributary of the Pawn River that is a major tributary of the Salween River, and one is on the Teng River in Shan State, namely the Kengtawng dam completed in 2009. Alongside the four completed projects, an additional nine large-dam projects are also at various stages of planning and construction on five major tributaries, with two at advanced stages of construction in Shan State on the Nam Teng and Baluchaung rivers.

Meanwhile, Memorandum of Agreement (MoA) have been signed for four large dams on the mainstream, not including the Mong Ton hydropower project (previously Tasang) which had reached an advanced stage of planning before the Myanmar government stated that it must be redesigned as a two-dam cascade in November 2016. Even as full-scale construction of mainstream Salween dams is yet to be initiated, preparatory work has been undertaken, including field site investigations, public consultations and, in some cases, preliminary construction work that has already resulted in some community resettlement and the exploitation of natural resources in project areas.

The tributary projects built in the Salween basin to date, while not much written about outside of Myanmar, are an important part of the contested history of conflict and displacement linked to development in the Salween basin. The Baluchaung 2 hydropower station (known locally as the Lawpita hydropower plant) was the first project, and was completed in 1974. It is fed by water diverted from the Mobye dam, and was the country's first large hydropower project constructed as part of a bilateral war reparation between Myanmar and Japan.[4] For its construction, between 1969 and 1972 an estimated 8,000 people from 114 villages were forced to

[3]Here we focus on the Lower Salween, but plans in the Upper Salween or Nu Jiang, in Tibet and Yunnan Province, China for a 13-dam cascade on the mainstream also emerged in the late 1990s (Magee 2006). At present however, these projects have been suspended as government policies promoting national parks and protected areas have gained momentum (Yu et al. 2018; Yu et al., Chap. 4, this volume), and there are high electricity reserve margins across China's regional power grids as the economy has restructured reducing demand for new supply (Lin et al. 2016). However, the Nu Jiang's tributaries have been extensively dammed (Ptak 2014). These plans and their assessments and processes have similarly involved a whole range of actors, not limited to local and state governments, conservationists, engineers, and scientists.

[4]Local leaders in Pekhon Township formed an Anti-Dam Construction Committee in 1963 to rally against the construction of the Mobye dam which began in 1962. Met with threats of arrest and finding no recourse to justice, the committee founded the Kayan New Land Party (KNLP) as an armed resistance in 1964 (KDRG 2006: 35; KWU 2008: 1). Construction work at the dam site resumed in 1966.

relocate from the Mobye dam's reservoir area (KDRG 2006; KWU 2008). The Baluchaung 1 hydropower plant, fed by water from the Dawtacha dam, was completed in 1992 during a period of heightened militarization and human rights violations as the Tatmadaw expanded its operations and bases in the area. Most recently, in 2015, construction work on the Baluchaung 3 hydropower plant was completed (MOEP 2011). Electricity from these projects is primarily transmitted to Yangon and Mandalay, while their local social and environmental legacy remains unaddressed.

Regarding specific projects proposed in Myanmar, there has been strong civil society concerns expressed about the social and environmental impacts. For instance, in the case of the proposed Mong Ton dam, affected residents protested the consultation meetings held in Shan State in 2015 by the Snowy Mountain Engineering Corporation (SMEC). Then, in 2016, for a variety of reasons, not all related to the stated public concerns, the Myanmar government requested that the dam be redesigned as two dams (EGATi 2016), leading to a team of Chinese experts restarting work at the site under military guard in 2018 (Yee 2018). There are similar concerns put forward about the impacts of the proposed Hatgyi dam, and, most recently the Kunlong and Naung Pha dams.

3.4 Transboundary Electricity Trade and the Salween Dams

The Salween River mainstream dams were conceived principally as cross-border power trade projects and were formulated within the context of the ADB GMS program for economic integration. Commissioned by the ADB in 1994, the Norwegian consultants Norconsult prepared an initial plan for regional power trade heavily based on large hydropower dam construction, including on the Salween River (Norconsult 1994). As plans for regional power trade evolved over the 1990s, the Tasang (now Mong Ton) dam in particular was identified as one to be incorporated into the regional power grid, exporting its power to Thailand. In 2002, the leaders of the GMS countries signed the Inter-Government Agreement on Regional Power Trade in the Greater Mekong Sub-Region Countries, which was a high-level endorsement to further plans for cross-border power trade in the region, and since then several further intergovernmental agreements have been signed (ADB 2012b). More recently, the GMS power trade plans have also been incorporated into a vision for an ASEAN-wide power grid (Affeltranger 2008). As plans have been furthered for the basin, the multi-scalar politics of these processes have been evident, as seen in their relationship to the regional power trade and grid, as well as through and across further scales – from the subnational to the bilateral and international – as well as across a range of actors, from states to conglomerates.

At the bilateral level, following several exchanges between government representatives in July 1996, an MoU was signed for 1,500 MW of power exports from Myanmar to Thailand, which expired in 2010. In May 2005, the governments of

Thailand and Myanmar signed an MoU for the joint development of hydropower dams along the Salween River, which mandated EGAT and Myanmar's Department of Electric Power to jointly study in detail the locations and generating capacity of four large projects: Tasang (Mong Ton), Hatgyi, Weigyi (referred to as "Upper Salween") and Dagwin (referred to as "Lower Salween").

Some of the proposed Salween projects are intended to export their electricity to China, namely the Nongpha and Kunlong Dams, and potentially the Ywathit Dam. China already imports electricity from the Shweli 1 Dam on a tributary of the Irrawaddy River (completed in 2008) and the Dapein hydropower plant, while the controversial Myitsone Dam in Myanmar, suspended in 2011, was also designed principally as a power export project to neighboring Yunnan Province, China. Yet, with China now facing power surplus (Lin 2016), commitment to further power imports on first impression makes little sense. However, as highlighted by Ptak/Hommel (2016), these planned power trade projects are not only intended for the purposes of Yunnan's and more broadly China's national electricity security. They also function to deepen regional interconnection and economic interdependence between China and Myanmar that overall serve China's geopolitical interest of closer ties with neighboring Southeast Asia, as well as securing opportunities for investment. More recently, however, plans have also been proposed for China to export electricity to Myanmar from Yunnan Province (Straits Times 2017).

Thailand's most recent Power Development Plan for 2015 to 2036 details hydropower imports from unspecified projects in Myanmar cumulatively increasing from 2027 onwards to a total of 6,300 MW, which overall would reflect a shift from 7 to 20% of Thailand's total electricity consumption met by imports (EPPO 2015). Thailand's electricity planners have long argued that Thailand needs to secure more electricity because current supplies are insufficient to meet its future demand and renewable energy and energy efficiency are insufficiently reliable, although these claims are also contested by civil society (Greacen/Greacen 2004; Delina 2018). The sector is heavily influenced by EGAT and large independent power producers, and has resulted in Thailand's highly centralized electricity system (Middleton/Dore 2015). Large-scale power import projects, including the proposed hydropower dams on the Salween River, are meant to ensure Thailand's electricity security through diversifying sources (while not overly depending on power imports), reducing carbon emissions and diversifying fuel types away from gas.

In Myanmar, at present, there is no definitive plan for electricity sector development in the public domain.[5] The current NLD-led government has sent mixed

---

[5]Since the transition to a semi-civilian government in 2010, there have been various plans prepared for meeting the country's electricity needs and these have been a priority for both the USDP-led and NLD-led governments. The Ministry of Electric Power, now the Ministry of Electricity and Energy (MOEE) has received significant technical and financial support from bilateral aid and multilateral financial institutions. Recently developed masterplans include: a National Electrification Plan by the World Bank published in 2015, a Myanmar Energy Master Plan (for primary energy sources) by the Asian Development Bank published in 2015, and ongoing support from the Japan International Cooperation Agency for a Power Sector Master Plan.

messages on plans for hydropower on the Salween River. On the one hand, the NLD's election manifesto made a commitment not to build new large dams. On the other hand, as noted above, shortly after the NLD came to power, in August 2016 the Permanent Secretary of the Ministry of Electricity and Energy declared the government's commitment to construct large hydropower dams on the Salween River (Myanmar Eleven 2016). More recently, it seems that discussion has turned to the use of natural gas for domestic electricity generation, which some commentators have suggested should weaken the claim that the Salween dams are needed (Pollard 2018). At the same time, it has also been reported that the Minister for Electricity and Energy is still actively supportive of hydropower construction (Kean 2018).

Domestic conglomerates in Myanmar are also influential actors in hydropower. Three of Myanmar's largest conglomerates have won significant construction contracts in 11 out of 14 Salween hydropower projects and their associated infrastructure, including the Shwe Taung Group, International Group of Entrepreneurs (IGE), and Asia World Company (Table 3.1: Proposed dams on the Salween River mainstream in Myanmar). These companies were established under the previous military governments, and maintained close relationships with them thus becoming regarded as crony companies. While playing important roles in ceasefire negotiations between the Myanmar government and EAOs in the Salween basin, they have also used their political capital in business negotiations for hydropower contracts, winning lucrative deals in construction and concessions for natural resource extraction in the project areas.

Thus, the range of state and non-state private actors active to seek the construction of the Salween Dams is diverse and working across scales in Thailand, China and Myanmar. Drawing on their range of influences and resources, they have produced studies and formulated discourses rationalizing the need for the hydropower dams for cross-border and domestic electricity supply. While some tributary projects have already been built, to date the mainstream Salween in Myanmar remains free flowing. We now turn to focus on violent conflict in the Salween basin, and its implications for the hydropolitics of the basin.

3.5 Myanmar's Peace Negotiations, Conflict and Hydropower Dams

In November 2010, highly-flawed elections paved the way for the military-backed Union Solidarity and Development Party (USDP) to form a semi-civilian government. With Myanmar's transition from decades of military rule there have been signs of democratic reform, including greater freedom of movement, the easing of rules on protests and media censorship, a degree of greater civil and political freedoms, and the opening of limited political space through the establishment of a parliamentary system and the granting of elections (Jones 2014; Farrelly et al. 2018). Local communities and civil society groups have mobilized campaigns at a variety of political scales, targeting a wide range of issues related to human rights, land and natural resource governance, and legislative reform. A number of ethnic civil society organizations continue to attach these calls to the broader movement for the establishment of a democratic federal union. However, comprehensive and deep-rooted democratic reforms are impeded by the continued significant role of the Tatmadaw in Myanmar's political system, enshrined within the 2008 constitution (Egreteau 2016; KPSN 2018b; Pedersen 2018).

Between September 2011 and April 2012, the USDP government renewed and signed new ceasefires with 14 EAOs—nine of which are based in the Salween River basin area—while also initiating a multilateral peace negotiation process. Nine formal rounds of multilateral negotiations between 2013 and 2015 involving representatives of the Tatmadaw, EAOs, political parties, government, parliament and other stakeholders resulted in a Nationwide Ceasefire Agreement (NCA) in October 2015, although not all groups signed the agreement. The NCA represents the first attempt to bring every EAO under one common ceasefire text and agreement, while also containing political provisions to guide the formation of a future democratic federal union (Box 3.1 Signatories and non-signatories to the NCA).

Box 3.1: Signatories and non-signatories to the NCA. *Source* The authors.

Much of the negotiations over peace, conflict, and authority are still very much in process. Significantly, NCA non-signatory EAOs command 70–80% of the country's EAO armed forces, including the United Wa State Army (UWSA), the Shan State Progressive Party (SSPP), the Ta'ang National Liberation Army (TNLA), the Myanmar National Democratic Alliance Army (MNDAA) and the National Democratic Alliance Army of Eastern Shan State (NDAA-ESS), all based or controlling territory in the Salween River basin in Shan State. In the Salween Basin, two of the largest EAOs—the Karen National Union (KNU) and the Restoration Council of Shan State (RCSS)—signed the NCA in October 2015 alongside six smaller EAOs. However, the KNU and RCSS ceasefires have been put under severe strain due to persistent Tatmadaw operations in their territories and restrictions on political space for genuine political dialogue. Thus, aside from the on-paper agreements, distinctions between ceasefire and non-ceasefire groups has become increasingly opaque in the Salween basin.

As of the time of writing, three Union-level peace conferences have been held under the name of the 21st Century Panglong, most recently in July 2018. However, limited progress has been made towards sustainable nationwide peace. The ongoing "21st Century Panglong" peace process appears increasingly uncertain (TNI 2017) and the present level of armed conflict has escalated to its most intense since the

1980s (Lintner 2016; Jolliffe 2018). Armed conflict is concentrated in the Salween basin in northern Shan state, and beyond the basin in Rakhine and Kachin states. In Karen state alone, since 2016, more than 8,500 civilians have been forced to flee Tatmadaw military operations (KPSN 2018a).[6]

Restrictions imposed through existing legislation, including the 2008 constitution, continues to present a road-block in peace negotiations, particularly regarding questions of territorial administration and land and natural resource governance (Bauer et al. 2018). The Burma Environmental Working Group, a coalition of ethnic CSOs, has argued that, for sustainable peace to be realized, a moratorium on new and incomplete large-scale natural resource investments must be enforced until devolved federal structures and policies can be operationalized under new federal institutions, guided by a federal constitution (BEWG 2017). Yet, legislation and policies established under the USDP-led and now NLD-led Union governments have further centralized state ownership and control over land and natural resources. For example, in 2018, revisions to the Vacant, Fallow and Virgin Land Management Law received wide criticism from civil society groups that it weakened customary land tenure rights, as well as small-holder farmers land tenure rights in general, and could undermine commitments made by the Union government within the peace negotiations (Gelbort 2018). National water policy has also increasingly defined a role for the Union-level government (Götz, Chap. 6, this volume; Sect. 3.7: Transboundary water governance below).

A fundamental issue regarding dams in the Salween basin is their relationship to contested territories in Myanmar, where there is a complex history of conflict and multiple associated claims for political authority and legitimacy (Stokke et al. 2018). There are no less than ten non-state EAOs, 18 Border Guard Forces (BGFs) and 28 militias active in the Salween Basin area (Buchanan 2016; Burke et al. 2017). South (2018: 52) differentiates between "relatively small and mostly quite remote areas controlled exclusively by EAOs and more extensive areas of mixed administration, where authority is exercised variously by one or more EAOs and the government or various Myanmar Army-backed militias." Thus, there are significant areas of the Salween basin where the central government has limited ability to project its authority (Callahan 2007; Jolliffe 2015). This includes all of the proposed and under construction projects on the Salween River basin, with the mainstream dams having the largest territorial impact often being the most contested.

Numerous analysts have linked the current plans for Salween dams to the risk of armed conflict and in some cases ongoing cases of fighting (BEWG 2017; Burke et al. 2017; Karen News 2014; Suhardiman et al. 2017). For instance, around the Hatgyi dam site, fighting between a splinter faction of the Democratic Karen Buddhist Army (DKBA) and a Tatmadaw-allied BGF has sporadically broken out, including in 2014 (KRW 2014) and 2016 (Hein 2016), and has been linked by CSO networks to plans for the project (KRW 2016; KPSN 2018a). Similarly, a

[6]Well over 900,000 civilians have been forcibly displaced since 2011, including close to 800,000 Rohingya who have fled from Rakhine state.

representative of the Shan Nationalities League for Democracy has suggested the Mong Ton dam could threaten the peace process and national reconciliation, a feeling echoed by CSOs (SWAN 2015; Hein 2016).

Contested political authority over territory in the Salween basin requires a different way of looking at water governance, as the assumption of a single sovereign political authority does not hold. From a hydropolitical perspective, it means that multiple claims for legitimacy to govern must be given serious regard. Yet, there are also a large number of predominantly pro-government armed actors, or militias, operating in the Salween River basin area, with a far small number allied with EAOs. Pro-government militias tend to be largest and most active in areas where the Tatmadaw has engaged in combat with the EAOs, and although they may receive direct and indirect support from the Tatmadaw many also raise additional revenues via business activities across a range of sectors, from natural resource projects such as agro-industry, logging and mining, to illicit activities such as narcotics (Buchanan 2016).

The ongoing political uncertainty around the peace negotiations, which have been stated as working towards defining a federal union in Myanmar, adds additional complexity to understanding how water is to be governed, as the foundational political agreement between the national and subnational levels of administration is not yet established. These contestations are also operating at, and generative of, various scales of mobilization and governance in terms of the peace process and with regard to water governance (Götz, Chap. 6, this volume).

3.6 Civil Society Work Across Scales

While above we detailed the broader political economic context and the various actors and scales invoked in the electricity trade, the evolving politics of hydropower development are also explicitly linked to the work of civils society actors (Lamb et al., Chap. 7, this volume). Momentum for the projects has been stalled as riverside communities in Thailand and Myanmar, sometimes in collaboration with a range of local, national and international CSOs, have contested and challenged the plans for large plans (Magee/Kelley 2009; Simpson 2013). Strategies have ranged from undertaking different types of research to celebrating the river and organizing protests and lodging complaints with the Myanmar Government and the National Human Rights Commission of Thailand. For example, in February 2007, a Global Day of Action Against the Salween Dams in Myanmar took place in 19 cities globally, when a petition letter endorsed by 232 organizations was also released (International Rivers 2007). These actions have continued to present (Deetes, Chap. 16, this volume).

It is important to recognize that CSOs and community-based organizations (CBOs) have sought different visions from those of large dam proponents that have emphasized more local, community-centered forms of development. For example, eighty kilometers downstream of the proposed Hatgyi dam the community-based NGO KESAN has been working in collaboration with five village communities around the Salween's Daw Lar Lake. The initiative, known as Community Based Water Governance, focuses on advancing local people's rights to land, natural resources and environmental conservation, and to expand civic space for community engagement in resource governance (KESAN 2018; Bright, Chap. 5, this volume). A Daw Lar Lake committee has been formed among the five villages around the lake that is now documenting the lake watershed's natural resources and seasonal livelihoods, mapping village boundaries and conservation zones, and detailing customary governance arrangements. This bottom-up initiative aspires to empower community rights "… to self-determination, which includes the right to manage and govern their own lands, water, forests and natural resources through community based institutions" (KESAN 2018: 16), given that only in 2013 the state government had tried to auction Daw Lar Lake as a commercial fishing concession to a private company without informing the communities.

Upstream of the proposed Hatgyi dam, in Karen State's Mutraw district, another grassroots-led initiative, promoting peace and self-determination, environmental integrity, and cultural survival has been steadily building momentum. This initiative is known as the Salween Peace Park. Founded on a longstanding partnership between local communities, Karen civil society and the Karen National Union (KNU), this indigenous conservation initiative has sought to

> … expand the concept of "Water Governance" beyond just the water in the river itself, to include the land, forest, biodiversity, upland shifting cultivation, customary land systems, and cultural and sacred sites along the Salween River Basin. (KESAN 2017: 1)

This work has been generative of much debate and attention. Significantly, it has introduced new scales for analysis and action: a Peace Park operating with the collaboration of Karen civil society, communities and the KNU.

3.7 Transboundary Water Governance

Finally, we turn to the transboundary scale of the Salween River and its relationship with national and subnational scale water governance. Unlike the Mekong River, there has not yet been a significant commitment by governments to a comprehensive transboundary water cooperation agreement for the Salween River. As discussed above, existing intergovernmental agreements have focused on transboundary electricity trade rather than transboundary water cooperation. Following the aforementioned scholars' insights into 'cooperation' and what it produces (Bakker 1999; Sneddon/Fox 2006), as well as who it includes and excludes, and what it 'forgets,' we consider the extent to which transboundary arrangements have emerged to date beyond a 'comprehensive transboundary water cooperation agreement', the lack of which is itself significant.

In Thailand, concerns over the transboundary impacts of the Salween dams in Myanmar on communities in Thailand highlight the range of actors involved, across both countries. Concerns about the cross-border impacts were, for instance, formalized and in part formed the basis for a National Human Rights Commission of Thailand (NHRCT) investigation towards the Salween dams, and in particular the Hatgyi Dam (Lamb 2014). Meanwhile, similar concerns have been raised regarding plans for cross-border water transfers from the Salween River to the Chaophraya River. Despite a series of detailed studies, including one most recently initiated in 2015 (Wipatayotin 2015), it is not certain there is consensus from Myanmar that water transfer would be acceptable (Affeltranger 2008). The impacts of the project still need to be fully assessed and made public. Meanwhile, for trans-basin water transfer projects proposed to be located within Thailand, for example those on the Yuam River or Moei River, it is not clear how Thailand would proceed with informing or negotiating with Myanmar (Interview with retired Myanmar government official 2016).

These cross-border ventures are likely to be re-assessed now, as water governance in Thailand is undergoing significant reform. A new water law was passed on 4 October 2018 that aims to consolidate 38 water-related departments across ten ministries under the Office of National Water Resources (ONWR) housed in the Prime Minister's Office. This holds uncertain implications for transboundary water governance in practice, and who would be involved or consulted in 'cooperation' efforts, and in what ways. For example, many of the international functions of the Department of Water Resources within the Ministry of Natural Resources and Environment, including the Thai National Mekong Committee, have been transferred to the ONWR.

Regarding national water governance in Myanmar, a National Water Resources Committee (NWRC) was established by presidential degree in 2013 and reconstituted in March 2016 after the elections (Aye 2016). The NWRC is comprised of 20 members from water-related state agencies, chaired by the Vice President and also has an Advisory Group made up of water experts (Aye 2016). In 2014, the NWRC published Myanmar's National Water Policy and subsequently a National Water Framework Directive that broadly adopts the principles of Integrated Water Resources Management (NWRC 2014; Nesheim et al. 2016). While national in scope, in practice a principle focus of the NWRC has been the Irrawaddy River. On the Salween River, the limited capacity of the central government to project authority across the basin has constrained the implementation of the National Water Policy. There has, however, recently been some discussion on the preparation of a Salween policy and potentially a Salween River Basin Organization. Yet, it would be important to heed concerns around bilateral or MRC-like cooperation that have been critiqued for their apolitical nature and ability to obscure or depoliticize the concerns of affected peoples (see for instance Salween University Network Meeting 2016; Sneddon/Fox 2006).

3.8 Salween Cooperation Through the Lens of Hydropolitics

In this section, we ask what does 'cooperation' mean, who is included or excluded, what is foregrounded versus forgotten, and what are the implications for the future of the basin. We suggest that a hydropolitics lens helps reveal and analyze the means through which water becomes politicized.

One mode of cooperation has emerged around the pursuit of large hydropower projects in the basin. Here the chapter has sought to draw attention to the long history of these projects, and the role played by various iterations of bilateral and regional inter-government political and economic cooperation, as well as the role of domestic and international project developers. This mode of cooperation which assumes common pursuit of large hydropower has been backed up by commissioned technical studies that focus on project feasibility in terms of engineering and economic considerations. Cooperation for the projects in Myanmar is largely framed around cross-border power trade relations, either with Thailand or China. At the same time, this mode of cooperation foregrounds national and regional development which often privilege particular terms of economic growth. Largely forgotten from these formal processes are affected communities and non-state actors such as electricity consumers, and in this particular case, Myanmar's EAOs and CSOs.

However, through various strategies, these actors have shaped and challenged the projects and their terms, for example through the NHRCT in the case of the Hatgyi dam. This case flagged, among other issues, the social impacts of the project, including from a human rights perspective, and their security implications. Meanwhile violent conflict in project areas appears linked at least partly to ensuring territorial control of project sites, and has created severe impacts on communities who live within the contested areas. Natural resource governance and 'resource sharing' is a theme within the ongoing peace negotiations, but to date has not been thoroughly discussed as political space to address even more fundamental political questions remains foreclosed by the Tatmadaw's influence over the process that is enshrined within the 2008 constitution.

A very different mode of cooperation has focused on community-based initiatives, such as the Salween Peace Park. Here, viewed through a hydropolitics lens, a different set of actors have sought to produce a vision for the future of the Salween River. Significantly, it is also an initiative emerging from a sense of history, but one that is narrated from territorial claims linked to the Karen people's calls for self-determination – a call shared by myriad ethnic groups in the Salween River basin and across the country. Inevitably, the violent conflict also shapes this sense of history and has in part forged the notion that there should be a *peace* park. Cooperation is between CSOs, the KNU and communities, while recognition of legitimacy is sought from a wider array of actors that range from the international community to the Union government. In this mode of cooperation, foregrounded are the Karen communities' claims for political authority, alongside CSOs and the KNU, who are organized through community-based institutions. In contrast to the first mode of cooperation, it is the Union government's political authority to govern resources such as water and land that is questioned. Local ecological knowledge is also emphasized, as well as a range of other studies produced by groups such as KESAN that addresses issues such as biodiversity and suitable forms of governance.

3.9 Conclusion

This chapter has sought to analyze the Salween basin through the lens of hydropolitics, with a focus on Myanmar. It has drawn attention to the multi-scaled processes and multiple actors involved, including for water governance and electricity governance, which intersect in the planning and materialization of large hydropower dams. These projects are advanced by consortiums formed of transnational corporations from Thailand and China working with Myanmar companies, and backed by various Myanmar and Thai government agencies via bilateral and regional agreements and through national planning processes.

We have also sought to draw out how scale itself is produced through these contested processes. Proponents of large dams have framed them as a "development solution" for the Salween basin via promoting regional connectivity, industrialization, electrification and associated poverty alleviation. Yet, this frame has also been contested at different scales, in particular by EAOs, CSOs and INGOs across a range of issues including on human rights, environment and social impacts, and the ongoing prevalence of conflict in the context of the ongoing peace negotiations and assertions by some actors for democratic federalism. Meanwhile, civil society collaborations for the Salween Peace Park also reconceptualize and decentralize water governance. These efforts not only position the actors involved as agents in water governance, but also rethink the scales of governance across local and subnational arenas.

Finally, it is the history of conflict in the Salween Basin, which remains unresolved to this day, that must be foregrounded as a key issue in how plans for large dams on the Salween River have unfolded. In the technical documents of planners, the issue of conflict is not really acknowledged, or is only recently so (i.e., ICEM 2017), but on the ground conflict and security fundamentally determine project outcomes including for potentially impacted communities. Given the technical, legal, and political complexity of these large dams and the great uncertainty in Myanmar's ongoing peace process, there is a strong argument that Myanmar's peace negotiations need to be concluded before such projects are discussed as a part of a broader discussion on resource governance in the context of federalism.

References

ADB (Asian Development Bank). (2012a). *The Greater Mekong Subregion at 20: Progress and prospects*. Mandaluyong City, Philippines: Asian Development Bank. Retrieved from: https://www.adb.org/sites/default/files/publication/30064/gms-20-yrs-progress-prospects.pdf.

ADB (Asian Development Bank). (2012b). *The Greater Mekong Subregion power trade and interconnection: 2 decades of cooperation*. Mandaluyong City, Philippines: Asian Development Bank.

ADB (Asian Development Bank). (2015). *Myanmar 2015 Energy Master Plan (EMP)*. Mandaluyong City, Philippines: Asian Development Bank.

Affeltranger, B. (2008). Inter-basin water transfers as a technico-political option: Thai–Burmese projects on the Salween River. In N.I Pachova, M. Nakayama & L. Jansky (Eds.), *International water security: Domestic threats and opportunities* (pp. 161–179). Tokyo, Japan: United Nations University Press.

Aye, S.P. (2016, July 11). Re-launched committee to manage water resources. *Myanmar Times*. Retrieved from: https://www.mmtimes.com/national-news/21308-re-launched-committee-to-manage-water-resources.html.

Bakker, K. (1999). The politics of hydropower: developing the Mekong. *Political Geography*, *18* (2): 209–232.

Bauer, A., Kirk, N., & Sahla, S. (2018). *Natural resource federalism: Considerations for Myanmar*. Natural Resource Governance Institute. Retrieved from: https://resourcegovernance.org/sites/default/files/documents/federalism-considerations-form-myanmar.pdf.

BEWG (Burma Environment Working Group). (2017). *Resource federalism: A roadmap for decentralized governance of Burma's natural heritage*.

Buchanan, J. (2016). *Militias in Myanmar*. San Francisco, CA: The Asia Foundation.

Burke, A., Williams, N., Barron, P., Jolliffe, K., & Carr, T. (2017). *The contested areas of Myanmar: Subnational conflict, aid and development*. San Francisco, CA: The Asia Foundation.

Callahan, M. (2007). *Political authority in Burma's ethnic minority states: Devolution, occupation and coexistence*. Washington, DC: East-West Center.

Deetes, P. (2016, September 29). Salween dams threaten river communities. *Bangkok Post*. Retrieved from: https://www.bangkokpost.com/opinion/opinion/1097621/salween-dams-threaten-river-communities.

Delina, L.L. (2018). Whose and what futures? Navigating the contested coproduction of Thailand's energy sociotechnical imaginaries. *Energy Research & Social Science*, *35*, 48–56.

EGATi (EGAT International). (2016). EGATi Upper Thanlwin Hydropower Project (Mong Ton). Retrieved from: http://www.egati.co.th/en/investment/en-mongton.html.

Egreteau, R. (2016). *Caretaking democratization: The military and political change in Myanmar*. Oxford University Press.

EPPO (Energy Policy and Planning Office). (2015). *Thailand Power Development Plan 2015–2036* (PDP2015). Bangkok, Thailand: EPPO, Ministry of Energy.

Farrelly, N., Holliday, I., & Simpson, A. (2018). Explaining Myanmar in flux and transition. In A. Simpson, N. Farrelly & I. Holliday (Eds.), *Routledge handbook of contemporary Myanmar* (pp. 3–12). London and New York: Routledge.

Gelbort, J. (2018, December 10). Implementation of Burma's Vacant, Fallow and Virgin Land Management Law: At Odds with the Nationwide Ceasefire Agreement and Peace Negotiations. *Transnational Institute*. Retrieved from: https://www.tni.org/en/article/implementation-of-burmas-vacant-fallow-and-virgin-land-management-law.

Greacen, C.S., & Greacen, C. (2004). Thailand's electricity reforms: Privatization of benefits and socialization of costs and risks. *Pacific Affairs*, *77*(3), 717–541.

Hein, K.S. (2016, October 20). Activists damn Salween plans. *Frontier Myanmar*. Retrieved from: https://frontiermyanmar.net/en/activists-damn-salween-plans.

ICEM (International Center for Environmental Management). (2017). *Baseline assessment report hydropower: Strategic environmental assessment of the hydropower sector in Myanmar*. Washington, DC: International Finance Corporation.

International Finance Corporation (IFC) (2018). *Strategic Environmental Assessment (SEA) of the Myanmar hydropower sector*. IFC. Washington, DC.

International Rivers. (2007, February 18). *Worldwide protests against Salween Dams in Burma*. Retrieved from: https://www.internationalrivers.org/resources/worldwide-protests-against-salween-dams-in-burma-3900.

IRN (International Rivers Network). (1999). *Power struggle: The impacts of hydro-development in Laos*. Berkeley, CA: International Rivers Network.

Jolliffe, K. (2015). *Ethnic armed conflict and territorial administration in Myanmar*. Yangon, Myanmar: The Asia Foundation.

Jolliffe, K. (2018). Peace and reconciliation. In A. Simpson, N. Farrelly & I. Holliday (Eds.), *Routledge handbook of contemporary Myanmar* (pp. 359–370). London and New York: Routledge.

Jones, L. (2014). The political economy of Myanmar's transition. *Journal of Contemporary Asia*, *44*(1), 144–170.

Karen News. (2014, October 5). Fighting 'directly linked' to Hat Gyi Dam Project – Claim Karen leadership. *Karen News*. Retrieved from: http://karennews.org/2016/10/fighting-directly-linked-to-hat-gyi-dam-project-claim-karen-leadership/.

KRW (Karen Rivers Watch). (2014). *Afraid to go home: Recent violent conflict and human rights abuses in Karen State*.

KRW (Karen Rivers Watch). (2016). *Karen State September 2016 Conflict: The real motivations behind renewed war*.

KDRG (Karenni Development Research Group). (2006). *Damned by Burma's generals: The Karenni experience with hydropower development - From Lawpita to Salween*.

KWU (Kayan Women's Union). (2008). *Drowning the green ghosts of Kayan Land: Impacts of the upper Paunglaung Dam in Burma*.

Kean, T. (2018, August 9). Hydropower is back. *Frontier Myanmar*. Retrieved from: https://frontiermyanmar.net/en/hydropower-is-back.

KESAN (Karen Environmental and Social Action Network). (2017). *The Salween Peace Park: A Scoping Study (unpublished)*. Yangon, Myanmar: KESAN.

KESAN (Karen Environmental and Social Action Network). (2018). *Community based water governance: A briefing report on Daw Lar Lake*. Yangon, Myanmar: KESAN. Forthcoming.

KPSN (Karen Peace Support Network). (2018a). *The nightmare returns: Karen hopes for peace and stability dashed by Burma Army's actions*.

KPSN (Karen Peace Support Network). (2018b). *Burma's dead-end peace negotiation process: A case study of the land sector*.

Lamb, V. (2014). Making governance good: The production of scale in the environmental impact assessment and governance of the Salween River. *Conservation and Society*, *12*(4), 386–397.

Lebel, L., Garden, P., & Imamura, M. (2005). The politics of scale, position, and place in the governance of water resources in the Mekong region. *Ecology and Society, 10*(2), 18.

Lin, J., Liu, X., & Kahrl, F. (2016). *Excess capacity in China's power systems: A regional analysis*. Commissioned by the Energy Foundation through the U.S. Department of Energy. Berkeley, CA: Ernest Orlando Lawrence Berkeley National Laboratory.

Lintner, B. (2016, October 11). Burma's misguided peace process needs a fresh start. *The Irrawaddy*. Retrieved from: http://www.irrawaddy.com/opinion/commentary/burmas-misguided-peace-process-needs-a-fresh-start.html.

Magee, D. (2006). Powershed politics: Yunnan hydropower under great western development. *China Quarterly*, *185*, 23–41.

Magee, D., & Kelley, S. (2009). Damming the Salween River. In F. Molle, T. Foran & M. Käkönen (Eds.), *Contested waterscapes in the Mekong Region: Hydropower, livelihoods and governance* (pp. 115–140). London, United Kingdom: Earthscan and USER (Chiang Mai University, Thailand).

Middleton, C. (2018). *Policy brief: Reciprocal transboundary cooperation on the Lancang-Mekong River: Towards an inclusive and ecological relationship*. Bangkok, Thailand: Center for Social Development Studies, Faculty of Political Science, Chulalongkorn University.

Middleton, C., & Allouche, J. (2016). Watershed or powershed? Critical hydropolitics, China and the 'Lancang-Mekong Cooperation Framework'. *The International Spectator*, *51*(3), 100–117.

Middleton, C., & Dore, J. (2015). Transboundary water and electricity governance in mainland Southeast Asia: Linkages, disjunctures and implications. *International Journal of Water Governance*, *3*(1), 93–120.

MOEP (Ministry of Electric Power). (February 2011). *Environmental Impact Assessment (EIA) and Social Impact Assessment (SIA)*. Baluchaung No.3 Hydropower Project. Retrieved from: https://www.shwetaunggroup.com/wp-content/uploads/2017/03/Baluchaung-No3-Hydropower-Project-ESIA-Report.pdf.

Molle, F. (2007). Scales and power in river basin management: The Chao Phraya River in Thailand. *The Geographical Journal*, *173*(4): 358–373.

Myanmar Eleven. (2016, August 14). Myanmar pushes for more hydropower projects. *The Nation*. Retrieved from: http://www.nationmultimedia.com/business/Myanmar-pushes-for-more-hydropower-projects-30292893.html.

Nesheim, I., Wathne, B.M., Ni, B., & Tun, Z.L. (2016). Myanmar: Pilot introducing the national water framework directive. *Water Solutions*, *1*, 18–27.

Norconsult. (1994). *Promoting subregional cooperation among Cambodia, Lao PDR, Myanmar, Thailand, Viet Nam and Yunnan Province of the People's Republic of China: Subregional energy sector study for the Asian Development Bank*. Mandaluyong City, Philippines: Asian Development Bank.

NWRC (National Water Resources Committee). (2014). *Myanmar National Water Policy*. Yangon, Myanmar: Ministry of Transport.

Paoletto, G., & Uitto, J.I. (1996). The Salween River: is international development possible? *Asia Pacific Viewpoint*, *37*(3): 269–282.

Pedersen, M.B. (2018). Democracy and human rights. In A. Simpson, N. Farrelly & I. Holliday (Eds.), *Routledge handbook of contemporary Myanmar* (pp. 371–280). London and New York: Routledge.

Pollard, J. (2018, February 11). Myanmar power plans could spare the Salween. *Asia Times Online*. Retrieved from: https://burmariversnetwork.org/news/myanmar-power-plans-could-spare-the-salween.html.

Ptak, T. (2014). Dams and development: Understanding hydropower in Far Western Yunnan Province, China. *Focus on Geography*, 57(2), 43–53.

Ptak, T., & Hommel, D. (2016). The trans-political nature of Southwest China's energy conduit, Yunnan Province. *Geopolitics*, *21*(3), 556–578.

Rogers, S., & Crow-Miller, B. (2017). The politics of water: a review of hydropolitical frameworks and their application in China. *Wiley Interdisciplinary Reviews: Water* 4(6): e1239.

Salween University Network Meeting. (2016, January 29–31). Shared Vision. Retrieved from: https://www.academia.edu/29283801/Meeting_Report_TOWARDS_A_SHARED_VISION_FOR_THE_SALWEEN-THANLWIN-NU_.

Salween Watch Coalition. (2016). *Current status of dam projects on the Salween River*.

Simpson, A. (2013). Challenging hydropower development in Myanmar (Burma): cross-border activism under a regime in transition. *The Pacific Review*, *26*(2), 129–152.

Sneddon, C. (2002). Water conflicts and river basins: the contradictions of co-management and scale in northeast Thailand. *Society and Natural Resources*, *15*(8), 725–741.

Sneddon, C., & Fox, C. (2006). Rethinking transboundary waters: A critical hydropolitics of the Mekong basin. *Political Geography*, *25*(2), 181–202.

South, A. (2018). Hybrid governance and the politics of legitimacy in the Myanmar peace process. *Journal of Contemporary Asia*, *48*(1), 50–66.

Stokke, K., Vakulchuk, R., & Øverland, I. (2018). *Myanmar: A political economy analysis*. Oslo: Norwegian Institute of International Affairs.

Straits Times. (2017, August 5). China in talks to sell electricity to Myanmar. *Straits Times*. Retrieved from https://www.straitstimes.com/asia/se-asia/china-in-talks-to-sell-electricity-to-myanmar.

Suhardiman, D., Rutherford, J., & Bright, S.J. (2017). Putting violent armed conflict in the center of the Salween hydropower debates. *Critical Asian Studies, 49*(3): 349–364.

SWAN (Shan Women's Action Network). (2015). *Naypyidaw must cancel its latest plans to build the Upper Salween (Mong Ton) dam in Shan State* [Press release].

TERRA (Towards Ecological Recovery and Regional Alliance). (2006). *Chronology of Salween dam plans (Thailand and Burma)*. Bangkok, Thailand: TERRA.

TNI (Transnational Institute). (2017). *Beyond Panglong: Myanmar's national peace and reform dilemma*. Amsterdam, The Netherlands: TNI.

Vatcharasinthu, C., & Babel, M.S. (1999). *Hydropower potential and water diversion from The Salween Basin. Workshop on trans-boundary waters: The Salween Basin.* Chiang Mai, Thailand.

Wipatayotin, A. (2015, July 18). RID pushes two new water diversion bids. *Bangkok Post*, Retrieved from: https://www.bangkokpost.com/print/626428/.

Yee, T. H. (2018, May 7). Work on Myanmar dam goes on under heavy guard. *The Straits Times*. Retrieved from: https://www.straitstimes.com/asia/se-asia/work-on-myanmar-dam-goes-on-under-heavy-guard.

Yu, X., Chen, X., Middleton C., & Lo, N. (March 2018). *Charting new pathways towards inclusive and sustainable development of the Nu River Valley*. Kunming and Bangkok: Green Watershed and Center for Social Development Studies, Faculty of Political Science, Chulalongkorn University.

Chapter 4
From Hydropower Construction to National Park Creation: Changing Pathways of the Nu River

Yu Xiaogang, Chen Xiangxue and Carl Middleton

4.1 Introduction

The Nu River, also known as the Upper Salween, is one of China's sixteen major cross-border rivers. It originates from the Tibetan Plateau, and runs through Yunnan Province of China, before entering Myanmar. Well-known for its rich biodiversity, the Nu River and surrounding areas are home to more than 4,300 plants species, of which 24 species fall under the first and second national protection classes (Nu Prefecture Forestry Bureau 2015). In acknowledgement of this, the Nu River is one of the "Three Parallel Rivers" (along with the Lancang and Jinsha Rivers), which is the largest World Heritage site in China. Furthermore, many ethnic groups live within the basin, including Lisu, Nu, Dulong, Bai, Pumi, Yi, Tibetan, Naxi and Jingpo.

There has been considerable debate in China over the past two decades about the fate of the Nu River. In contrast to most major rivers in China, the mainstream of the Nu River is largely free flowing, with the exception of two hydropower dams near the headwater. Yet, over the past two decades, many of its tributaries have been dammed for hydropower and irrigation (Ptak 2014). This chapter explores the range of visions for the Nu River and the extent to which they have materialized through exploring five 'pathways', namely (see also Yu et al. 2018):

- The *hydropower construction pathway* that details how China's government developed a strategy for promoting hydropower dam construction in the 1990s as a means of accelerating national economic growth. It incorporates projects across two scales, namely (a) large dams under the West-East Power Transfer (WEPT) regional development scheme that led to plans for a 13-dam cascade on

---

Yu Xiaogang, Director, Green Watershed; Email: greenwatershed@163.com.

Chen Xiangxue, Project Officer, Green Watershed; Email: chenxx417@163.com.

Dr. Carl Middleton, Director, Center of Excellence for Resource Politics in Social Development, Center for Social Development Studies, Faculty of Political Science, Chulalongkorn University; Email: carl.chulalongkorn@gmail.com.

C. Middleton and V. Lamb (eds.), *Knowing the Salween River: Resource Politics of a Contested Transboundary River*, The Anthropocene: Politik—Economics—Society—Science 27, https://doi.org/10.1007/978-3-319-77440-4_4

the Nu River mainstream; and (b) medium- and small-scale dams for provincial or prefecture-level initiatives.

- The *civil society river protection pathway* that explores the significance of changing society-state-nature relations in China. This pathway is intertwined with the other pathways, including resistance to the hydropower construction pathway, and in support of environmental protection via national parks and energy reform. It emphasizes how democratic space has – to an extent – been produced and utilized to engage high-level government policy objectives on "Ecological Civilization."
- The *energy reform pathway* that details the implications of China's current electricity surplus and shifting policy positions on climate change and renewable energy, in particular for the hydropower dam pathway.
- The *national park pathway* that documents how the local government is now pursuing the establishment of national parks, which will give priority to ecological protection. This pathway to a certain extent also reflects the transformation of development thinking over the recent decade in China.
- The *water conservancy pathway* that seeks to reframe earlier large hydropower dam plans as multi-use projects for irrigation and flood/drought management. This plan has only recently emerged, and few details are in the public domain.

4.2 Methodology

The method adopted in this chapter is that of development pathway research. Within this approach, development pathways are defined as "… the particular directions in which interacting social, technological and environmental systems co-evolve over time" (Leach et al. 2010: xiv). The study of a development pathway entails analyzing how different actors' visions for the basin are being formulated and acted upon, including the narratives produced, and the decision-making processes invoked. Overall, the analysis of pathways is a useful heuristic tool to render visible options that exist – or that are proposed – and hence to evaluate their implications for a range of policy goals for the future.

We reviewed relevant policies, legal frameworks and operational guidelines relating to the Nu River Basin's development and environmental protection at the national and local levels (see also Yu et al. 2018 for further details). We also conducted institutional and actor analysis, drawing on an extensive literature review, as well as in-depth interviews with government officers, experts and environmental NGO staff (Table 4.1: In-depth interviews conducted). Interviews were also conducted in two communities in Nu Prefecture and the Baoshan Area. Drawing on their role as leaders in the Nu River protection campaign, Green Watershed also contributed their firsthand experience to inform this research.

Table 4.1 In-depth interviews conducted

Government agencies	– National Energy Agency (NEA) – National Development and Reform Commission (NDRC) – Ministry of Environmental Protection (MEP)
Experts	– Expert from the Energy Research Institute, NDRC – Expert from the Department of International Economic Cooperation, NDRC – Expert from the China Hydropower Engineering Consulting Group Co. – Expert from the Research Center of Environment and Development, Chinese Academy of Social Sciences – Expert from the China Institute of Water Resources and Hydropower Research – Expert from the Center for the Study of Contemporary China
NGOs	– Green Earth Volunteers – Friends of Nature – International Rivers – Hengduan Mountain Research Institute

Source The authors

4.3 The Transformation of China's Development Thinking

Government plans for the Nu River have been defined by several critical turning points. From the initiation of plans for large hydropower dam construction in the late 1990s to the decision to suspend these plans in February 2004 by Premier Wen Jiabao (2003–2013), the local government has played a key role. Overall, key elements of the local government have maintained a commitment to hydropower construction in collaboration with project developers, and this obstructed the exploration of other options for the basin over the 2000s. However, in November 2012, the central government upgraded its "Ecological Civilization" policy to become a national strategy, and proposed the establishment of a national park system. Shortly afterwards, the Yunnan Provincial government and Nu Prefecture government began to advance new plans accordingly.

More broadly, China has witnessed significant policy reorientation. In 1978, China halted its "class struggle as core" approach, and reoriented towards a measured policy of "opening up" the country to private investment. Over the following three decades, economic growth was a central priority, and socio-economic development of overriding importance, with the goal of meeting the growing material and cultural needs of China's population. However, the pursuit of rapid economic growth led to a long-term pattern of development that could be described as high-input, high-pollution, high-energy-consumption and low-efficiency, and which did not attain a broad-based sustainable mode of development. This not only resulted in a series of social problems, such as an increasing gap between the rich and the poor, capital rent-seeking and political corruption, but also led to environment degradation, including consumption of vast amounts of natural resources, serious pollution and the degradation of ecological systems.

For several reasons, these challenges urged China's leaders to redirect the country's strategy of economic development towards achieving a more comprehensive form of development of the economy, society and the environment. Firstly, China has a low amount of natural resources per capita, and the growing volume of resource utilization together with low energy efficiency has gradually led to an economic growth bottleneck (partly-addressed by China's growing resource imports). Secondly, with living standards improved for many, there is a growing demand for a better environment. Rising public unrest trigged by environmental pollution has raised questions about government accountability and challenged social stability. Thirdly, China's economic growth has been paralleled by a growing geopolitical role, in which China is playing a more important role in international cooperation and global governance. China is seeking to conform with green low-carbon development trends that are emerging globally, and wishes to promote its competitiveness and influence in the international arena (Ronghua 2012).

The Central Committee of the Communist Party of China (CPC) gradually accepted the concept of sustainability, with the 15th Party Congress held in 1997 adopting a strategy of sustainable development and the 16th Party Congress held in 2002 establishing principles of resource savings and an environment-friendly society. The 17th Party Congress held in 2007 recognized a good ecological environment as one of the important requirements for achieving a well-off society by 2020. However, even as the concept of sustainability has gradually gained acceptance, economic development still remains the first priority. Environmental protection policies are subordinate to (economic) development needs, and thus the old development model has not fundamentally changed.

The 18th Party Congress held in 2012, however, appears to have been a turning point. The report that came out of the meeting included a whole chapter highlighting the construction of an "Ecological Civilization" for the people's wellbeing, and solidified a long-term national development strategy towards attaining it. This meeting adopted the revised Constitution of the CPC, declaring: "The CPC will lead the Chinese people to build a socialist ecological civilization." In the fifth plenary session of the Congress, the General Secretary of the Communist Party of China, Xi Jinping, put forward five new development concepts: innovation, coordination, greening, openness and sharing. Ecological Civilization and the five development concepts formed the development guidelines for China's 13th five-year plan (2016–2020).

4.4 The Hydropower Development Pathway

In the late 1990s, Premier Li Peng's (1988–1998) retirement left behind an economic bubble in China's coastal areas and a legacy of underdevelopment in the Western regions. The Asian Financial Crisis of 1997 resulted in the stagnation of

China's foreign trade and the shrinking of its export-oriented economy, while major flooding in 1998 also harmed the economy. Premier Zhu Rongji (1998–2003) led the new State Council to address these challenges by expanding domestic demand to create an economic "soft landing" (Yonglin 2006). The country's development pattern shifted from an export-oriented economy alone to one that also prioritized investment in undeveloped regions of Western China.

In 2000, the central government launched the Western Region Development Strategy (2000–2020). Within this approach, the "West East Power Transfer" (WEPT) strategy supported large dam construction in Southwest China to transmit electricity to the industrialized Southeastern seaboard where power demand was high and resources to produce electricity limited (Magee 2006; Dore et al. 2007). The WEPT strategy was favored by China's leaders because designing plans for the development of China's western water resources allowed the WEPT strategy to promise to not only meet the electricity demands of Eastern China, but also stimulate economic growth in the West of the country (Peiyan 2010). In this policy discourse, hydropower was considered a clean and renewable energy. Yet, Southwest China, where the projects were to be built, is also the area with China's best-preserved ecology, most abundant biodiversity, and largest ethnic minority population, thus raising the risk of ecological and social harms due to the projects.

4.4.1 Dissolution of the State Power Corporation

China's continued economic liberalization under Premier Zhu Rongji led to the reform of state-owned enterprises (SOEs). The power generation section of the monopoly SOE State Power Company was split into five companies (Huaneng, Datang, Huadian, Guodian and China Power Investment Corporation), while the power grid section was split into the State Grid and the Southern Power Grid (Xiaobing 2003). Following this reform, the five power generation companies began competing to increase their profits through bids to build hydropower dams on domestic rivers mainly in Southwest China, which became known as the "river-enclosure movement" (Xiaohui/Baiying 2010). The companies claimed that these hydropower projects were key to China's national energy strategy.

While reform of SOEs was needed to improve their economic performance, under WEPT the central government also unleashed market forces to drive hydropower development. Despite a strong emphasis on planning and environmental impact assessment (EIA), the government could not control the SOEs. Furthermore, the local and national government themselves were often captured by the benefits of strong capital forces, including the generation of revenues from taxes and various fees. Moreover, SOEs often relied on corrupt government officials to obtain project approval.

4.4.2 Local Government Response to the WEPT Strategy and Nu River Hydropower Plans

There are six rivers flowing through Yunnan Province, with the most profitable for hydropower development being the Nu, Lancang and Jinsha Rivers. Yunnan Province thus became the main focus of WEPT. The Nu River has 36,400 megawatts (MW) of potential hydropower resources and accounts for about one-third of the total hydropower capacity in Yunnan. It was viewed by the government and project developers as one of China's largest potential hydropower resources yet to be developed. The total investment of the proposed thirteen dam cascade proposed for the Nu River was 89.65 billion RMB. When completed, the total annual income from the 13 hydropower dams was anticipated to be more than 30 billion RMB. Meanwhile, annual state revenue would increase by 5.20 billion RMB and annual local revenue for Yunnan Province would increase by 2.72 billion RMB, of which the Nu Prefecture Government would obtain 1 billion RMB of tax revenue (Yaohua/Jiankun 2004).

In November 2000, the Yunnan Provincial government responded to the strategy of WEPT by affirming its intention to speed up the development of the hydropower industry (Hao 2000). On 31st January 2003, the Yunnan Provincial government signed an "Agreement on promoting the cooperation of power development in Yunnan" with the Huadian Group. Subsequently, on 10th July 2003, the Yunnan Huadian Nu River Hydropower Company was jointly established by Huadian Group (51%), Yunnan Energy Investment Group (20%), China Resources Power Holdings (19%) and Yunnan Power Investment (10%).

Yet, in the same month of the creation of the Yunnan Huadian Nu River Hydropower Company, on 3rd July 2003, the UNESCO World Heritage Committee voted to list the "Three Parallel Rivers" as a World Nature Heritage site. This listing offered a different development option for both the Nu River and the Nu Prefecture government. However, the local government quickly initiated a propaganda campaign to ensure that hydropower development was the option to be prioritized. For example, the Secretary of the Nu Prefecture government, Xie Yi, requested each government department to collectively write an article explaining that hydropower was "the only way to facilitate Nu River development" (later to be published as "An inevitable choice for the Nu River"). The aim of the articles was to influence civil servants' views on hydropower development. Additionally, the local newspaper and radio began reporting daily on the significance of hydropower to various industries. When Cha Chaoou, a government official of Lisu origin in charge of the prefecture's poverty alleviation program, pointed out that there were many ways to alleviate poverty and that hydropower development was not the only option, he was quickly removed from his position (Bizhong 2004).

4.4.3 Challenges from the State Environmental Protection Administration

The State Environmental Protection Administration (SEPA) played a key role in the debate over the Nu River's development pathway. With the implementation of China's EIA Law on 1st September 2003, SEPA gained the authority to question EIA reports and processes, and subsequently challenged the EIA of the Nu River hydropower development plan. SEPA's main criticism centered on the fact that the EIA report was based on the "Regulations for River Basin Environmental Impact Assessments" issued in 1992. As there were technical differences between these regulations and the new EIA Law, SEPA questioned the overall credibility of the Nu River Hydropower Development Planning Unit and the EIA Unit of the Beijing Engineering Corporation (Yikun 2004).

Following a review meeting held by the National Development and Reform Commission on Nu River hydropower planning in Beijing on 12th–14th August 2003, SEPA held two expert forums on ecological environment protection and hydropower development in the Nu River Basin on September 3rd and October 21st 2003. Following objections raised around the EIA report, the media flagged the issue as a public concern. Yang Chaofei, Director of the Natural Ecological Protection Division of SEPA, began to question the hydropower boom and the ecological impact of dams (Chaofei 2003). In this debate, he also introduced the report of the World Commission on Dams (WCD 2000) into the discussion in China about hydropower. These studies informed the cautious attitude and discourse of SEPA toward large hydropower dams on rivers in China's Southwest.

The Nu River hydropower planning also stimulated public debate. Chinese environmental NGOs cooperated with the public, media and experts to conduct a series of advocacy campaigns (see below).

4.4.4 Suspension and Consequence of the Nu River Hydropower Plan

In February 2004, Premier Wen Jiabao (2003–2013) officially suspended the Nu River hydropower development plan. However, project proponents continued to support the projects and decision-making on the future of the Nu River became complicated. Supporters of hydropower construction applied three new strategies: (1) form an alliance with the media to influence public opinion; (2) create a hydropower nationalism discourse; and (3) cultivate pro-hydropower academics.

For hydropower developers, the media had become a challenge to them as it tended to report civil society opinions against the dams. Hydropower developers thus saw it necessary to create hydropower-friendly media to cultivate a discourse endorsing dam development. This strategy manifested itself in the emergence of a group of on-line writers, including Zhang Boting, Fang Zhouzi, and Sima Nam, who positioned themselves against the civil society movement seeking to halt large dam development.

Relatedly, a 'hydropower nationalism' discourse emerged that centered on framing river protection as a Western country-induced plot against China. The 'hydropower nationalism' discourse presented hydropower development as a means to protect China's public and national interests. Under this discourse, the transnational environment NGO movement questioning hydropower development and advocating for indigenous people's rights were portrayed as a Western scheme to sow seeds of dissent within China. Rumors such as "Premier Wen Jiabao was coerced by NGOs into impeding hydropower development" (Xiaolan 2007) or that SEPA officials were against hydropower because they wanted to increase their own decision-making power were commonly used in this discourse.

Regarding cultivating pro-hydropower academics, in 2004, the Yunnan Power Company set up a "Yunnan Power Prize" worth 800,000 RMB to support research by the Yunnan Academy of Social Science. He Yaohua, former President of the Yunnan Academy of Social Science and Chairman of the Southwest National Research Association, wrote, collected and published papers in a book that supported hydropower development entitled "Nu, Lancang and Jinsha Rivers: Research on the exploitation of hydropower resources and the protection of the environment" (Jiangkun/Yaohua 2004). The publication was disseminated as policy input for decision-making by the Yunnan Province government and its senior leaders. In promoting hydropower development, this academic strategy reduced space for public debate on hydropower by framing hydropower decision-making as purely an academic and technical issue. Qin Guangrong, Vice Secretary of the Yunnan Provincial Party Committee and Executive Vice Governor, praised the role of the academy in supporting Yunnan's development (Guangrong 2004).

4.4.5 Current Status of WEPT in Yunnan Province

Although large hydropower development has been questioned by the public since the early 2000s, hydropower in Southwestern China continues to be built under the direction of various government agencies responsible for energy and the local government. By the end of 2014, the total installed capacity of hydropower in China had reached over 300 million kilowatts, accounting for a quarter of the world's total installed capacity (Hui 2015). However, as China's economy enters a new norm of less rapid economic growth and a restructuring of the economy away from energy intensive industries, hydropower developers have been left with an excess of power generation capacity (Lin et al. 2016). In response, as discussed in the energy reform pathway below, the government has proposed to expand electricity exports into Southeast Asia, and domestically to shift from WEPT to increasing demand in Yunnan Province itself.

4.4.6 Small Hydropower on Nu River Tributaries

Following the official suspension of large hydropower dam plans in 2004, the Nu Prefecture government reoriented towards promoting small- and medium-scale hydropower. These plans, however, have also been backed by various Central Government policies since 1983, when the Ministry of Water Resources endorsed small hydropower projects for rural electrification. Investors from eastern areas of China, such as Shanghai and Zhejiang provinces, rushed into Southwestern China attracted by small hydropower's low initial investment and running costs.

There was, however, a lack of binding regulations, especially regarding small hydropower's environmental impacts. SEPA, which in 2008 was upgraded to the Ministry of Environment Protection, issued a "Notice on the Orderly Development of Small Hydropower to Effectively Protect the Ecological Environment" in June 2006. The Nu Prefecture government has also issued measures on small hydropower development and utilization management in 2007 and 2015 respectively. However, given the fact that the Nu Prefecture government was also promoting small hydropower, it tended to be flexible with regulations to encourage and enable development. In contrast to large hydropower projects, small hydropower projects are rarely reported on in the media, so there is a lack of public awareness. In the absence of policy constraints and public awareness, small hydropower has expanded rapidly in just a few years, with numerous legal violations and some dams even being built within the World Heritage site.

While accurate data is difficult to ascertain, according to some media reports, there were 45 companies developing small hydropower as of 2008, with agreements to build 85 power stations on 65 branches. WLE-Mekong (2018) document 21 hydropower dams of 15MW and above, and 10 irrigation dams with reservoir surface areas larger than 0.5 square kilometers. In a case study of 12 villages along the Dimaluo tributary river in Nu Jiang Prefecture, Ptak (2014) highlights that there are benefits and challenges. For instance, small-scale hydropower projects have brought some limited material benefits to relatively inaccessible rural ethnic minority communities; mainly this is linked to electricity production, but also is evidenced during the construction phase. Yet, the process of creating the projects have lacked participation and did not addressed multidimensional needs, such as access to employment opportunities nor improved access to education and health services. In his case study, Ptak found that the projects had also created difficulties for those who had been resettled (Ptak 2014).

Small hydropower is often presented as a means to help local communities increase their household income (as low electricity prices from small hydropower can help farmers reduce their household burden) and to protect forests by decreasing reliance on firewood for cooking and heating. However, uncoordinated small hydropower dam construction has also resulted in significant environmental degradation. The dams, canals, and diversion tunnels built by small hydropower stations can alter river channels and even cause the original river to sometimes dry up, leading to the disappearance of aquatic species that rely on specific niches or that cannot migrate. According to Kibler/Tullos (2013), small hydropower in the Nu River basin has resulted in a huge cumulative impact on the ecological environment, particularly with regard to habitat and hydrologic change. Furthermore, a significant proportion of the small hydropower dams are linked to mining projects nearby within the Nu River basin. In 2007, the local government had encouraged mining projects near to the hydropower stations to use their power, as there was insufficient transmission grid infrastructure to export the electricity farther away.

As investors built these power stations one by one, power generation grew incrementally, and generation overcapacity became an increasingly serious problem. In general, there is sufficient water for more power generation, but due to the lack of transmission grid infrastructure, most of the small power stations cannot run at full capacity. By the end of 2015, the Yunnan provincial government had halted further small hydropower construction on the Nu River.

4.5 Civil Society River Protection Pathway

Chinese environmental NGOs have played an important role in protecting the Nu River from large dam development, and are one of the major forces pushing for greater access to information and public participation. Key organizations have included: Green Watershed (GW), Green Earth Volunteers, and Friends of Nature. Pursuing existing legal means to mobilize public attention and support, these NGOs formed alliances through different networks to influence hydropower decision-making. They also drew on scientific assessments from renowned experts and field investigations to present the potential social, environmental and geological problems posed by the Nu River hydropower development plans to top-level officials to urge them to make informed decisions.

4.5.1 Civil Society Strategies and Activities

In early 2002, GW conducted a social impact assessment (SIA) in the area around the Manwan dam on the Lancang River, documenting the poverty that resettled migrants were living in. In June 2002, following the state-run Xinhua News' coverage of the Manwan dam's adverse impacts, Premier Zhu Rongji ordered the Yunnan Provincial government to verify the situation, resolve the dam-induced social issues, and compensate the Manwan dam-affected community for their losses. This research laid a strong foundation for GW's credibility to participate in the Nu River decision-making processes.

From November to December 2003, GW visited Beijing to establish a collaboration with other NGOs there, including Friends of Nature, Green Earth Volunteers, the Global Village, Conservation International, Wild China, Green Island and Home Watch Volunteers. In 2004, these environmental groups formed a loose coalition called the China River Network, later renamed the River Watching Network. The China River Network worked with member organizations to share resources, take collective action, and put pressure on groups promoting hydropower. They signed several petition letters to protect the Nu River, increasing social pressure on the government and the hydropower developers by mobilizing the general public in large numbers.

One of the most successful methods adopted by NGOs was building alliances with public figures, environmental officials, the China Democratic Party, and various experts and scholars in order to influence decision-making. While environmental officials, technical experts and scholars joined forces to promote protecting the Nu River, Wang Yongchen, a reporter for China National Radio and founder of Green Earth Volunteers, also liaised with a number of well-known writers, literary figures, and movie stars to influence the public. In October 2003, at the second Congress of the China Environmental Culture Promotion Association, Wang Yongchen mobilized 62 celebrated individuals from the fields of science and culture, along with media and civil society groups, to sign their names in opposition to the Nu River hydropower dams.

The China Democratic Party Yunnan Branch (DP) also disagreed with the positioning of hydropower as a main pillar of Yunnan's economy, especially with regards to the development of the Nu River. GW and the DP co-wrote a proposal submitted to the party's political consultation session in early February 2004. While the DP had always supported government decisions in the past, this time they were against Nu River hydropower development. These advocacy campaigns played a significant role in influencing Premier Wen Jiabao to suspend plans for Nu River hydropower development in 2004.

The NGO network's ability to garner support for a free-flowing Nu River cannot be viewed in isolation from the network's relationship with the media. From 2003 to 2006, there were hundreds of news reports on the environmental and social impacts of hydropower. Media in Yunnan had been warned that the phrase "Green Watershed" should not appear in local media. Despite this warning, media representatives poured in from other provinces to interview GW about the Nu River. In 2004, the "Nu River" was listed as the ninth most widely covered issue in China. Thus, the media amplified the voice of civil society. The People's Daily, Youth Daily, Economic Observer, 21st Century Economic Report, International Herald Tribune, South Reviews, Southern Weekend, and Beijing News all reported on the Nu River campaign at great length. Some reports were released at critical times to influence decision-making within the central government. It placed decision-making processes around hydropower development on the Nu River in the public spotlight.

Central to the perspective of civil society was that local communities should be positioned as policy actors rather than mere recipients of development policy. Linking policy-making at the national level with the need for representation of affected communities, GW arranged for 14 villagers from two potential resettlement communities in the Nu River basin to visit the Manwan Dam in May 2004 (Hongwei 2004). After seeing how local communities affected by the Manwan Dam made their living by collecting garbage, the trip participants shared their experiences with other villagers living along the Nu River. A documentary film maker, Shi Lihong, produced a video documenting the visit entitled "The Voice of the Nu River," which was shared widely. While arranging the visit held serious negative repercussions for GW and its ongoing work, the activity did push the government to solve the resettlement problems at the Manwan Dam and ensured that a new resettlement compensation policy was developed in Yunnan province. As a result, resettlement compensation rates were raised in 2005 from 5,000 RMB (in the case of the Manwan Dam) to 80,000 RMB per affected person, which includes the cost of rebuilding houses as well as community infrastructure such as new roads, schools and markets.

In October of 2004, a resettled community from the Tiger Leaping Gorge dam made history when five community representatives participated in a UN Conference on Dams and Sustainable Development co-organized by the UNDP, World Bank and the State Council in Beijing. At the conference, resettled people took advantage of the rare opportunity to engage in open dialogue with officials, including the Director General of the National Energy Agency (NEA), the Director of the Yunnan Resettlement Department, and the CEOs of dam companies. Based on this dialogue, the focus of the conference shifted from showing the magnificent achievements of hydropower development to exploring alternative means to reduce social and environmental impacts (Liangzhong/Ying 2005).

4.5.2 The World Heritage Committee and Nu River Protection

Following the listing of the Three Parallel Rivers as a World Natural Heritage site, Chinese NGOs and NGO representatives from 60 countries signed a letter during a conference held by the International Rivers Network in Thailand in November of 2003 requesting the World Heritage Committee (WHC) to give serious consideration to the potential harms of Nu River hydropower development. In July 2004, at the 28th World Heritage Conference, Chinese NGOs signed a joint letter to the WHC regarding dam development on the Nu River, and in April 2006, experts from the IUCN and UNESCO conducted an investigative trip to Yunnan (Kejia 2006). Their report pointed out that the main threats to World Natural Heritage sites on the Three Parallel Rivers include: hydropower development, mining, and development of the tourist industry (UNESCO 2006). It also identified that proposed adjustments would cause a 20% reduction in the original heritage site area. In general, if a heritage site is reduced by 20% it will likely be removed from the World Heritage list.

Based on the report, the 30th session of the WHC General Assembly held in July 2006 stated that the impact of exploratory activities for Nu River hydropower development is obvious, and that if the dam construction starts, it would have a very large impact to the Nu River's aesthetic value, and the naturally flowing river will be transformed into a series of reservoirs (Wang 2007). The WHC General

Assembly asked the Chinese government to submit a status report on heritage protection to the WHC by February 2007.

In January 2007, the Chinese government submitted the status report, explaining that hydropower plans for the middle and lower reaches of the Nu River are currently the subject of ongoing research and debate. Their report noted, "there are strict laws and regulations in China to protect heritage sites. These regulations prohibit dam construction and mining activities within heritage sites." In June 2007, the Chinese government was required by the WHC to conduct a comprehensive scientific assessment of the impacts of dams and other projects on the environment of the "Three Parallel Rivers" in the middle reaches of the Nu River. According to the WHC, if the "Three Parallel Rivers" cannot be governed properly by 2008, the area will lose its World Heritage status.

The WHC specifically required that each project near the 'Three Parallel Rivers' disclose information to the public in a timely and transparent manner. Similarly, while the local government benefits from the site's World Heritage status, they must also be aware of their responsibility to protect the site, and must not over-develop tourism to increase income. The Ministry of Housing and Urban-Rural Development issued a warning to the local government that if they fail to protect the site, this will badly impact China's overall image. In the four years that followed, the 'Three Parallel Rivers' were monitored by the WHC.

In early 2016, GW, Green Earth Volunteers and Friends of Nature visited the NEA as concerned civil society organizations to call upon the administration to support the establishment of a national park, protect the World Heritage site, and prioritize ecological protection in the context of hydropower development. After this dialogue, seven NGOs (Friends of Nature, Green Earth Volunteers, GW, Green Han River, Hengduan Mountain Institute, Green Zhejiang and Chengdu Urban Rivers Association) collectively released an open letter appealing to the NEA to emphasize EIAs and public participation in the 13th five-year energy plan, particularly disclosure of hydropower plans on the Nu River. The dialogue and open letter were effective, as in the latter part of 2016, hydropower construction on the Nu River was excluded from the 13th Five-year Energy Development Plan, the 13th Five-year Renewable Energy Development Plan and the 13th Five-year Hydropower Development Plan.

4.6 The Energy Reform Pathway

As discussed above, the hydropower development pathway in Yunnan Province, including on the Nu River, was fundamentally shaped by the WEPT strategy. At the same time, China's electricity industry was undergoing profound reform. As noted above, in 2002, the state monopoly of the State Power Corporation was restructured into two state monopoly grid companies, and five state-owned power generation companies (the "Big Five") that were intended to compete alongside private and other state-owned investors in power generation. Wang/Chen (2012) highlight the slow pace of market-orientated reform after the initial creation of the "Big Five," suggesting that in contrast to the earlier monopoly in China, a "relative monopoly" now exists (see also Ngan 2010). Thus, these large state-owned enterprises have enjoyed the benefits of both market and planned economies (Dore et al. 2007). Meanwhile, China's 1996 Electricity Law encouraged the government to support renewable and clean energy for power generation, which incentivized the construction of large-scale hydropower dams for nation-wide electrification, as well as the construction of small hydropower stations to promote rural electrification. As noted before, the increasingly competition between the "Big Five" power companies accelerated the exploitation of Yunnan's hydropower resources (Han 2013).

Over the past couple of years, electricity markets in China's eastern industrial areas are now facing a power surplus (Solomon 2016). This is due to the global economic slowdown, as well as over-investment in coal and nuclear power generation, and China's trend towards less energy intensive industries. The installed capacity of Yunnan Province's hydropower is at present not fully utilized (Xiao n.d.), and the government's slogan has shifted since 2016 from 'WEPT' to "Yunnan Hydropower for Yunnan Use." Meanwhile, the Yunnan Provincial government, hydropower companies and power grid companies are also exploring opportunities to export power to neighboring countries such as like Laos, Vietnam, Bangladesh, Thailand and Myanmar (Yaqin 2016). For example, in 2015, China was already exporting approximately 70 MW of electricity to Laos, with an agreement to export up to 3000 MW (RPTCC 2015).

Overall, a profound transformation of China's electricity sector is increasingly underway, incentivized by the need to reverse environmental degradation, meet international climate change obligations, clean up corruption, and address China's power surplus (Hao 2016). Technical solutions to meeting the country's energy demands are currently being proposed. The 13th Five-year Plan (2016–2020) responded to demands to reduce dependence on coal, develop clean and renewable energy, and enhance energy efficiency. China is also committed to achieving 15% non-fossil energy production by 2020. Hydropower, however, in this context is considered essential to attaining this non-fossil energy target.

Yet, China's energy surplus has cooled down the fever to develop the Nu River dams, at least for the time being. Yunnan's high level of hydropower development can either be seen as over-exploitation of the provinces' natural resources, or as one of many steps towards a less carbon-intensive energy future (Magee/Henning 2017). Civil society groups argue that due to their social and ecological impacts large dams cannot be considered as clean energy. They argue that governance reform is required, namely: Firstly, (re)balancing the interests of various stakeholders, including restraining special interest groups and protecting public interests, especially the interests of disadvantaged groups. Secondly, clean relationships between government and enterprises must be fostered within the hydropower sector. Thirdly, the full and true costs of hydropower must be assessed, including impacts on watershed ecosystems and the rights of resettled communities and indigenous people. Lastly, hydropower governance must be improved, with increased transparency, public participation and accountability in decision-making (Xiaogang 2016).

4.7 The National Park Pathway

Now, over ten years since the suspension of hydropower construction, the Yunnan Provincial government and Nu Prefecture government is proposing the establishment of a national park. This pathway is still in the decision-making stage, and it is not yet clear how proposals for the Nu River Grand Canyon National Park will unfold given that the full plans are yet to be disclosed to the public.

In November 2012, the CPC Central Committee made the creation of an "Ecological Civilization" in China a national strategy, and subsequently issued a "Decision on Several Major Issues Concerning Comprehensive Deep Reform" in November 2013. The Central Committee State Council also issued "Opinions on Accelerating the Development of Ecological Civilization" and a "General Plan for the Reform of the Ecological Civilization System" in 2015, which supported the establishment of a national park system and the implementation of stricter protection measures within national parks. Such protection measures include prohibiting any construction and development that damages the ecosystem and affects indigenous people's homelands. Furthermore, these policies convey the importance of protecting the natural ecology and the authenticity and integrity of natural and cultural heritage.

This "national park pathway" was affirmed in the 13th Five-Year Plan (2016–2020). Although the outline of the plan proposed to coordinate hydropower development and ecological protection, the plan places high priority on ecological protection, and dictates scientific considerations for hydropower development in Southwestern China. The plan also mentioned establishing a national park system. However, despite this new direction in the national 13th Five-Year Plan, Yunnan's 13th Five-Year Plan Outline still makes mention in a couple of sentences to plans for the Nu River hydropower dams and suggests doing more follow-up work on it.

The Policy Research Office of the Yunnan Provincial government put forward the idea of a Nu River Grand Canyon National Park in 2007 (Yihua/Yaqi 2007). However, this pathway had not been fully considered within the context of the regional hydropower boom. Even as plans for large hydropower dams were shelved in 2004, the Nu Prefecture government still implemented an "electricity combined with mining" economic strategy, and spared no effort to promote small hydropower projects to this end. This situation continued until 2013, when the Yunnan Provincial Party Committee appointed a new Nu Prefecture Secretary. The new Secretary, Zhiyun Tong, led a delegation to visit national parks in the United States, and subsequently began the development of national parks in Yunnan Province in 2014. The Secretary of the Yunnan Provincial Party Committee declared on 25th January 2016 that all small hydropower and mining development on the Nu River would be stopped to protect the river's ecology and to support the Nu River Grand Canyon National Park proposal.

In May 2016, the Yunnan Provincial government approved the establishment of the Nu River Grand Canyon National Park. Later, the Secretary of Nu Prefecture stressed that national park construction was a new concept, and a step forward in promoting Nu River tourism development and ecological construction. He deemed this as the most suitable approach for Nu River development, incorporating both protection and utilization, in line with Premier Xi Jinping's statement: "clean waters and green mountains can bring us prosperity and wealth" (Jufen 2016).

The Nu River Grand Canyon National Park plan was approved by experts in August 2016, but as of the time of writing (January 2019) has not yet been publicly disclosed. Furthermore, the public, including indigenous people in local communities and NGOs, did not have the opportunity to give direct input on the Nu River Grand Canyon National Park plan. This despite the fact that in 2015, the Yunnan Provincial government had issued a National Park Management Regulation that encourages local community participation in national park management.

A number of problems have emerged in the pilot phase of establishing national parks in Yunnan Province, the first being ecological protection. For example, in Pudacuo National Park, which was established in 2007 in Shangri-La County, in order to increase income, the core area of the nature reserve was developed for tourism. Furthermore, the government's National Park Service became a liaison between tour companies, communities and government agencies. The fact that the National Park Service is supervising tourism development complicates the incentives for the National Park Service, as its funds are provided by tour companies. Meanwhile, in terms of community participation, local residents can only access low level employment opportunities, mainly cleaning work. Compensation for community members who lost their livelihoods with the establishment of the park consists mainly of cash compensation rather than sustainable livelihood capacity-building and reconstruction of their livelihoods.

The experience to date raises the question will the Nu River Grand Canyon National Park follow the old development model of government and capital domination and the neglect of minority community rights to participate in decision-making and to share benefits? Could it be possible that the development and management of the Nu River Grand Canyon National Park will ensure minority communities enjoying full access to information, inclusive participation and benefit sharing?

Regarding ecological conservation, the planned Nu River Grand Canyon National Park area does not cover the Nu River itself, the riverbanks and some small hydropower areas. This is because the National Park's boundaries, like those of the Three Parallel Rivers Protected Areas and existing Nature Reserves, cut off at a certain elevation. The Secretary of Nu Prefecture has stated that "National park construction does not affect the Nu River hydropower development or other industry development, so do not put them against each other" (Meng 2016). The argument posed by the Yunnan Province Forestry Department is that if the national park's boundaries include the Nu River itself, large number of people on both sides of the river would have to be resettled. The Yunnan Provincial Forestry Department said, "There is no direct relationship between the Nu River Grand Canyon National Park and halting Nu River hydropower development. If the national park covers the Grand Canyon, this means that the people on both sides of the river will be resettled, leading to a resettlement cost that is too high" (Meng 2016).

4.8 Water Conservancy Pathway

The final pathway we address is the Nu River water conservancy pathway, although very few details are available in the public domain. Interviews undertaken with government agencies for this research revealed that the Nu River is considered a very important freshwater resource, possibly even of higher value than its power generation capacity. This perspective has recently been incorporated into government planning documents.

In Yunnan Province's 13th Five-Year Plan Outline, there are three large- to medium- scale hydropower comprehensive utilization projects planned on the Nu River, namely Fugong, Lushui and Saige, with Lushui also being detailed as a key water conservancy infrastructure project. Furthermore, in the Nu Prefecture 13th Five-Year Water Conservancy Plan Recommendations, the Water Resources and Hydropower Planning and Design General Institute proposed that in order to achieve flood control, irrigation, water supply, power generation, and other benefits, the Nu River hydropower development plans should be adjusted for comprehensive utilization of water resources. Furthermore, more focus should be placed on the important role of comprehensive water resource utilization in poverty alleviation, ecological protection, and livelihood improvement. It proposes that to promote the comprehensive development and utilization of water resources on the Nu River, priority should be given to water conservation projects such as Maji, Yabiluo, Lushui and Saige, which it suggests have great potential benefits, with little environmental impact. The Lushui water conservancy project, it states, can not only provide irrigation water to more than 300,000 mu of farmland downstream (1 mu = 666 m^2), but can also supply water to households, as well as contribute to industrial and drinking water supplies in rural areas for Lushui County and some townships in the Baoshan Prefecture area. In a meeting between the Nu Prefecture government and the Ministry of Water Resources (MWR) in December 2016, the Secretary of the Nu Prefecture government proposed that comprehensive development of the Nu River water resource poses a fundamental solution to ecological protection and is the key to poverty alleviation in Nu Prefecture.

In summary, it seems that the rationality of the hydropower development pathway of the Nu River is moving towards that of water conservancy projects. Dams will likely shift their function from solely power generation to more comprehensive functions, including water supply, irrigation, flood control and drought resistance. So far, the local government and MWR are the main proponents of this pathway. There were initial indications that the Nu River Basin Comprehensive Plan may be approved in the first half of 2017, but as of the time of writing (January

2019) there has been no further details disclosed publicly. GW had applied to the Yangtze River Water Resources Commission for information disclosure in 2016, but received a response that the information involved state secrets so they were unable to provide access.

4.9 Conclusion

The analysis of pathways discussed above reveals that there are many possible alternatives for the Nu River. However, not every pathway has been given equal consideration.

There is no doubt that in the context of the national strategy of Western Development and WEPT, the hydropower development pathway has been given more consideration. Hydropower proponents have more power, money, resources, technology and knowledge, and include the NDRC and the NEA in positions of authority, local government, hydropower companies, and academics who use professional knowledge to promote hydropower development. Also important is that decision-making processes about the Nu River hydropower projects to date have proceeded without complete information disclosure and public participation, and have not fully taken account of the objections to the plans.

The civil society river protection pathway led by China's environmental NGOs was originally in a marginalized position, but the NGOs' ongoing actions introduced civil society voices to China's central leadership. The history of the Nu River protection since 2003 shows that civil society can participate in and promote more inclusive, informed and responsible decision-making processes. As a result of the tireless efforts of China's environmental NGOs, the Nu River avoided the fate of other dammed rivers, and its natural heritage and ecosystems remain largely intact.

Over more than 10 years of disagreement between the proponents of big hydropower development versus those seeking natural heritage protection, the local government consistently waited for decisions to be handed down by the central government, instead of pursuing other development pathways. Economic development and living standards in the Nu River basin are behind those of other regions in Yunnan, but the ecological environment has been maintained. When central authorities proposed the creation of an Ecological Civilization in China, the Nu Prefecture government only then made a choice to work on poverty alleviation and approved the national park's creation as a development pathway.

In recent years, China's leaders have promoted new development concepts, including innovation, coordination, and green, open and shared development. The national park pathway for the Nu River has the potential to provide more opportunities for local people and nature to coexist sustainably. However, the future of the Nu River remains uncertain, as the local government and the energy administration of the central government have not completely abandoned the Nu River hydropower plans, and there remains much resistance to public participation and information disclosure which are necessary for sustainable development. One of the biggest indicators of this resistance to public participation is the fact that the Nu River basin comprehensive plan has been kept confidential. Therefore, most importantly, decision-making about which development pathway is chosen for the future for the Nu River, should be inclusive, informed and accountable to ensure that the rights and entitlements of ethnic communities living along the river are recognized.

References

Bizhong, C. (2004, July 17). In addition to hydropower, is there any other choice for the Nu people out of poverty? *21st Century Economic Report*. Retrieved from: http://www.china.com.cn/chinese/difang/612599.htm.

Chaofei, Y. (2003). Examining the ecological mire of dam construction: a new decision-making framework. *Water Resources and Electric Power. 4*, 38–42.

Dore, J., Yu, X., & Li, K. (2007). China's energy reforms and hydropower expansion in Yunnan. In L. Lebel, J. Dore, R. Danial & Y.S. Koma (Eds.), *Democratizing water governance in the Mekong region* (pp. 55–92). Chiang Mai, Thailand: Mekong Press.

Guangrong, Q. (2004). Opening ceremony of the seminar on hydropower resources exploitation and environmental protection of Nu River, Lancang River and Jinsha River. In F. Jiangkun & H. Yaohua (Eds.), *Nu River, Lancang River and Jinsha River: Researches on the exploitation of hydropower resources and the protection of environment*. China: Social Sciences Academic Press.

Han, H. (2013). China's policymaking in transition: a hydropower development case. *Journal of Environment & Development*, *22*(3), 313–336.

Hao, F. (2016, April 7). China puts an emergency stop on coal power construction. *The Diplomat*. Retrieved from: http://thediplomat.com/2016/04/china-puts-an-emergency-stop-on-coal-power-construction/.

Hao, W. (2000, November 20). Accelerating the development of pillar industry in Yunnan. *Xinhua Net*. Retrieved from: http://www.people.com.cn/GB/channel3/26/20001120/319639.html.

Hongwei, Y. (2004, June 2). Dam or not? Indigenous people from the Nu River visited Manwan. *South Reviews*. Retrieved from: http://finance.sina.com.cn/roll/20040602/1121791662.shtml.

Hui, L. (2015, May 19). Hydropower installed capacity of China accounted for one quarter of the world. *Xinhua Net*. Retrieved from: http://news.xinhuanet.com/finance/2015-05/19/c_1115339094.htm.

Jiangkun, F., & Yaohua, H. (Eds). (2004) *Nu, Lancang and Jinsha Rivers: Research on the exploitation of hydropower resources and the protection of the environment*. China: Social Sciences Academic Press.

Jufen, Y. (2016, June 7). The Nu prefecture government held national park construction meeting. Retrieved from the Forestry Department of Yunnan Province website: http://www.ynly.gov.cn/yunnanwz/pub/cms/2/8407/8415/8477/107222.html.

Kejia, Z. (2006, June 18). World Heritage 'three parallel rivers' is facing threats. *China Youth Daily*. Retrieved from: http://zqb.cyol.com/content/2006-07/18/content_1449846.htm.

Kibler, K.M., & Tullos, D.D. (2013). Cumulative biophysical impact of small and large hydropower development in Nu River, China. *Water Resources Research, 49*(6), 3104–3118.

Leach, M., Scoones. I., & Stirling, A. (2010). *Dynamic sustainabilities: Technology, environment, social justice*. Abingdon, United Kingdom: Earthscan.

Liangzhong, X., & Ying, X. (2005). The rational appeal of Jinsha River farmers. *China Society Periodical, 15*, 16–19.

Lin, J., Liu, X., & Kahrl, F. (2016) *Excess capacity in China's power systems: A regional analysis.* Commissioned by the Energy Foundation through the U.S. Department of Energy. Berkeley, CA: Ernest Orlando Lawrence Berkeley National Laboratory.

Magee, D. (2006). Powershed politics: Yunnan hydropower under Great Western Development. *China Quarterly, 185*, 23–41.

Magee, D., & Henning, T. (2017, April 27). Hydropower boom in China and along Asia's rivers outpaces electricity demand. *China Dialogue*. Retrieved from: https://www.chinadialogue.net/article/show/single/en/9760-Hydropower-boom-in-China-and-along-Asia-s-rivers-outpaces-electricity-demand.

Meng, Z. (2016, September 19). The Nu River Grand Canyon National Park will not include the grand canyon. *The Paper News*. Retrieved from: http://www.thepaper.cn/newsDetail_forward_1530425.

Ngan, H.W. (2010). Electricity regulation and electricity market reforms in China. *Energy Policy, 38*(5), 2142–2148.

Nu Prefecture Forestry Bureau. (2015). Brief introduction of the Nu prefecture forestry.

Ptak, T. (2014). Dams and development: Understanding hydropower in far Western Yunnan Province, China. *Focus on Geography, 57*(2), 43–53.

Ptak, T., & Hommel, D. (2016). The trans-political nature of Southwest China's energy conduit, Yunnan Province. *Geopolitics, 21*(3), 556–578.

Peiyan, Z. (2010). WEPT: to create a new pattern of China's power development. *CPC History Studies, 3*, 5–13.

Ronghua, W. (2012, December 3). Interpretation of ecological civilization. *Wenhui Daily*. Retrieved from: http://theory.people.com.cn/n/2012/1203/c40531-19774896-1.html.

RPTCC. (2015). *Summary of Proceedings: 19th Meeting of the GMS Regional Power Trade Coordination Committee (RPTCC-19).* Bangkok, Thailand: Asian Development Bank.

Solomon, D. (2016, June 16). China's impressive clean energy progress confronted by emerging challenges. *East by Southeast*. Retrieved from: http://www.eastbysoutheast.com/chinas-impressive-clean-energy-progress-confronted-by-emerging-challenges/.

Tingzhen, L. (2010). Investment attraction but the local government was captured by central enterprises. *Economy & Nation Weekly, 25*, 32–33.

UNESCO. (2006). *Convention concerning the protection of the world cultural and natural heritage. Thirtieth session. Vilnius, Lithuania 8–16 July 2006*. Retrieved from: https://whc.unesco.org/archive/2006/whc06-30com-7b.addE.pdf.

Wang, Q., & Chen, X. (2012). China's electricity market-oriented reform: From an absolute to a relative monopoly. *Energy Policy, 51*, 143–148.

Wang, Y. (2007). A yellow card for the Three Parallel Rivers. China Dialogue. Retrieved from: https://www.chinadialogue.net/article/show/single/en/1200-A-yellow-card-for-the-Three-Parallel-Rivers.

WCD. (2000). *Dams and development: A new framework for decision-making.* The Report of the World Commission on Dams. London, United Kingdom: Earthscan.

WLE-Mekong. (2018). Greater Mekong Dams Observatory. Retrieved from: https://wle-mekong.cgiar.org/changes/our-research/greater-mekong-dams-observatory/.

Xiao, F. (n.d.). China's energy demands do not need such a large-scale hydropower development. Green Earth Volunteers. Retrieved from: http://www.chinagev.org/index.php/greenpro/environmentview/5392-zhonggbuxuyaorucidaguimoshuidiankaifa.

Xiaobing, W. (2003, February 20). The whole story of power reform program. *Finance & Economy.*

Xiaogang, Y. (2016). *What is the future of hydropower governance?* REEI energy system transformation workshop.

Xiaolan, Y. (2007, November 6). Zhang Boting: preacher of hydropower science popularization in China. *Yunnan Electric Power News*. Retrieved from: http://www.hydropower.org.cn/showNewsDetail.asp?nsId=112.

Xiaohui, L., & Baiying, X. (2010, August 2). Re-emergence of enclosure boom in hydropower market. *China Securities News*.

Yaohua, H., & Jiankun, F. (2004). The scientific outlook on development and the hydroelectric exploitation of Nu Jiang. In F. Jiangkun & H. Yaohua (Eds.), *Nu River, Lancang River and Jinsha River: Researches on the exploitation of hydropower resources and the protection of environment*. China: Social Sciences Academic Press.

Yaqin, J. (2016, February 23). Hydropower overcapacity in Yunnan is expected to alleviate, the key is the national market and the international market. *China Energy News*.

Yihua, M., & Yaqi, Z. (2007, February 25). Yunnan province will use 10 years to build a 'national park' system. *Spring City Evening News*. Retrieved from: http://news.sohu.com/20070225/n248338027.shtml.

Yikun, H. (2004, April 11). The Nu River hydropower planning need to do re-EIA. *The Economic Observer*. Retrieved from: http://finance.sina.com.cn/b/20040411/1446712692.shtml.

Yonglin, T. (2006, March 20). 1996-2000: Macro-economic control and soft landing during 9th five-year period. *China Youth Daily*. Retrieved from: http://zqb.cyol.com/content/2006-03/20/content_1338340.htm.

Yu, X., Chen, X., Middleton, C., & Lo, N. (2018, March). *Charting new pathways towards inclusive and sustainable development of the Nu River Valley*. Kunming and Bangkok: Green Watershed and Center for Social Development Studies, Faculty of Political Science, Chulalongkorn University.

Chapter 5
Rites, Rights, and Water Justice in Karen State: A Case Study of Community-Based Water Governance and the Hatgyi Dam

Saw John Bright

5.1 Introduction

The issue of justice lies at the heart of debates over water governance in Myanmar. Nowhere is this more apparent than the contested plans for five large hydropower dams in politically contested areas on the Salween River's mainstream, including the Hatgyi Dam in Hpa An District, Karen State (Salween Watch 2016). Initial preparations for the Hatgyi Dam have already been linked to violent armed conflicts and human rights violations, and further harm to livelihoods and cultural values of communities could ensue if construction proceeds (Magee/Kelley 2009; KRW 2016; Middleton et al., Chap. 3, this volume). The dam site is located in an area of "mixed administration" where the Union Government, the Karen State Government, the Karen National Union (KNU), and the Democratic Karen Benevolent Army (DKBA; before 2010 known as the Democratic Karen Buddhist Army) compete for political authority, legitimacy, and territorial control (South 2018). For over 60 years, the Karen National Liberation Army (KNLA) and its political wing the KNU have been in conflict with Myanmar's Union Government and the Myanmar armed forces (the *Tatmadaw*) demanding for the right to self-determination and to govern its own natural resources (Jolliffe 2016). The dam site itself is partly control by the DKBA, which is an ethnic armed organization (EAO) that splintered from the KNLA in 1994 and that was first allied with the *Tatmadaw*, but following the 2010 elections became independent from them.

Eighty kilometers downstream from the Hatgyi Dam site on the Salween River is Kaw Ku Island and the nearby Daw Lar Lake that connects to the Salween River mainstream (KESAN 2018). Five villages are located around Daw Lar Lake with a combined population of more than 8,000 people. In this area, the State-level Karen State Government has a stronger presence compared to the Hatgyi Dam site

Saw John Bright, Water Governance Program Coordinator, Karen Environmental and Social Action Network (KESAN); Email: hserdohtoo@gmail.com.

C. Middleton and V. Lamb (eds.), *Knowing the Salween River: Resource Politics of a Contested Transboundary River*, The Anthropocene: Politik—Economics—Society—Science 27, https://doi.org/10.1007/978-3-319-77440-4_5

upstream, and claim formal authority over resource management. At the same time, the government's authority is a relatively weak presence in the area on a day-to-day basis, and the five villages have governed the Daw Lar Lake and seasonally-flooded Kaw Ku Island through evolving local customary arrangements as a form of local authority over water resources. More recently, with support from the NGO Karen Environment and Social Action Network (KESAN), a project has sought to consolidate Community-Based Water Governance (CBWG) in a quest to gain recognition from the Karen State Government on the right to govern these resources principally through a local committee formed of the community themselves (see also, Götz, Chap. 6, this volume).

Researchers of water issues in Southeast Asia have begun to address the concept of justice in water governance, arguing that there is a need to identify the winners and losers in decision-making towards water (Lazarus et al. 2011; Middleton/Pritchard 2016). The purpose of this chapter is to examine the competing claims for justice in water governance on Kaw Ku Island and around Daw Lar Lake and work undertaken to establish CBWG, in the context of plans for the Hatgyi Dam located upstream. Conceptually, I draw on the work of Badenoch/Leepreecha (2011) who studied watershed governance in Northern Thailand through the lens of 'Rights' and 'Rites' to analyze claims to legitimacy and the power relations. Rites-based approaches are understood as locally-defined natural resource management arrangements that center upon cultural norms and local knowledge. Meanwhile, Rights-based approaches are formalized and legalistic approaches to resource governance that are normally recognized by the state and codified in laws and policies (see also Dore 2014). There is a complex relationship between the central government and other actors in the Hatgyi Dam area and downstream at Kaw Ku Island and the Daw Lar Lake. The concepts of Rights and Rites enables me to explore the formal and informal means by which a range of actors (central-level and state-level government; KNU, DKBA, and communities) interact around the issue of water governance, the relative influence of each actor and power relations between them, and ultimately the issue of justice in water governance.

The interviews for this research were principally undertaken between November 2015 and July 2016. I conducted in-depth interviews, informal dialogues and participant observation at various types of meetings (policy dialogues; conferences; workshops, etc) at the national level, state level; and at the village/community level. My interviews with those who are working on water governance issues in Myanmar included representatives of State Government (n = 2), the National Water Resource Committee (n = 1), and research institutes (n = 2). I held three focus group discussions in Mikiyan Village with representatives from all five villages around Daw La Lake, and five in depth interviews with community leaders. Many of the meetings and dialogues that I have joined both during the interview period and more recently related to my work at the time with KESAN as Water Governance Program Lead.

In this chapter, I explore how 'Rights' and 'Rites' have long shaped the interaction between state and community in resource governance in Karen State. I argue that to ensure inclusive decision-making processes and therefore social justice in water governance in Myanmar, formal policy and institutional arrangements must reflect Rights but also accommodate Rites. The chapter is structured as follows. In the next section, I analyze CBWG at Daw La Lake and Kaw Ku Island through a Rites and Rights lens. I then discuss the decision-making process to date for the Haygyi Dam and the role of the civil society movement challenging the project. In the subsequent discussion, I synthesize these two aspects of water governance on the Salween River in Karen State to discuss water justice from the perspective of Rights and Rites.

5.2 Community-Based Water Governance Arrangements in Karen State

To better manage local resources, and in response to large-scale development projects such as the proposed Hatgyi Dam upstream, the Karen local community living along the Salween River near Daw La Lake and Kaw Ku Island have undertaken a local initiative known as CBWG. CBWG is intended to build community members' capacity to lead and implement their own vision of governance and management of natural resources. Within the CBWG process and collaborating with the NGO KESAN, community members are center to decision-making. Throughout the learning process, traditional resource use practices are explored as a basis to develop a community resource management plan. The intention is that the process of CBWG will ensure recognition and support from the Karen State Government for the community's approach. Both Rights and Rites are reflected in the CBWG model and are the basis of the community claiming legitimacy for local authority over the Salween River's governance in their area.

In the next section, I detail the existing local knowledge and practices towards managing water resources in Daw La Lake and on the Kaw Ku Island. I then show how the CBWG arrangements have been undertaken through a community learning process, leading to the establishment of a local governance institution.

5.2.1 Community's Rites at Daw La Lake and Kaw Ku Island

The ecosystems of the Salween River, Daw Lar Lake and Kaw Ku Island are intimately connected (KESAN 2018). Daw La Lake is approximately 26 square kilometers when fully flooded and is located on the western bank of the Salween River, 12 km upstream from Hpa An town. It is connected to the Salween River via a large natural stream. Kaw Ku Island is approximately 1.2 km^2 when fully exposed, and is located within the Salween River's mainstream, nearby to the lake. The seasonal fluctuations of the Salween River, to which people have long adapted, shape the ecosystems and livelihoods around Daw Lar Lake and on Kaw Ku Island.

When the river rises and floods in the rainy season (July–September), Daw Lar Lake expands, while Kaw Ku Island is submerged beneath the water. This period is good for fishing. As the Salween River drops and the flood water recedes in the summer (October–December), Daw Lar Lake diminishes in size and Kaw Ku Island appears again. During this period, cash crops such as mung bean and peanuts, are planted on Kaw Ku Island (Ko Thout Kyar Interview, 11 November 2015) (Table 5.1: Seasonal farming calendar for Kaw Ku Island and Daw La Lake). With the complex land/waterscape that annually transitions between being flooded and dry, the villages in this area have locally created systems (i.e. non-written rules) for governing and using the productive, seasonally-changing natural resource base.

The seasonally-flooded Kaw Ku Island is shared by three villages located on both banks of the Salween River, namely: Mizan Village, Motekadi Village and Kaw Ku Village. When the Salween River's floodwaters recede at the beginning of October highly fertile soil becomes exposed that is well-suited to seasonal plantings. During this time, between 70 and 80 families farm on the island, and each family earns approximately US$3,000 per season (U Myint Sein Interview, 25 January 2016). This income is generated over three to four months when the island is used for growing seasonal crops. As a seasonally flooded island, from the perspective of formal law its land ownership is ambiguous. Some parcels of land on the island have a land registration certificate issued by the Land Department of the General Administration Department (GAD) under the Ministry of Interior. However, the majority of land use on the island is managed through informal means. The heads of the villages usually lead management for sharing the use of the island including dispute resolution, although they also have to consult with the monks, village elders, and informal community leaders including women representatives to facilitate decision-making (Ko Nyan Win Interview, 12 November 2015). However, as the farmers here are not sure whether the government land use policy on the island will change, they are working with the CBWG project towards defining more clearly the village-led management for the island to reduce the perceived risk of this external threat.

Regarding Daw La Lake, there are five villages located around it, namely: Mikayin Village, Motekadi Village, Kangyi Village, Kankalay Village, and Kedauk Village. During the rainy season, when the Salween River level rises, Daw La Lake floods and fish migrate from the river to breed and grow in the lake (Ko Kyaw Hla Interview, 12 November 2015). At this time, the main occupation of the community members is fishing. However, fishing is conducted all year round and the largest fish are actually caught when the water becomes low in the lake during March and April. The flood also nourishes the soil, which is important for farming once the water recedes (Daw Mya Nyut Interview, 4 February 2016). When the water recedes, the exposed land is used for seasonal crops and vegetables. The farmers main cash crop in these villages is rice. The farmers also plant vegetables together with beans. After harvesting the agricultural crops, the land is used for grazing buffalos and cows (Elders Group Interview, 26 February 2016).

Around Daw La Lake, the community members have created informal rules for lake and land use. Rather than use official maps to define resource use boundaries,

Table 5.1 Seasonal farming calendar for Kaw Ku Island and Daw La Lake

Month (Burmese and Karen names in brackets)	Farming practice			
	"Muyin Rice" (mixed land and water) grown both around Daw La Lake and on Kaw Ku Island	"Farm Rice" (on land) grown around Daw La Lake	"Patat Rice" (in water) grown around Daw La Lake	Beans and other vegetables grown mostly on Kaw Ku Island
January (Burmese: Pyatho, Karen: Lar Plu)	Prepared and planted			
February (Burmese: Taboetwal, Karen: Tha Lae)				
March (Burmese: Tabaung, Karen: Htay Ku)				Harvested
April (Burmese: Tagu, Karen: Thway Kaw)	Harvested			
May (Burmese: Kason, Karen: Day Nyar)				
June (Burmese: Nayon, Karen: Lar Ku)		Prepared and planted		
July (Burmese: Waso, Karen: Lar Nwei)				
August (Burmese: Wakhaung, Karcn: Lar Khoe)				
September (Burmese: Tawthalin, Karen: Lar Khut)			Prepared and planted	
October (Burmese: Thadingyut, Karen: Sie Mu)		Harvested		Prepared and planted
November (Burmese: Tazaungmone, Karen; Lar Nor)			Harvested	
December (Burmese: Nadaw, Karen: Lar Plue)				

Source Interviews and fieldwork conducted by the author
Note Burmese/Karen calendar months should be referred to as the primary timing for this seasonal calendar, as the Burmese/Karen calendar months do not always correspondent precisely with the days of the months in the Roman Calendar

Table 5.2 Community learning process towards CBWG

	Step	Activities
1	Laying the ground work	Mobilize community members and build up capacity to facilitate the CBWG process. Form an interim committee to assist the learning process until the management plan is established
2	Documentation of natural resources	Map out resource use, seasonal calendar and livelihood cycle. Investigate resource boundaries and facilitate a negotiation process between various users of resources
3	Strengthening community governance for natural resource management	Develop a management plan with community regulations and institutions to enforce it
4	Preparing community action plan for livelihood support	Integrate livelihood support into management plan and ensure equitable regulations
5	Monitoring and evaluation	Create community platform for resource management and advocacy, and a permanent village committee. Ensure the sustainability of resource management plan

Source Interviews and fieldwork conducted by the author

informally agreed boundaries have been agreed between the villages using landmarks such as big trees, mountains and small streams, and there are no territorial disputes between the five villages (Ko Nyan Win Interview, 12 November 2015). Regarding disputes over fishing practices, informal dispute handling mechanisms between the five villages exist that have been further developed in the CBWG process (see next section and Table 5.2: Community learning process towards CBWG). This has led to practices and rules to avoid overfishing and threats to fish species, including restrictions on the types of equipment used for fishing. Small-mesh fishing nets and electric shock devices, for example, are not permitted. If someone is found using them, the first time they are given a warning, and if there is repeated use, the equipment is confiscated (but only if the equipment has a low monetary value) (Youth Groups Interview, 26 February 2016). It is also not permitted to catch fish during the hatching period because it can harm the breeding of the fish species; for example, in Mikayin Village traditionally fishing is not permitted during the flooded period from Waso (July) to Tawthalin (September). Meanwhile, the month of Nadaw (December) is a religious period. At this time, there are meditation retreats and no one can go fishing as it is not permitted to kill during this period according to Buddhist teaching (Elders Group Interview, 26 February 2016).

Until now, similar to land management at Kaw Ku Island, representatives of the five villages meet together to make decisions on setting rules and regulations and disseminating them based on a general consensus. However, as there is at present no government recognized legal protection to acknowledge their traditional ownership and management of these natural resources, they are concerned about their long-term entitlement to access the lake.

5.2.2 A Rights Approach to the Daw La Lake

Daw La Lake is crucial to the food and water security of villagers living around it. Yet, according to community leaders interviewed, the Karen State Government does not act in a way that acknowledges and respects this. In 2013, for example, during the Union Solidarity and Development Party (USDP) government period (2011–2016), the Karen State Government Fisheries Department sought to grant Daw Lar Lake as a concession to a private company for commercial fishing (KESAN 2018). The communities surrounding the lake were not consulted on this decision, and once they learned of the plan they resisted the plan. They organized a public signature campaign and submitted a letter to the Karen State Chief Minister claiming that the lake had been owned by their communities since their ancestors and is a key source of livelihood for fishing and farming.

Realizing the potential threat from such initiatives of the Karen State Government, community leaders and members settled on a plan to establishment a CBWG model in Daw Lar Lake that would enable them to advocate for legal recognition of the village's claim to manage the lake and thus reduce the likelihood that the Lake would be sold as a concession to private companies. Table 5.2: Community learning process towards CBWG details the five main steps in the community learning process for the CBWG model. Through this process, local institutional arrangements have emerged that reflect the traditional practices of resource governance and are a claim to the community's authority over the management and ownership of Daw Lar Lake.

The local histories in these Karen areas are complex and have been in one way or another impacted by armed conflict for over half a century (South 2018). In laying the groundwork for CBWG, confidence and trust had to be carefully built between the communities and the NGO KESAN that partnered with the communities to guide the CBWG process. A series of events towards a cement factory that was proposed nearby to Mikayin Village in 2013 coincidently led to deepened understanding and trust between the community and KESAN. The cement factory would have had a significant negative impact on Mikayin Village's culture and livelihood. Community members from the village networked with other villages nearby and cooperated with KESAN to stop the cement factory's construction. This was an experience that led to the realization amongst the community members that they could influence policy if they mobilized themselves together. Following this, community members from Mikayin Village proposed to work with KESAN to develop a conservation plan for Daw La Lake with the ambition to gain legal recognition from the Karen State Government.

Subsequently, staff from KESAN worked with local community leaders to facilitate discussions and meetings in the five villages around the lake to establish village committees in late 2015. However, it was not an easy task as these communities have long been living under the threat of conflict and an authoritarian regime that had influenced their confidence to establish community-led committees. Various community members were concerned that they would be perceived by the government as breaking the law if they participated in the created village committee, as it is not a registered organization. However, a core group of community leaders were confident in their knowledge of how to utilize political space and seek legal opportunities in the context of the new semi-civilian government. They therefore established an interim committee formed of members from the five villages in early 2016.

The purpose of this interim committee is to facilitate the CBWG process that is working towards the creation of a lake conservation and management plan (KESAN 2018). In the longer term, a permanent management committee will be established. Members of the interim committee were selected democratically by representatives of the existing village committees, which themselves are democratically elected by the community members. Gender equality was also carefully accounted for. The interim committee facilitates dialogue between the different users of the lake that include both fishers and farmers across the five villages. Their role includes to help resolve disputes related to resource management; for example, in some areas of the lake, fishers want to retain flood water for fishing, while farmers want the water to rapidly recede to grow rice and other crops.

Traditionally, in Karen communities, men and women share responsibilities for livelihood activities such as fishing and farming. For example, in some families women go fishing, while in other families women make fishing nets and baskets while men fish. However, overall men have a greater say in resource management decisions in the community. Women are keen to – and do – take part in some village meetings, but do not normally hold administrative duties in the village. There are exceptions, however. For example, in Mikayin Village there are four women "ten-household" leaders who take responsibility for keeping minutes of village meetings related to the management of the lake. Thus, despite strong prevailing cultural norms, there is some flexibility for women to take leadership roles (Women's Group Interview, 26 February 2016).

The CBWG initiative around Daw La Lake is a work in progress. With the interim committee now established, community members and KESAN are engaged in activities to document natural and man-made resources and their use around the Daw Lar Lake, and to agree upon boundaries and associated resource uses within them. To this end, the interim committee selected community learning facilitators (CLFs) from the five villages to assist them. Most are youths aged between 25 and 35, who were selected with the intention to encourage the younger generation to actively engage in decision-making processes around the management and governance of natural resources and environmental conservation in their community. The CLFs have worked closely with KESAN to document activities on the use of natural resources and facilitate the preparation of a resource management plan with livelihood support. They also conduct village meetings and an annual forum, facilitate monitoring and evaluation activities, and mobilize and encourage the community members.

While the CBWG initiative has made important progress towards rendering visible the communities' claims to govern local natural resources in support of their livelihoods, there are a number of threats. They range from the proposal of the

Karen State Government mentioned above to auction the Daw Lar Lake as a private fishing concession, to other threats such as plans for rock quarries and expanding rubber plantations, as well as the proposed Hatgyi Dam upstream that I discuss in more detail in the next section. These threats reveal the centralized nature of decision-making that exclude communities' Rites – and often Rights – in the process, and that have motivated the community to seek legal recognition of their local CBWG institution and livelihood systems.

5.3 Hatgyi Dam: Centralization of Decision-Making and Civil Society Response

Plans for large dams on the Salween River first emerged in 1979 when the Electricity Generating Authority of Thailand (EGAT) initiated a series of feasibility studies for water diversions and later hydropower dams on tributaries and the mainstream (see also, Middleton et al., Chap. 3, this volume). Initial studies gained political momentum when a Memorandum of Understanding was signed between the Thailand and Myanmar (military junta) governments in July 1996 for Thailand to purchase 1,500 MW of hydroelectricity from Myanmar by 2010 (TERRA 2006). The initial design of the Hatgyi Dam, prepared in 1999, proposed a 300 MW "run-of-river" dam. However, a subsequent study published in November 2005 redesigned and significantly enlarged the project to 1,200 MW with a 33-m-high dam. In December 2005, EGAT signed a Memorandum of Agreement with the Myanmar military junta government's Department of Hydroelectric Power to proceed with the Hatgyi Dam (TERRA 2006). The entire process was shrouded in secrecy (Magee/Kelley 2009), with no consideration of local communities concerns in Myanmar neither from a Rights nor Rites perspective.

Military activities have occurred around the Hatgyi dam site creating fear and insecurity for communities living in the area (see also TERRA 2006, 2014; Magee/Kelley 2009). In June 2009, for example, a major military attack by the *Tatmadaw* and the at-the-time allied DKBA occurred on the KNU, in an area only 17 km away from the proposed Hatgyi Dam site. The intention of the attack was to secure the wider area around the dam site, which the KNU controlled. As a result, 3,500 villagers were forced to flee across the Thai border (KHRG 2009). Three years later, on 13 January 2012, the KNU and the *Tatmadaw* signed an initial ceasefire agreement that opened the door to political dialogues on peace and a federal settlement between the Myanmar Union Government and the KNU. However, even despite this agreement, contradictory activities occurred on-the-ground, as the number of Myanmar military troops deployed into Karen areas increased (Wade 2012) and military attacks have continued until present with major incidents in 2014, 2015 and 2016 (KRW 2014; KHRG/KRW 2018).

Hence, during the period of conflict, and during the ceasefire, the Myanmar Union Government has tried to advance plans for dams on the Salween River (Salween Watch 2016). Civil society groups, in response, have sought to challenge these plans, raising a range of concerns related to the projects' environmental and social impacts, human rights violations, and that the projects are being pushed forward even as conflict in the areas continue and obstruct the peace negotiations (KPSN 2015). For example, in 2014, 80,000 local people from Shan, Karenni, Karen and Mon States and more than 130 civil society organizations and political parties in Myanmar signed a petition urging an immediate halt to the six dams planned by the Union Government on the Salween River. Furthermore, a network of community leaders and community-based organizations have organized under the name "Save the Salween Network" who have regularly issued statements stating that the Salween River is the lifeblood of local people, providing livelihood opportunities and links with cultural identity (e.g. SSN 2016).

Eighty kilometers downstream of the dam site, in the villages at Daw La Lake and Kaw Ku Island, community leaders have expressed their serious concern about the government's plans to build the Hatgyi Dam. One community leader from Kaw Ku Village said:

> If the dam is built, the natural balance will be completely destroyed. The river's flows will be blocked by the dam, and there will be no seasonal lake or island, and no chance for fishing and farming. (Ko Thout Kyar Interview, 11 November 2015)

These communities have received no official information from the government, but just hear about the project from the media and information received from civil society organizations, including KESAN, Karen Rivers Watch, Save the Salween Network and Burma Rivers Network. These organizations have conducted research and undertake advocacy. Community members receive information both directly or indirectly from CSOs publications and media coverage talking about Salween dams. Some also participate in public events and meetings organized by CSOs to mark the international day of action for rivers on March 14 annually. They are concerned who will take the responsibility for their lives if the dam is built.

The Myanmar Union Government does not have clear policies or legal mechanisms to address conflict-sensitive water disputes such as on the Salween River neither domestically nor at the transboundary-level. The Union-level National Water Resources Committee (NWRC), created in 2013, is chaired by the Second Vice President, and co-chaired by the Union Minister of the Ministry of Transport, with representatives from different government ministries. Reflecting the political character of the committee, the NWRC as formed under the USDP of President U Thein Sein was reformed in June 2016 under the subsequent National League for Democracy (NLD) government with a revised membership (Phyu 2016). The NWRC is seeking to define a central role for itself in policies related to the water sector in Myanmar. To date, the committee has developed a National Water Policy, published in March 2014, and at the time of writing is working on a Water Law. However, the details of the process to develop the Water Law are not publicly available (U Cho Cho Interview, 25 July 2016).

Despite the absence of a comprehensive Water Law, there have been many investments related to water resources in Myanmar, including in hydropower projects. The most relevant water-related legislation at present is the Conservation of Water Resources and Rivers Law (2006), which addresses three main themes: navigation on rivers; water infrastructure, in particular ports and constructions alongside rivers; and water pollution. The scope of this law, which excludes hydropower and irrigation infrastructure, mandates the Ministry of Transportation and Communications, which also explains the scope of the law. Another related legislation regards Environmental Impact Assessment (EIA). The EIA Procedure – approved by the cabinet in November 2015 – was drafted with support from the Asian Development Bank, and although it was said to be based on guidelines from the International Finance Corporation it fails to cover complaint mechanisms and only poorly recognizes community rights as it does not, for instance, mention "Free, Prior and Inform Consent," both of which are important community rights for ethnic nationalities.

Which government department leads the decision-making over the Salween dam projects is also ill-defined. During the USDP administration (2010–2015), the Ministry of National Planning and Economic Development and the Ministry of Electric Power were the key ministries responsible for hydropower construction in Myanmar (Doran et al. 2014). Subsequently, under the NLD government (2016-present), following a consolidation of various ministries, the new Ministry of Electric Power and Energy takes a key role in hydropower policymaking (Parliamentarian Interview, 29 July 2016). The Ministry of Natural Resources and Environmental Conservation (MONREC), formed in March 2016, also takes a role in decision-making to enforce environmental regulations. However, there remains ambiguity on the precise role of each, together with that of other relevant ministries. Until now, decisions that have sought to push forward the Hatgyi Dam have been taken at the Union-level, with little role for the State-level Government.

Civil society groups concerned about plans for hydropower on the Salween River had hoped that the new NLD government would revise the government's position on the dams, as its 2015 Election Manifesto states:

> The construction of the large dams required for the production of hydropower causes major environmental harm. For this reason, we will generate electricity from existing hydropower projects, and repair and maintain the existing dams to enable greater efficiency.

However, in practice, since coming to power, the NLD government has not clearly removed the Salween Dams from the electricity-sector agenda. According to Myanmar's 2008 Constitution Article 445, it is stated that any new government will have to implement the projects agreed by the previous government. Therefore, the NLD government has had to navigate various legal issues in decision-making over the Salween River dams, even though the agreements of the previous governments were made without information disclosure or consultation with the public.

5.4 Federalism, the Peace Negotiation Process and the Hatgyi Dam

Underlying the issue of large dam decision-making in the Salween River basin are the issues of decentralization, federalism and the peace negotiation process, in particular with regard to resource governance and revenue sharing (see also Middleton et al., Chap. 3, this volume). The Karen State Chief Minister, in an interview stated that:

> We are aware of the negative impacts of the Hatgyi Dam on the environment and people, but we are not sure exactly how to deal with the Hatgyi Dam as the decision was made by the Union Government of the previous Government. (Daw Nan Khin Htway Myint Interview, 19 June 2016)

A Member of Parliament from the NLD at the Union-level, elected to the Lower House and who is Chairperson of the Lower House Committee for Natural Resources Affairs, also stated that State Governments should have decision-making powers for development projects in their areas (U Soe Thura Htun Interview, 29 July 2016). Reflecting this perspective, the Myanmar Times newspaper reported that the Shan State government also asserted that it planned to stop all hydropower projects in Shan State (Htwe 2016). However, the controversial 2008 constitution states under Schedule Two on the authority delegated to the State-level government that:

> 4. (a) Medium and small scale electric power production and distribution that have the right to be managed by the Region or State not having any link with national power grid, except large scale electric power production and distribution having the right to be managed by the Union.

Yet, given the complexity and implications of large dams, the actual role in decision-making of the State-level Government still requires further clarification. Beyond this, the legal entitlements of communities who would be affected by such projects if they proceed are even more ambiguous.

The rush towards foreign investment and development projects in conflict-affected areas are resulting in human rights abuses in ethnic areas in Myanmar. This is leading to distrust and questions by some ethnic leaders and the wider ethnic people's community towards the political process, including on its inclusiveness and commitment to decentralizing decision-making powers under a federal system versus affirming the authority of the Central Government over ethnic areas (KPSN 2018). In response, ethnic civil society groups and political parties at the local, state and national-levels have been demanding customary rights in national laws and policies, including in the Water Law now reportedly under preparation but not yet made publicly available (BRN 2017). Many have also called for a moratorium on mega-development projects, such as the Hatgyi Dam, until the peace negotiations are fully concluded, which would establish Myanmar's new federal political structure and accordingly a new Constitution with new rules on decentralized resource governance (BEWG 2017). For example, the KNU Vice Chairperson, Padoh

Naw Seporah Sein, stated in a documentary interview in 2016 that the KNU has a clearly stated policy that:

> … a moratorium for the Hatgyi Dam project and other Salween dams should be made and no discussion on this before peace has settled down in the ethnic areas and resource governance and federal issues are sorted out. (Fawthrop 2016)

Meanwhile, communities living alongside the Salween River near the Hatgyi Dam have been almost entirely excluded from the decision-making process to date. Only once the political negotiations are complete for peace, and it is detailed within a new Federal Constitution, can the conditions really be in place for meaningful participation and deliberation amongst river-side ethnic communities.

5.5 Towards a Rites and Rights Approach for Justice in Water Governance

In the recent literature on environmental justice, attention is paid to distributional justice, procedural justice, and recognitional justice (Walker 2012). Distributive aspects of justice refer to the fair distribution of environmental harms, risks and benefits. Procedural justice considers the ways in which decisions are made, who is involved, and who has influence. Justice as recognition addresses who is and is not valued, and incorporates social and cultural (lack of) recognition. Furthermore, there are various formal and informal arenas across multiple scales within which decision-making processes and access to justice take place (Middleton/Pritchard 2016). In this concluding section, I link these notions of justice to the situation faced by the river-side communities seeking to defend their claims for their Rights and Rites in relation to access, use and control of natural resources in the Daw La Lake and the Kaw Ku Island, and in responding to the threat to their livelihoods posed by the plans for the Hatgyi Dam upstream.

At the local-scale, the Karen communities around Daw La Lake and on Kaw Ku Island have sought to claim their Rites to access resources based on their customary practices and situational knowledge. Procedurally, they have sought to develop a democratized inclusive grassroots approach to resource management, working with the NGO KESAN in the CBWG project. The opportunity to undertake the CBWG activities reflects the changing political arenas in Myanmar since 2010 when the semi-civilian government was elected. It is an example of how communities and civil society are working together to undertake action research as a basis for influencing decision making and promoting both Rites and Rights (see also, Lamb et al., Chap. 7, this volume). The claims for Rites and Rights by the communities around Daw La Lake and on Kaw Ku Island along the Salween River are the basis of claims for political authority for local resource management.

The subnational level is a second scale around which communities Rights and Rites are being contested, and at which justice is sought. Under the current 2008 Constitution, a key challenge for the Karen communities around Daw La Lake and on Kaw Ku Island is to render their claims for community ownership Rights to be legally recognized by the Karen Regional Government. This recognition is important given the potential threats to their claims from other competing plans, ranging from granting the Daw La Lake as a private fishing concession, to the threat to their local resources if the Hatgyi Dam were to be built upstream.

Cutting across the national, subnational and local scales, the ongoing peace negotiation between the Union government and ethnic armed organizations, including the KNU, is a key arena of justice that relates to restructuring the current political system in Myanmar towards a federal system. Within the peace negotiations, the issue of resource sharing is a key issue to be addressed, and must address all three dimensions of justice: distributional; procedural; and recognitional (BEWG 2017). The importance of the outcome of these peace negotiations cannot be understated in terms of the implications for riverside communities along the Salween River.

Finally, decisions largely claimed to be national by the Union Government regarding the proposed Hatgyi Dam have led to a centralized and opaque formal decision-making process that has excluded communities and civil society groups to date. Despite this, riverside communities and civil society groups have challenged the proposed Hatgyi Dam project on the basis that it threatens both their Human Rights and cultural identity (i.e. Rites-based practices). This is not to suggest that there have not been significant improvements in civil and political freedoms in Myanmar overall since the transition from a military to a semi-civilian government, yet in the case of the Hatgyi Dam decision-making remains lacking in accountability.

In conclusion, I propose that water governance in an ethnically diverse area, such as in the Salween River basin, will create more positive outcomes when both a Rights-based and Rites-based perspectives are integrated. Thus, formal state policies and institutions in Myanmar must become better informed by a Rights-based approach while also accommodating the Rites-based perspectives of communities via creating policy platforms that enable inclusive decision-making in water governance and therefore social justice on the Salween River.

References

Badenoch, N., & Leepreecha, P. (2011). Rights and rites: Local strategies to manage competition for water resources in Northern Thailand. In K. Lazarus, N. Badenoch, N. Dao & B. Resurreccion (Eds.), *Water rights and social justice in the Mekong Region*. London, United Kingdom: Earthscan.

BEWG (Burma Environment Working Group). (2017). *Resource federalism: a roadmap for decentralised governance of Burma's natural heritage.*

BRN (Burma Rivers Network). (2017, March 14). *Countrywide gatherings on International Rivers Day to oppose large dams in Burma's conflict zones.* Statement by Burma Rivers Network (BRN), Save the Salween Network (SSN) and the Burma Environment Working Group (BEWG).

Doran, D., Christensen, M., & Aye, T. (2014). Hydropower in Myanmar: Sector analysis and related legal reforms. *International Journal on Hydropower and Dams, 21*(3), 87–91.

Dore, J. (2014). An agenda for deliberative governance arenas in the Mekong. *Water Policy, 16*, 194–214.

Fawthrop, T. (2016). *Mega Dams on Thanlwin.* [Documentary film]. Democratic Voice of Burma and Karen Environmental and Social Action (Producer). Retrieved from: https://www.youtube.com/watch?v=AEt6bagvt6w.

Htwe, C.M. (2016, July 8). Hydropower dams, major development projects suspended in Shan State: minister. *Myanmar Times*. Retrieved from: https://www.mmtimes.com/national-news/21276-hydropower-dams-major-development-projects-suspended-in-shan-state-minister.html.

Jolliffe, K. (2016). *Ceasefires, governance and development: The Karen National Union in times of change*. The Asia Foundation.

KESAN (Karen Environmental and Social Action Network). (2018). *Community based water governance: A briefing report on Daw Lar Lake*. Yangon, Myanmar: KESAN. Forthcoming.

KHRG. (2009). Over 3,000 villagers flee to Thailand amidst ongoing SPDC/DKBA attacks. Retrieved from: http://khrg.org/2009/06/09-3-nb1/over-3000-villagers-flee-thailand-amidst-ongoing-spdcdkba-attacks.

KHRG (Karen Human Rights Group) & KRW (Karen Rivers Watch). (2018, June). *Development or destruction? The human rights impact of hydropower development in villages in Southeast Myanmar*. Retrieved from: http://khrg.org/2018/07/18-1-cmt1/development-or-destruction-human-rights-impacts-hydropower-development-villagers.

KPSN. (2015, July 10). *Statement of Karen Peace Support Network.*

KPSN (Karen Peace Support Network). (2018). *Burma's dead-end peace negotiation process: A case study of the land sector.*

KRW (Karen Rivers Watch). (2014). *Afraid to go home: Recent violent conflict and human rights abuses in Karen State.*

KRW (Karen Rivers Watch). (2016). *Karen State September 2016 Conflict: The real motivations behind renewed war.*

Lazarus, K., Badenoch, N., Dao, N., & Resurreccion, B. (Eds.). (2011). *Water rights and social justice in the Mekong Region*. London, United Kingdom: Earthscan.

Magee, D., & Kelley, S. (2009). Damming the Salween River. In F. Molle, T. Foran & M. Käkönen (Eds.), *Contested waterscapes in the Mekong Region: Hydropower, livelihoods and governance* (pp. 115–140). London, United Kingdom: Earthscan and USER (Chiang Mai University, Thailand).

Middleton, C., & Pritchard, A. (2016). Arenas of water justice on transboundary rivers: A case study of the Xayaburi Dam, Laos. In D. Blake & L. Robins (Eds.), *Water governance dynamics in the Mekong* (pp. 57–88). Petaling Jaya, Malaysia: Regional Strategic Information & Research Development Centre.

Phyu, A.S. (2016, July 11). Re-launched committee to manage water resources. *Myanmar Times*. Retrieved from: https://www.mmtimes.com/national-news/21276-hydropower-dams-major-development-projects-suspended-in-shan-state-minister.html.

Salween Watch. (2016). *Current Status of Dams on the Salween River - February 2016.* Retrieved from: https://www.internationalrivers.org/resources/11286.

South, A. (2018). 'Hybrid governance' and the politics of legitimacy in the Myanmar peace process. *Journal of Contemporary Asia, 48*:50–66.

SSN (Save the Salween Network). (2016). *Statement of Save the Salween Network in Yangon – February 2016.*

TERRA (Towards Ecological Recovery and Regional Alliance). (2006). *Chronology of Salween dam plans (Thailand and Burma)*. Retrieved from: http://www.terraper.org/web/sites/default/files/key-issues-content/1305712134_en.pdf.

TERRA (Towards Ecological Recovery and Regional Alliance). (2014). *Chronology of Hatgyi Dam From 1996-October 2014*. Retrieved from: http://www.terraper.org/web/sites/default/files/key-issues-content/1421399932_en.pdf.

Wade, F. (2012, February 15). Myanmar: Ceasefire does not mean peace. *Aljazeera Opinion*. Retrieved from: https://www.aljazeera.com/indepth/opinion/2012/02/201221211854783599.html.

Walker, G. (2012). *Environmental justice: Concepts, evidence and politics*. London and New York: Routledge.

Chapter 6
Contested Water Governance in Myanmar/Burma: Politics, the Peace Negotiations and the Production of Scale

Johanna M. Götz

6.1 Introduction

It's a clear morning in early 2017. The sun has only come up a few hours ago, however, the heat of the day is already tangible. Villagers and activists from all over Myanmar/Burma,[1] media representatives, Karen National Union (KNU) members and non-state armed forces peacefully gather on an island in the Salween/Thanlwin river in Karen State. They talk about the river's importance to the livelihood of communities and about its beauty. They express their rejection of the construction of mega-dams across it and discuss these dams' influences on conflict and peace building. Prayers and poems, discussions and talks, pictures and music try to capture the river in all its dimensions. A couple of days after the event, a letter arrives from the Karen State Government addressed to the KNU condemning this 'unauthorized' gathering. It criticizes the KNU, stating *inter alia* that they had infringed the agreements of the National Ceasefire Agreement (NCA), including Article 25(c) that states:

> [...] government and the individual [...] Ethnic Armed Organizations shall coordinate the implementation of tasks that are specific to the areas of the respective Ethnic Armed Organization. (NCA 2015)

These criticisms are connected with the request towards the KNU to follow the NCA by controlling their armed members. What can this short insight tell about

[1] I use the terms Burma/Myanmar, and further names changed under the 2008 constitution, interchangeably. This is far from implying that naming is irrelevant. However, throughout a myriad of conversations, I realized that either term has its own limits. *For now*, I therefore choose to follow the advice of a colleague stating: "I think we have to move beyond that debate – there are more important issues at the moment."

Johanna M. Götz, Ph.D. Candidate, Department of Geography, University of Bonn & Development Studies, University of Helsinki; Email: johanna.goetz@helsinki.fi.

C. Middleton and V. Lamb (eds.), *Knowing the Salween River: Resource Politics of a Contested Transboundary River*, The Anthropocene: Politik—Economics—Society—Science 27, https://doi.org/10.1007/978-3-319-77440-4_6

water governance in Myanmar? Throughout this chapter, I demonstrate how the Salween River (cf. Fig. 6.1a: Myanmar's nation-state borders and the Salween River Basin) represents much more than just material water flowing downhill, and how the river is deeply embedded into a complex history and current socio-political struggles. Simultaneously, I show how highly contested power-relations play out in claims for decision-making around various scales, and its significance to the reshaping of water governance. To approach this, two substantial foci arise as fundamental conceptual frames for this chapter: (1) an understanding of water as hydrosocial (Linton/Budds 2014); and (2) how within that understanding the production of scale is historically embedded, highly political and socially contested. Overall, I show how the scales of water governance in Myanmar are currently being contested and argue that this constitutes a key battleground for future decision-making not only in terms of water governance, but also with implications for the peace negotiations and the federal structure of government that it is working towards.

Burma is at a unique historic moment as a key struggle over the future rules of water governance, and the role and relative influence of actors involved, is unfolding. With clear rules and laws around water yet to be established, different narratives, imaginaries, material practices on the ground and institutional arrangements around water governance become apparent. This becomes evident looking at two groups of actors with rather different visions: The National Water Resources Committee (NWRC) and its Advisory Group (AG) acting under the Union government, and a group of non-state actors connected to the Salween Peace Park (SPP) within the KNU-controlled, ethnic-minority dominated Mutraw District in Karen State. While at the Union level a rather centralized, expert-led water policy-making process is seeking to generate a unified Myanmar narrative, actors around the SPP propose a contrasting arena of community-led, bottom-up federalism.

This chapter builds on an explorative approach with qualitative interviews and informal conversations conducted in 2017 (Götz 2017). Actors consisted of informants connected to the NWRC (including: NWRC members, AG individuals) and the SPP (including: informants from local CSOs and the KNU; international supporters).[2] The research is further supported through media and document analysis.

The chapter is organized as follows: First, I outline the conceptual frame of the chapter before providing some relevant background on Burma's socio-political history as it relates to current claims around the scales of water governance. Recent developments regarding the peace negotiations and the struggle over the degree of decentralization are also highlighted. Subsequently, the two groups of actors with their specific claims around future rules of water governance and its implications for the production of scale are introduced. Finally, I draw those claims together to explore the arenas and processes through which scale is being contested, before concluding on the implications for the future rules of water governance in Myanmar.

[2]To protect the informant's anonymity all informants quoted in this chapter are coded in accordance with one of the two key groups they are related to (i.e. NWRC or SPP).

(a)

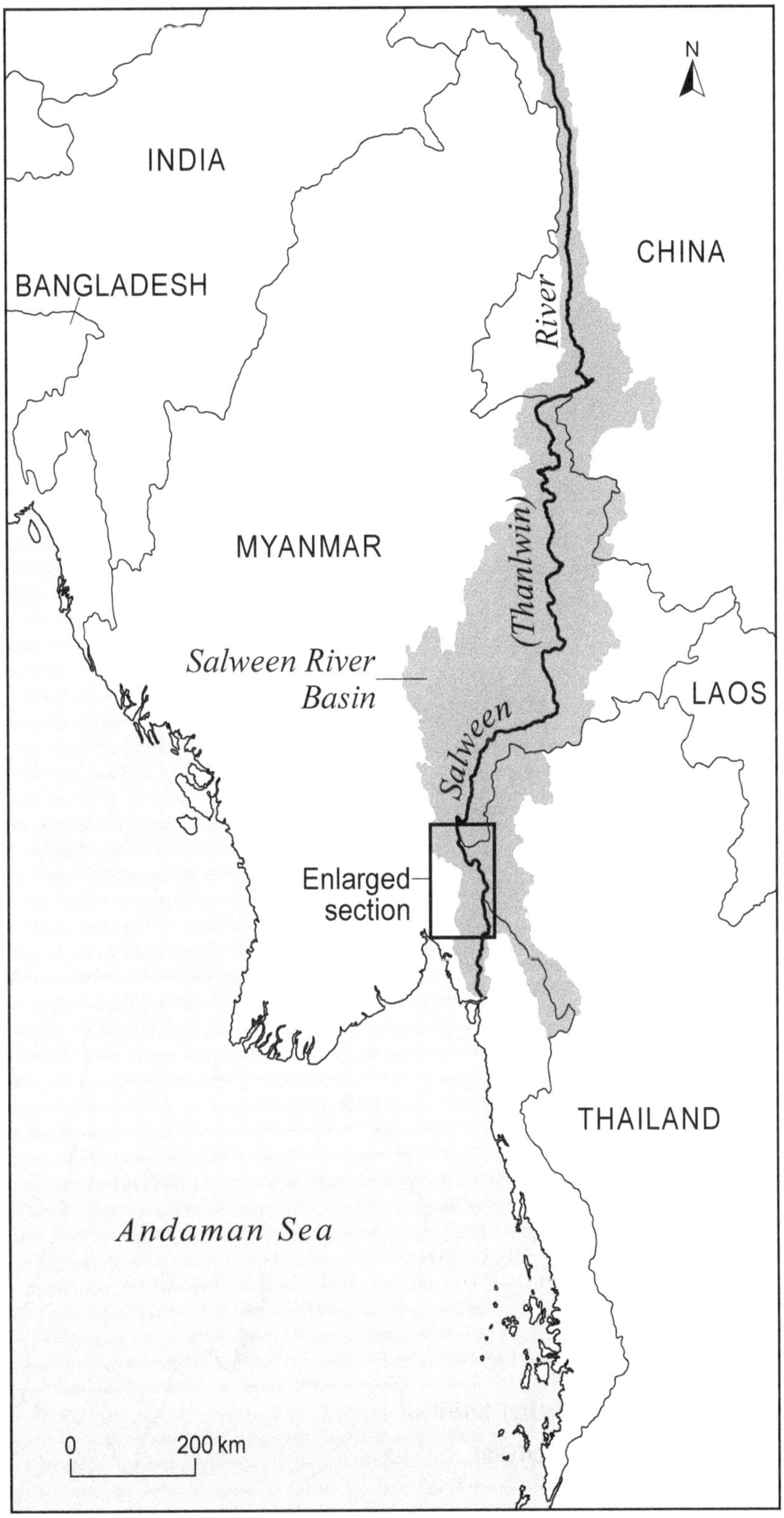

Fig. 6.1 Articulated scalar manifestations/ claims around water governance in Burma/ Myanmar. **a** Myanmar's nation-state borders and the Salween River Basin. **b** The Salween Peace Park within Karen State's/ Kawthoolei's Hpapun/ Mutraw District. *Source* Cartography by Chandra Jayasuriya, University of Melbourne, with permission

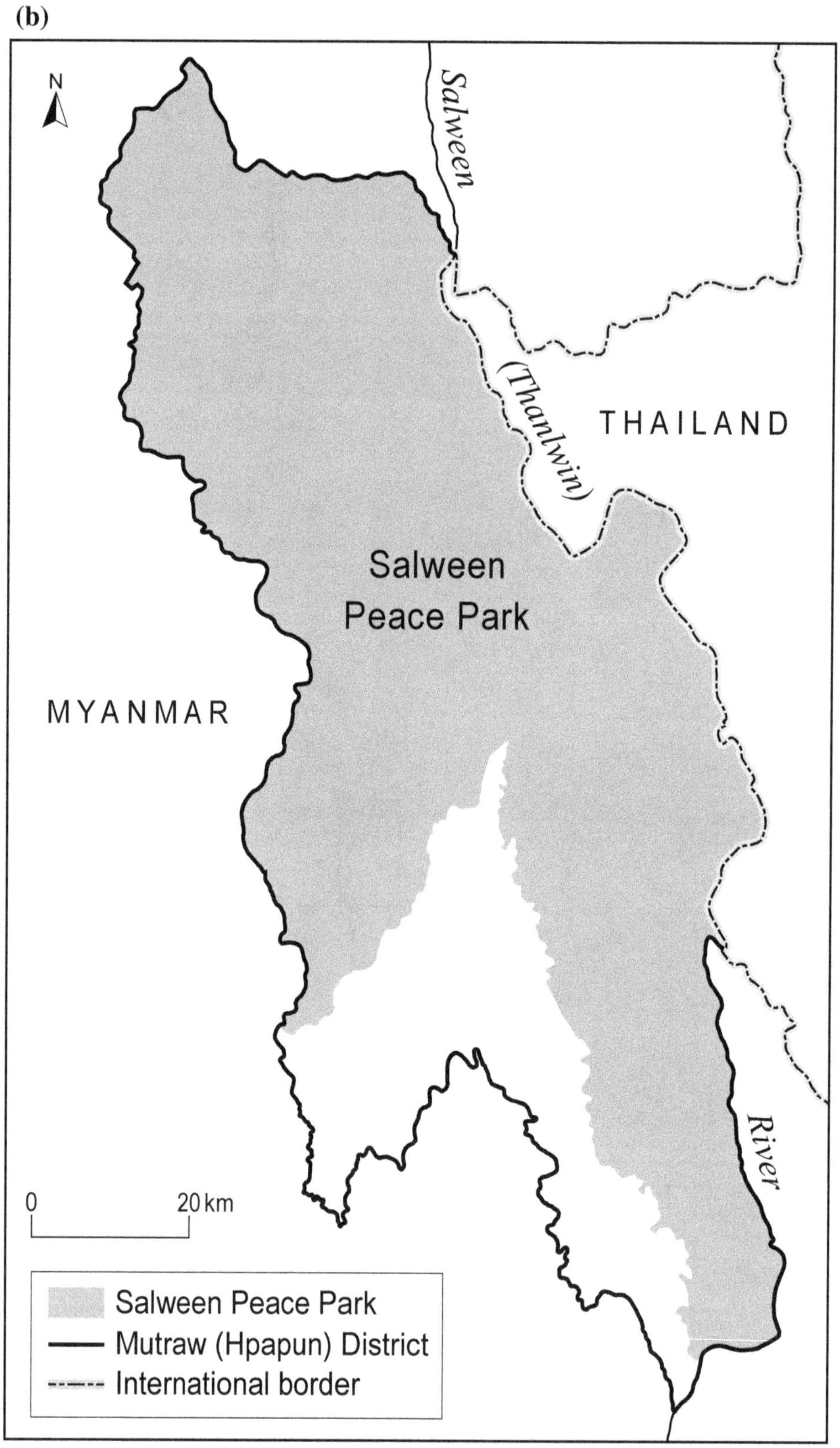

Fig. 6.1 (continued)

6.2 Water Through a Hydrosocial Lens and the Production of Scale

Considering water from a hydrosocial perspective holds major implications for understanding water and water governance (e.g. Boelens et al. 2016; Linton/Budds 2014). Through a hydrosocial lens, water is seen not as a mere material "natural resource", but rather as a "historical and relational-dialectic process" (Linton/Budds 2013: 10) wherein water and society constantly constitute themselves as "socionatural hybrids (cf. ibid.; Linton 2010). Viewing water and society not as two dualistic entities but as "hybrid nature" (Budds/Hinojosa 2012) reveals how water and society evolve through one another and consequently, water governance is understood as more than institutions and policies, but reflects the wider array of co-constituting human and more-than-human relationships. This leads to the insight that "water is not politically neutral, but instead both reflects and reproduces relations of social power" (Perreault 2015: 118).

Scale as a concept has been subject of heated debates (e.g. Brenner 2001; Herod 2011; Neumann 2015). While some call for eliminating 'scale' as a concept (Marston et al. 2005), others show the relevance of it when applied critically (e.g. Leitner/Miller 2007). Critical research has sought to redefine scale away from understanding it as hierarchical, fixed and apolitical towards an understanding of scale as:

> […] something that is produced; a process that is always deeply heterogeneous, conflictual, and contested. Scale becomes the arena and moment, both discursively and materially, where sociospatial power relations are contested and compromises are negotiated and regulated. Scale, therefore, is both the result and the outcome of social struggle for power and control. (Swyngedouw 1997: 140)

Within this understanding, debates around water governance and the politics and production of scale have gained momentum (e.g. Dore/Lebel 2010; Norman et al. 2015). Within a hydrosocial understanding, scalar relations have been addressed through concepts like 'waterscapes' (Budds/Hinojosa 2012) or "hydrosocial territories" (Boelens et al. 2016). Drawing from these debates, I focus on the *processes* that influence the (re)construction of scalar notions (i.e. their continuous (re)production, contestation and reconfiguration) and (temporal) manifestations related to water in its socio-material dimensions rather than looking at scale per se.

Considering the widespread understanding of socially-constructed scales, within water policy-making it can be observed that certain scales are given priority, while downplaying or marginalizing less obvious scalar networks that influence water. Budds/Hinojosa (2012: 119) emphasize the importance of overcoming the still prevalent practice to "take the hierarchical physical boundaries and administrative structures that characterize most instances of water governance as given" and instead focus on the politics and production of scale. This becomes evident when looking at the widespread use of the watershed or river basin as the seemingly pre-given, most suitable scale for water policy-making. Since the 1990s, the old concept of river basin management gained new relevance and became *the* scale for water governance in "the West" (and subsequently far beyond) (Molle 2015; Warner et al. 2008). This is embedded in a notion of hydrological boundaries of the watershed, which is assumed to be a 'natural' scale for governing. However, in fact this assumption renders both scale and water as fixed, and thus stands in stark contrast with a hydrosocial understanding. While it is clear that hydrological flows – as water's materiality – exist and are indeed relevant, taking the scale of the river basin for granted does leave out important social, cultural and political dimensions. As Blomquist/Schlager (2005: 103–5) argue, choosing the watershed as the appropriate scale for water governance is always a conscious choice – rather than a 'natural' given necessity – that brings political momentum with it. Ultimately, defining a scale for water governance (be it the watershed, administrative boundaries or other envisioned or drawn spaces) brings with it questions about which actors are involved in decision-making, affected by it and how power is distributed. As such, it is not about neglecting material flows within hydrologically defined watersheds, but to avoid considering these flows as independent from other hydrosocial arrangements.

6.3 Transforming Water Governance in Myanmar

The production of scale around water governance and associated claims for power are, as I will argue, deeply embedded within a wider historical context. Under British colonialism, the borders of what is known as Myanmar today were defined for the first time. Included in it were the areas of Bamar-dominated "Burma Proper" under the centralized authority of the British, which covered the central low-lying areas. The more remote "Frontier Areas" in the hilly border regions, meanwhile, were inhabited by different non-Bamar ethnic minority groups and were under less rigorous authority by the British, thus "allowing traditional local leaders to run the day-to-day affairs" (Callahan 2007: 12). After independence, a short attempt to find a peaceful federal solution as written in the 1947 Constitution ended abruptly through Ne Win's coup in 1962. The decades to follow were characterized by centralized authoritarian regimes (Ninh/Arnold 2016). However, as Callahan (2007: 12–13) states, there have been areas that:

> [t]hroughout the postcolonial era, […] have never come under anything approaching central control. Large stretches of territory […] and large numbers of people have been governed, administered, and exploited by armed state challengers, such as […] the Karen National Union.

Recent economic and political reforms in Myanmar since 2010–11 have led to an increased liberalization of economic policy, attracting growing volumes of foreign investment. However, Burma remains unique in the region as it is in the state of a fragile (partial) ceasefire rather than peace. Within Myanmar's emerging political system of National and Regional/State governance, not formally represented non-state actors remain highly influential and are seeking strong localized federalism. The question remains open as to what a future federal system will look like (e.g. Ninh/Arnold 2016). Given that water is highly political, especially if understood from a hydrosocial perspective, then water governance in Burma must be positioned in this wider historical and political context. With the recent change from military dictatorship to a quasi-civilian government, clear rules and laws around water, or the scales of water governance, are yet to be established, which will, amongst other things, depend on the degree of decentralization under a future federal agreement.

After decades of military dictatorship and with it the oppression of most critical thinking, the nascent state of literature on water governance in Myanmar is unsurprising.[3] However, there are a growing number of articles, reports and scholarly literature on water-related issues. This includes reports looking at the current water-related Union government institutions and respective legal frameworks (e.g. Kattelus et al. 2014; van Meel et al. 2014). While providing some key issues (lack of transparency; public participations), solutions are mainly proposed along the line of a globally hegemonic water management approach, namely Integrated Water Resources Management (IWRM). Thereby, most research taking a localized perspective raises a management approach rather than a hydrosocial perspective (ibid.). Alternative paradigms and framings of water are much less discussed in academic literature on water governance in Burma, with some notable exceptions (e.g. Middleton et al. 2017; Suhardiman et al. 2017).

Within the two subsequent sections, I aim to exemplify the importance of moving beyond a simplified understanding of water governance and emphasize the wider hydrosocial relations and their scalar politics. Figure 6.1: Articulated scalar manifestations/claims around water governance in Burma/Myanmar illustrates the examined articulated scalar manifestations and claims around water governance, including Fig. 6.1a: Myanmar's nation-state borders and the Salween River Basin, and Fig. 6.1b: The Salween Peace Park within Karen State's/Kawthoolei's Hpapun/ Mutraw District. While the visualization of this may carry political clout with it, the mappings illustrate the intertwined and overlapping scalar articulations, unveiling the complexity of the future rules of water governance in Burma.

6.4 National Water Policy Regime

With the transition towards a quasi democratic government system, a major restructuring of state-based institutions has unfolded in Myanmar. This has been accompanied by new discourses and visions around the future-rules of water governance, which will influence how hydrosocial relations unfold. Under the presidential degree of U Thein Sein, a working group called the National Water Resources Committee (NWRC) was established in 2013 with the goal to "take responsibility for the overall management of national water resources and to

[3]This may not include sources in Burmese/other, ethnic-minority languages and unpublished work.

facilitate for a more coordinated approach" (Nesheim et al. 2016: 21). After the dissolution of the committee at the end of U Thein Sein's presidential period in March 2016, the NWRC was reestablished by the new NLD-led government in June 2016 (personal communication, NWRC2).

The current NWRC and its Advisory Group (AG) includes a network of actors from water-related Union ministries, departments, representatives from the State and Regional level, the mayors of Yangon, Naypyidaw and Mandalay, and 'water experts' under the Chair of Vice President U Henry Van Thio. The NWRC Secretary is the Directorate of Water Resources and Improvement of River Systems (DWRI) of the Union Ministry of Transport and Communication (NWRC 2015, 2017). The makeup of the NWRC, with involved ministries amongst others including the Ministry of Industry, the Ministry of Electric Power and Energy, the Ministry of Agriculture, Livestock and Irrigation, and the Ministry of Natural Resources and Environmental Conservation, already hints towards the heterogenous interests existent within water-related decision-making. Reflecting the fragmented responsibilities across sectors and ministries, one objective of the NWRC is the "strengthening of inter-ministerial cooperation, communication and information sharing" (NWRC 2015, Art. 2.4(ii)).

A key mandate of the NWRC is to draft water-related rules and regulations. With the National Water Policy (NWP) first published in February 2014 (NWRC 2015) and approved by the cabinet in June 2015 (personal communications, NWRC1 and NWRC6) together with the National Water Framework Directive (NWFD), two documents have already been published while a National Water Law is currently being drafted but not publicly available at the time of writing (January 2019; See also, Bright, Chap. 5, this volume). This is meant to be an umbrella law for any water-related laws.

Key paradigms and concepts visible in the NWRC's work indicate a rather centralized, expert-led water-policy making that is shaping specific narratives around water in Burma. Rereading the analysis of Nesheim et al. (2016) on current institutional settings suggests that hegemonic global influences can be found in Myanmar's emerging policies. For instance, the NWFD does not only carry the name of but

> [...] is inspired by the EU WFD [European Union Water Framework Directive] in that it parallels several of the same principles as those in the European directive, including among others River Basin Management. (ibid: 22)

However, in contrast to the EU WFD, it is not set to be a binding law, but is instead "an umbrella statement of general principles governing the exercise of legislative and/or executive (or devolved) powers by the Union, the States and Regions, and the local governing bodies" (van Meel et al. 2014: 12). IWRM is communicated as *the* central paradigm. The NWP clearly states this in various places, for example:

> [T]here are inequities in distribution and lack of a unified perspective in planning, management and use of water resources, i.e. little or no knowledge about [IWRM] except in the domain of water professionals. The *whole country needs to be aware of IWRM principles and participatory approach.* (NWRC 2015: 2, emphasis added)

With IWRM as a central paradigm, various water-related sectors are to be 'integrated' under the auspices of the NWRC as a powerful apex body. In so-called 'international best practice', IWRM has tended to go together with a call for the river basin as the appropriate level of governance (cf. Molle 2015). In Burma, a call for river basin management can be observed in the narrative of several government documents and actors, even though it is not formulated as the only level of governance:

> All the elements of the water cycle […] are interdependent and the basic hydrological unit is the river basin, which should be considered as the basic hydrological unit for planning. (NWRC 2015: 15)

Ongoing projects further accelerate the focus on a combined IWRM approach at the river basin scale, such as the Ayeyarwady Integrated River Basin Management (AIRBM) project financed by the World Bank with the objective "to help Myanmar develop the institutions and tools needed to enable informed decision making in the management of Myanmar's water resources and to implement integrated river basin management of the Ayeyarwady Basin" (MoT 2017). It should, however, be emphasized that to date no transboundary river basin organizations or similar agreements exist for the country's rivers, even though some informants are proposing such institutionalizations (personal communication NWRC3, 2017). Rather, the 'natural' river basins seem to be bounded by the borders of the nation-state.

With this Union government-led decision of IWRM implementation at a river basin level for the whole country, as affirmed in the NWP and materialized in projects such as the AIRBM, the river basin as a specific scale for water governance is (consciously or not) being produced as an appropriate level for decision-making over water. Furthermore, by prescribing a hegemonic global concept at the level of seemingly 'natural' boundaries, IWRM is being raised as *the* appropriate basis for decision-making. This brings with it a centralizing tendency: this occurs strategically, by prescribing IWRM as an overall paradigm for the country from a Union level perspective. Thereby, other established water governance practices (e.g. customary rights) might be overwritten. It also occurs institutionally, by 'integrating' various institutions – with their distinct power relations – under the umbrella of water management whereby a (re)centralization of power through IWRM implementation has been predicted and observed in other contexts (e.g. Mehta et al. 2014). On paper, IWRM is supportive of decentralization thus intending to redistribute decision-making to 'lower' levels. For Burma, this would mean decentralization towards the Region or State governments. However, what sounds locally empowering in theory might prove to be centralizing in practice, as a lack of capacity and internal conflict at the Region/State level might redirect power back to the National level or to certain powerful factions (cf. Mehta et al. 2014 for a South African exploration). Moreover, current customary rights and established practices around water governance, as prevalent in some areas, would be overwritten by Union state-controlled IWRM practices.

The above-mentioned centralizing tendency becomes further consolidated when examining national-level water policy actors who, besides the river basin, place a strong emphasis on the Union government and, with this, a centralized level of water governance.

> The objective of the National Water Policy is to take cognizance of the existing situation, to propose a framework for creation of a system of laws and institutions and for a plan of action with a *unified national perspective*. (NWRC 2015, emphasis added)

> [I]f the country is going to federalization and [...] if there is a region and region conflict of interest, there will be *the national level coming in and decide*. And also, if it is a national-level concern, like the national/international, transboundary river, then the national government is concerned [...]. So [the National Water Law] will be the master of all the laws related to water [...] and everything is to be integrated [...]. *According to our Constitution, the right is under the [Union] state of handling water*. (personal communication NWRC3, 2017, emphasis added)

While universal rights are underlined, the power of decision-making is clearly given to the central government (at least in the second instance). The NWRC has repeatedly been referred to as an important apex body for the water governance of Myanmar *inter alia* by its members (personal communications, 2017) and envisions a future influential role. The NWP further underpins this intended central position of the committee by stating:

> [The] NWRC should be a legislative body in national water sector and should have authorization of the government to play a significant role in any national/state level water use [...] of consumptive or non-consumptive nature. (NWRC 2015: 46)

The centralization of power is also influenced by the relatively small number of individuals – namely a limited number of "water professionals" (personal communication NWRC1, 2017) – active within the central institutions.

The NWRC and its AG are a significant part of current efforts to pave the way for future water governance at a Union level by drafting policies and laws. However, in practice they still do not have the final say in whether those are adopted within the rather fragmented government arrangements of the country. At the same time, they further their claim to legitimacy through collaborating with actors like the World Bank in the AIRBM project and other International Financial Institutions (IFIs), NGOs and national governments (i.e. the Netherlands or Norway). Thus, relying on concepts with a hegemonic status can be seen as important legitimization of authority by the Union within a 'transnational network' (cf. Hensengerth 2015). Additionally, the NWRC has been rather successful in being visible thorough, for example, joint reports with partners which represent some of the little information that is available on water governance in Burma (e.g. van Meel et al. 2014 on IWRM). Yet, some civil society actors repeatedly challenge the Union government's cooperation with actors like the World Bank and question the degree of participation in associated projects.

Historically-grown power relations in the central Bamar-majority area are reflected in today's centralizing tendencies both institutionally and materially (e.g. within the AIRBM project). Under colonial rule, this centralized control proved true for "Burma Proper" while the "Frontier Areas" were administered indirectly. After independence, little changed in those highly centralized power relations (Ninh/Arnold 2016: 225–6). Simultaneously, however, as Callahan (2007: 12–13) notes, the following authoritarian regimes were at no time able to bring the whole country under central control. Rather, different Ethnic Armed Organizations (EAOs) set up their own 'state-like structures' and governance. Questions around decentralization and federalism date back to the post-colonial era and remain unresolved. While decentralization seems to be a rather universal goal, the degree of power redistribution remains one of the most crucial questions towards federalism in the current peace negotiation process (Ninh/Arnold 2016: 226). Decentralization – currently bound to the 2008 constitution – will create new scales of governance and redefine power relations, which also holds implications for water governance. While the NWRC seems to support decentralization of water governance, for now it claims and affirms its leadership role as an apex body:

> Federal States should take charges for water governance and decision making around water in their state. But, for the early years they should [be] under the umbrella of Control Government body like NWRC. (personal communication NWRC5, 2017)

Concluding, it can be claimed that through rather centralized, expert-led water policy-making a national-state-scaled water policy narrative framed in a river basin-scaled unit of governance is currently emerging. These spatiotemporal notions – built around a hegemonic, unified Myanmar narrative on water governance – reflect historically-grown power relations and highlight the hydrosocial relations within which water and society constantly (re)produce themselves (cf. Linton/Budds 2014).

6.5 Salween Peace Park: A Local Initiative

The recently officially proclaimed Salween Peace Park (SPP) (e.g. KESAN 2019), located in Mutraw District in Northern Karen State (Fig. 6.1b: The Salween Peace Park within Karen State's/Kawthoolei's Hpapun/Mutraw District), poses a quite different arena of governing water. A group of actors including community representatives, KNU local leadership, and members of CSOs push for their own vision within an area of over 5,000 km^2, which encloses a range of 'community' and reserved forests, wildlife sanctuaries and customary land (KNU and KESAN 2016; personal communications, 2017). These actors envision

> [a] grassroots, people-centered alternative to the previous Myanmar government and foreign companies' plans for destructive development in the Salween River basin. Instead of massive dams on the Salween River, we see small hydropower and decentralized solar power […]. Instead of megaprojects that threaten conflict and perhaps the resumption of war, we seek a lasting peace and a thriving ecosystem where people live in harmony with the nature around them. (KNU/KESAN 2016: 3)

Based on these visions, a clear narrative becomes prevalent that calls for a "bottom-up", "community-based" approach with traditional systems to be preserved and promoted. Thereby 'local' people shall have "the right to self-governance", natural resources and wildlife shall be protected, and "Karen indigenous people's culture and life" are to be preserved (personal communication SPP3, 2017). In contrast to 'traditional' conservation areas, which tend to relocate people outside of the area, the SPP is claimed as integrating communities into it. One involved individual argues that there are many connecting

> ...issues that need to be considered when we talk about water governance and water management issues, because water is very political. Water is life, and everybody uses water every day. (Personal communication SPP1, 2017)

Here, a considerably different narrative of water that is more overtly stated as political is voiced out, which is also antagonistic vis-à-vis the rather centralized views described above. As the recently released SPP Charter (SPP Steering Committee/KNU 2018) demonstrates, a strong focus is placed on the 'community' as a central scale for decision-making. Connected to 'natural' ecosystems – for instance "community forests" – this scale can be, analogous to the watershed, understood as a naturalized social scale with the 'community' being handled as seemingly given entity and scale (cf. Cohen/McCarthy 2014). Nevertheless, the SPP Charter recognizes that "the Salween River is a precious resource that sustains the indigenous Karen people's way of life, and people who live beyond the boundaries of the Salween Peace Park" (SPP Steering Committee/KNU 2018: 30), recognizing the fluid character of water and the wider hydrosocial network beyond the borders of the SPP.

Spatially the SPP is based on historically-rooted power relations. It is within this historic context that the proposed borders of the SPP and therewith the envisioned scales of water governance are actively co-produced. The British colonial rule brought with it some major implications for power-relations within Myanmar. The "Frontier areas" included a small area of today's Karen State, namely the so-called Salween District. As part of these historically more autonomous areas, today's Mutraw District – while far from uncontested – up to this day "[...] remains the most autonomous KNU-controlled region and has never been brought under centralized state rule" (Jolliffe 2016: 9).

After independence, power-relations shifted several times not only between the KNU and Union government, but also amongst different KNU districts (Brenner 2017). During the 1990s, Mutraw District under Brigade 5 gained power relative to other districts due to declining economic trade with Thailand and strong interventions by Myanmar's armed forces. Power struggles at this time, according to Brenner (2017), led to today's factions within the KNU. The proposed SPP is located exactly in this still very much autonomous region, covering a majority of today's Mutraw District. Although the historic context can merely be touched upon, it already shows that the extent of today's SPP is anything but coincidental. Historically grown power-relations (re)shaped specific spaces and enable claim-making for future envisioned scales.

In order to legitimize explicit visions of the SPP, different strategies can be observed. These include reinforcing the visions and scalar politics of the SPP by raising public awareness *inter alia* by initiating debates and campaigning through protests and media releases (cf. Dore/Lebel 2010 for similar strategies in the Mekong). Solidarity with other ethnic minority communities also connects the SPP to wider networks within and beyond Myanmar. Furthermore, actors working to further the SPP engage in protests against mega-dam development and thereby connect to a network of a localized – but also global – anti-dam movement (e.g. SSN et al. 2017). A range of articles, short documentaries and meetings further campaign for the idea of the SPP contributing to the contestation and production of scale (e.g. KESAN 2017).

Another way of consolidating claims to the right to govern are formalized in KNU policies related to "natural resources governance" (i.e. land and forest policy). The comprehensive KNU Land Policy was first ratified in 1974 (KNU 2015) and most recently revised in May 2016 (TBC 2016). While the KNU is also planning on its own water policy (Personal communications SPP1 and SPP2, 2017), the KNU Land Policy also already includes clear visions about water governance, stating for instance

> The ethnic nationalities are the ultimate owners of all lands, forests, water, water enterprises and natural resources. (KNU 2015: 5)

This stands in direct contrast to the 2008 Constitution which defines the Union as "the ultimate owner of all lands and all natural resources above and below the ground, above and beneath the water and in the atmosphere in the Union" (Union of Myanmar 2008: 10).

The existence of autonomous laws within KNU-controlled areas represents a clear call for self-determination and different realities of current and future power structures:

> This is KNU-controlled area, the Government is not able to come and set up their administration. KNU has its administration, […and] operations, KNU collects taxes, KNU, even though they are not able to provide services, but in some way, they protect the people in their controlled territories, so the people believe and trust them […]. And we are still in the […] stage of negotiation with the Government that 'you have to recognize our policies and laws (and) administrations (while) we recognize your administration'. (Personal communication SPP1, 2017)

Hong (2017) within a legal pluralism frame looks at those "autonomous laws" within Burma and how the creation of law – may it be drafted by the state or, as in the case of the KNU, by an ethnic de-facto government – not only tries to consolidate power of the relevant actors, but through it contributes to the reshaping of scales. Referring to the KNU Land Policy, Hong (2017: 11) shows how law may serve "as a multi-scalar multi-temporal connecting tissue between the practice of autonomy in the here and now, and a long-term desire for self-determination and recognition." Looking at the SPP shows how scalar claims are being voiced out and how those are always closely connected to other (materialized/envisioned) scalar articulations – as with the land of the Karen people: Kawthoolei.

However, it is important to recognize that a unified, self-determined Kawthoolei is not self-evident. In early 2017, elections were held to vote for the new KNU central government. Power-relations shifted heavily towards Saw Mutu Say Po's faction, pushing aside the opposition under Naw Zipporah Sein. While the former is known to be rather pro-business and open to the NCA and adjoining processes in their current form, the latter tends to be more nationalistic and has its strongest support within autonomous areas such as Mutraw District (Jolliffe 2017). While striving for peace is omnipresent, the ways of achieving it vary greatly,[4] and the ideas around peace and (water) governance represented in the SPP are not feasible for all KNU-influenced territories. This is especially true as visions range from 'pro-development' to "conservation of traditional ways of life" (Jolliffe 2016). How 'appropriate' arrangements of governance will look like, are, however, up for debate, bringing the focus back to the heart of the scale debate and its highly political notion. Because, as Jolliffe (2017: n.p.) states, while Mutraw District oversees one of "the most powerful and well-supported KNU military units [, it has] barely been engaged in negotiations in the peace process so far." This limited representation might further fuel the SPP's narrative of the 'right' level of decision-making with a clear stance against any centralized approach of water governance.

The degree of power held by the central KNU administration seems to be at least twofold. Firstly, it seems to serve as a source of legitimization to get official recognition

> So, to make sure this [(SPP charter)] is official, we need to bring that document to the KNU decision-makers, to make sure it is a kind of official document, an official regulation, official policy to rule the Peace Park. (Personal communication SPP2, 2017)

Secondly, one informant emphasized that "it's not that the KNU central is establishing [the SPP] at the KNU level", rather "the local people, they're the ones who are working to establish the Salween Peace Park" in a process that is happening "gradually, from the bottom up". Concluding that, as it is being established, it is "then eventually going up [...] to get recognized at the central [KNU] level" (personal communication SPP3, 2017). As such, internal power relations and their contestation become revealed.

Talking to one informant involved in the SPP about the national water law currently being drafted, it becomes quite evident that a more or less centralized national scale is not being recognized:

> No, [the National Water Law would not have any influence within the SPP area] because the Salween Peace Park, as we say, is a demonstration to self-determination. It is not about the Union Government saying, 'you have to do this and that' – it's independent. As this is exactly what we are calling for: the right to self-determination. (Personal communication SPP2, 2017)

[4]This has become evidence once again in the recent (temporary) suspension of the KNU from the peace negotiations, demonstrating discrepancies both within the KNU and in relation to the central government.

With this call for self-determination, the scalar visions around the SPP emphasize local values of water and the right of local uses to benefit. To sum up, it can be argued that a narrative of a more localized arena of community-led, bottom-up decision-making within a federalism of self-determination can be observed.

6.6 Contesting Scale: A Battleground of Water Governance in Burma

As laid out in the previous sections, water governance in Myanmar is far from being undisputed or apolitical, but rather reveals a battleground around different scales at which water is to be decided upon in the future. This chapter represents only a small fraction of claims available. Nevertheless, it shows the significance of looking beyond both a naturalized understanding of water as well as an oversimplification of 'appropriate' scales of water policy-making. Following an understanding of scale as being hydrosocially produced inevitably opens up questions around actors involved in (re)producing scales of water governance.

The sections above took a closer look at two sets of key actors, namely the NWRC and its AG, which can be seen to represent a centralized, national scale under the Union government, and KNU members, CSOs, and activists closely related to the SPP in Mutraw District, Karen State as one example of a more localized approach to water governance. Interestingly, both set of actors legitimized their respective claims by "rescaling to ecosystem spaces" (e.g. watershed or community forest) and "rescaling to jurisdictional scales" (e.g. States/Regions under the Union or Kawthoolei under a strong KNU and SPP) (Cohen/McCarthy 2014: 19). Figure 6.1: Articulated scalar manifestations/claims around water governance in Burma/Myanmar exemplifies these intertwined and overlapping scalar projects. Those groups of actors are only two among many intersecting networks contesting scales of water governance and the observed scalar claims are just an extract of many interconnected and overlaying visions out there. However, it becomes evident by introducing those two different networks of actors that diverse visions and paradigms around water governance exist, which aim at different scalar articulations at which decisions are to be made at.

Regarding *how* scales are being produced, different strategies are evident. Amongst them are legal strategies of actors drafting their own respective policies under differently recognized institutional arrangements and power-relations (i.e. NWRC vs. KNU/SPP) that are to regulate specific arenas (nation-state vs. Kawthoolei). While the NWRC aims at integrating the various claims and actors related to water governance on a unified Myanmar scale, the actors around the SPP are instead focusing on their own regulations at more localized scales. Thereby, the seemingly 'fixed' national scale is being contested.

To support their claims, both groups draw on different concepts that connect them to different local, national and transnational networks (cf. Boelens et al. 2016;

Hensengerth 2015). With its claim for IWRM and its implication for water governance at the river basin, the NWRC and its AG focus on a dominant concept recognized by other governments, IFIs and INGOs. In contrast, while Peace Parks are also an internationally recognized concept, the actors around the SPP distance themselves from a pure conservationist approach and instead focus on community-based decision-making with rather strong 'indigenous' notions (e.g. "Karen traditional lifestyle") thus relating themselves to a network of national and international non-state actors as grassroots activist and CSOs. They also engage in wider protests against mega-dam development and, with it, against centralized infrastructure construction conducted by state-related actors (IFIs, global investors, etc.). Discursively, actors draw their own specific visions and "scalar narratives" (Swyngedouw 1997: 140), legitimizing respective ideas about the 'right' scales of decision-making around water. Intending to communicate (largely) within their respective networks both sets of actors increase their visibility by releasing documentation on their envisioned scales (i.e. rather techno-scientific reports vs. politicizing press releases), both at times supported by maps of the contested scales (e.g. van Meel et al. 2014 or KNU/KESAN 2016). Looking at the proposed scales for future rules of water governance within a hydrosocial framing – while considering the highly political dimension water carries – puts the current peace negotiation process at the very heart of the debate. Claims around the appropriate scale for water governance closely connect to wider claims around decentralization and federalism, which will inevitably produce new scales at which decisions will be made at, and with it the forms of power (re)distribution. This contestation emphasizes current politics of scale

> We are considering for governance and management of water resources in [a] federal system, but to form this system […] depend[s mainly] on the political change and the changes of the governance system. Without having federal states and regions, it is impossible to have federal system for Myanmar's water [governance]. (Personal communication NWRC6, 2017)

> The new Myanmar government has promised to lead the country toward a devolved, federal democracy. The Karen are not waiting idly for this: the Salween Peace Park is federal democracy in action. (KNU/KESAN 2016: 3)

The *purpose* of claiming certain scales for water governance for all informants seems to be a genuine will to reach some sort of "change for the better." How that change is to look like, however, differs tremendously. Different framings – depending on varying priorities and goals – then have an influence on how scale is understood, constituted and (re)produced (cf. Sneddon/Fox 2012). Underlying power-relations will contribute significantly in defining which visions will prevail.

Although the enormous power of the Union government as an actor should not be neglected, the complexity of power relations in Burma should not be disregarded either. The NWRC consists of members from the Union government and is backed by it and a range of international players, but does not have unlimited authority. The SPP arena is located in an autonomous region where, up to this day, the Union government has limited influence. This autonomy tries to be maintained in a time of radical political change through repeated calls for self-determination. In turn, the SPP is also connected to a wider network of actors supporting their claims and thereby strengthens their proposed scales for future decision-making around water.

6.7 Towards Future Rules of Water Governance

Let us move back to the introductory incident around the Salween River. I will highlight how, by moving beyond a simplified understanding of water as an apolitical resource to be managed, towards emphasizing the wider array of co-constituting human and more-than-human spatiotemporal relationships, the contested water governance in Myanmar gets illuminated. As such, water is ultimately bound to wider, historically-grown processes and power-relations as visible within the ongoing peace negotiations. Looking at different networks of actors within the hydrosocial arena of Burma, I have argued that the production of scale is currently being contested and constitutes an important battleground about the future rules of decision-making and the actors involved in it, as well as the power relations within the wider network of actors. Scales are not a priori given – neither as a centralized Union state with hierarchal scales nor as independent localized factions evolving around the 'community' – but are subject to continuous contestation.

Understanding scalar articulations as intertwined and overlapping with other scalar imaginations and manifestations, as present along the Salween, brings to the fore the complex hydrosocial relations. This is apparent when considering how the envisioned SPP both connects and opposes a range of other scalar articulations: as part of the land of the Karen – Kawthoolei – and as such connects to KNU-based ways of organization; or simultaneously, as located in the 'official' administrative demarcation of Karen State which are to underlie the Union-based water governance paradigms. In addition, the SPP is part of the Salween River basin, both bound by the nation-state borders and as a transboundary river that connects to neighboring China and Thailand. Similarly, the NWRC is ultimately bound to the nation-state's (internal and external) boundaries, but simultaneously favoring international hegemonic IWRM and the river basin scale with its 'ignorance' to socially constructed borders. These represent just a few of the multiple overlapping scalar articulations, structured by certain rules and normative frameworks, embedded in wider hydrosocial networks, and subject to distinct power relations.

Political authority in Myanmar, and in particular in the Salween basin, is highly fragmented with intensely contested political power-relations, which are built on complex historical accounts (e.g. Callahan 2007). Looking at Myanmar's current politics, the degree of decentralization is highly controversial with a clear unified vision lacking. This further complicates questions about the future rules of water governance. Even with federalism established and institutional structures changed, decentralization processes do not automatically redistribute power-relations or empower local actors, as demonstrated in other contexts (e.g. Marks/Lebel 2016; Norman/Bakker 2009). Looking at the "incomplete decentralization" (Marks/Lebel

2016) in neighboring Thailand shows how the degree of power redistribution is dependent on much more than an institutional restructuring. Within an arena of an emerging political system of National and Regional/State governance, as well as existing grey areas with quasi-/non-state actors seeking self-determination under a strong federalism, rules and regulations around water are yet to be (re)established with claims on the 'appropriate' scales being made by a range of actors with their respective narratives. Within such an understanding, the politics of scale constitutes a key battleground around which water governance is currently being (re)defined.

Acknowledgements My uttermost gratitude goes to my informants for their time and trust. I further thank the editors for all their work while compiling this volume; the HBF-Team, Yangon for their continued assistance; and everyone supporting my Master thesis process upon which this chapter draws – they know.

References

Blomquist, W., & Schlager, E. (2005). Political pitfalls of integrated watershed management. *Society & Natural Resources, 18*(2), 101–117.

Boelens, R., Hoogesteger J., Swyngedouw E., Vos, J., & Wester, P. (2016). Hydrosocial territories: A political ecology perspective. *Water International*, *41*(1), 1–14.

Brenner, D. (2017). Inside the Karen insurgency: Explaining conflict and conciliation in Myanmar's changing borderlands. *Asian Security*, *14*(2), 1–17.

Brenner, N. (2001). The limits to scale? Methodological reflections of scalar structuration. *Progress in Human Geography*, *24*(4), 591–614.

Bright, J. (2016). Rights and rites for water justice: A case study for the proposed Hatgyi Dam on the Salween river. In Center for Social Development Studies, the Mekong Sub-region Social Research Center & Vietnam Academy of Water Resources (Eds.), *Mekong, Salween and Red Rivers: Sharing knowledge and perspectives across borders* (pp. 732–765). Bangkok, Thailand.

Budds, J., & Hinojosa, L. (2012). Restructuring and rescaling water governance in mining contexts: The co-production of waterscapes in Peru. *Water Alternatives*, *5*(1), 119–137.

Callahan, M.P. (2007). *Political authority in Burma's ethnic minority states: Devolution, occupation, and coexistence.* Washington, DC: East-West Center.

Cohen, A., & McCarthy, J. (2014). Reviewing rescaling. *Progress in Human Geography*, *39*(1), 3–25.

Dore, J., & Lebel, L. (2010). Deliberation and scale in Mekong region water governance. *Environmental Management*, *46*, 60–80.

Götz, J. (2017). *The politics of water governance in Myanmar - a hydrosocial approach.* (Unpublished Masters thesis), University of Bonn, Germany.

Hensengerth, O. (2015). Where is the power?: Transnational networks, authority and the dispute over the Xayaburi Dam on the Lower Mekong Mainstream. *Water International, 40*(5–6), 911–928.

Herod, A. (2011). *Scale.* London, United Kingdom: Routledge.

Hong, E. (2017). Scaling struggles over land and law: Autonomy, investment, and interlegality in Myanmar's borderlands. *Geoforum*, *82*, 225–236.

Jolliffe, K. (2016). *Ceasefires, governance and development: The Karen National Union in times of change*. The Asia Foundation.

Jolliffe, K. (2017, April 26). A new chapter for the Karen movement. *Frontier Myanmar*. Retrieved from: https://frontiermyanmar.net/en/a-new-chapter-for-the-karen-movement.

Kattelus, M., Rahaman, M.M., & Varis, O. (2014). Myanmar under reform: Emerging pressures on water, energy and food security. *Natural Resources Forum, 38*(2):85–98.

KESAN (Karen Environmental and Social Action Network). (2017). *The Salween Peace Park: A place for all living things to share peacefully*.

KESAN. (2019). *Celebrating the Salween Peace Park proclamation*. Retrieved from: http://kesan.asia/celebrating-the-salween-peace-park-proclamation/.

KNU (Karen National Union). (2015). *Karen National Union - KNU Land Policy*.

KNU (Karen National Union)., & Karen Environmental and Social Action Network (KESAN). (2016). *Salween Peace Park: A vision for an indigenous Karen landscape of human-nature harmony in southeast Myanmar*. Yangon, Myanmar: KNU and KESAN.

Leitner, H., & Miller, B. (2007). Scale and the limitations of ontological debate: A commentary on Marston, Jones and Woodward. *Transactions of the Institute of British Geographers, 32* (1):116–125.

Linton, J. (2010). *What is water? The history of a modern abstraction* (1st ed). Vancouver, Canada: UBC Press.

Linton, J., & Budds, J. (2014). The hydrosocial cycle: Defining and mobilizing a relational-dialectical approach to water. *Geoforum, 57*, 170–180.

Marks, D., & Lebel, L. (2016). Disaster governance and the scalar politics of incomplete decentralization: Fragmented and contested responses to the 2011 floods in Central Thailand. *Habitat International, 52*, 57–66.

Marston, S., Jones, J.P., & Woodward, K. (2005). Human geography without scale. *Transactions of the Institute of British Geographers, 30*(4), 416–432.

Mehta, L., Alba, R., Bolding, A., Denby, K., Derman, B., Hove, T., Manzungu, E., Movik, S., Prabhakaran, P., & Van Koppen, B. (2014). The politics of IWRM in southern Africa. *International Journal of Water Resources Development, 30*(3), 528–542.

Middleton, C., Thabchumpon, T., Lian, V.B., & Pratomlek, O. (2017). *Water governance and access to water in Hakha Town, Chin State, Myanmar: Towards addressing water insecurity:* Bangkok.

Ministry of Transport and Communication (MoT). (2017). *Ayeyarwady Integrated River Basin Management (AIRBM) Project*. Retrieved from: https://www.airbm.org/.

Molle, F. (2015). Examining scalar assumptions: Unpacking the watershed. In E. S. Norman, C. Cook, & A. Cohen (Eds.), *Negotiating water governance* (pp. 17–24). London and New York: Ashgate.

National Water Resources Committee (NWRC). (2015). *Myanmar National Water Policy: NWP*.

National Water Resources Committee (NWRC). (2017). *National Water Resources Committee: Overview and activities 2016-2017*.

Nationwide Ceasefire Agreement (NCA). (2015). *The Nationwide Ceasefire Agreement between the Government of the Republic of the Union of Myanmar and the Ethnic Armed Organizations*.

Nesheim, I., Wathne, B.M., & Tun, Z.L. (2016). Myanmar: Pilot introducing the National Water Framework Directive. *Water Solutions, 1*, 18–28.

Neumann, R.P. (2015). Political ecology of scale. In R.L Bryant (Ed.), *The international handbook of political ecology* (pp. 475–486). Cheltenham, United Kingdom: Edward Elgar Publishing.

Ninh, K.N.B., & Arnold, M. (2016). Decentralization in Myanmar: A nascent and evolving Process. *Journal of Southeast Asian Economics, 33*(2), 224–241.

Norman, E.S., & Bakker, K. (2009). Transgressing scales: Water governance across the Canada-U. S. borderland. *Annals of the Association of American Geographers, 99*(1), 99–117.

Norman, E.S, Cook. C., & Cohen, A. (Eds.). (2015). *Negotiating water governance: Why the politics of scale matter*. London and New York: Ashgate.

Perreault, T. (2015). Beyond the watershed: Rescaling decision-making: Introduction to part II. In E.S, Norman, C. Cook & A. Cohen (Eds.), *Negotiating water governance* (pp. 117–124). London and New York: Ashgate.

Save the Salween Network (SSN), Burma Rivers Network (BRN), Burma Environmental Working Group (BEWG). (2017). *Countrywide gatherings on International Rivers Day to oppose large dams in Burma's conflict zones*.

Sneddon, C., & Fox, C. (2012). Water, geopolitics and economic development in the conceptualization of a region. *Eurasian Geography and Economics*, *53*(1), 143–160.

SPP Steering Committee, Karen National Union (KNU). (2018). *Charter of the Salween Peace Park*.

Suhardiman, D., Rutherford J., & Bright, J. (2017). Putting violent armed conflict in the center of the Salween hydropower debates. *Critical Asian Studies*, *3*(4), 1–16.

Swyngedouw, E. (1997). Neither global nor local: 'Glocalization' and the politics of scale. In K.R Cox (Ed.), *Spaces of globalization* (pp. 137–166). New York and London: Guilford Press.

The Border Consortium (TBC). (2016). *KNU recognizes rights to land restitution for displaced communities*. Retrieved from: http://www.theborderconsortium.org/news/knu-land-policy/.

Union of Myanmar. (2008). *The Constitution of the Republic of the Union of Myanmar*.

van Meel, P., Leewis, M., Tonneijck, M., Leushuis, M., de Groot, K., de Jongh, I., Laboyrie, H., Graas, S., Shubber, Z., Klink, T., Douma, D., & Nauta, T. (2014). *Myanmar integrated water resources managements. Strategic study: Research and analysis, strategies and measures*.

Warner, J., Wester, P., & Bolding, A. (2008). Going with the flow: River basins as the natural units for water management? *Water Policy*, *10*(S2), 121.

Chapter 7
A State of Knowledge of the Salween River: An Overview of Civil Society Research

Vanessa Lamb, Carl Middleton, Saw John Bright, Saw Tha Phoe, Naw Aye Aye Myaing, Nang Hom Kham, Sai Aum Khay, Nang Sam Paung Hom, Nang Aye Tin, Nang Shining, Yu Xiaogang, Chen Xiangxue and Chayan Vaddhanaphuti

7.1 Introduction

The Salween is a transboundary river supporting the livelihoods of more than ten million people in the basin (Johnston et al. 2017). Known by many names (see Lamb, Chap. 2, this volume)—*Gyalmo Ngulchu* in the Tibetan region where it originates, *Nu Jiang* in Yunnan, *Salawin* in Thailand, *Nam Khone* in Myanmar's (Burma) Shan State, and *Thanlwin* more generally throughout the country of

Vanessa Lamb, Lecturer, School of Geography, University of Melbourne, Melbourne, Australia; Email: Vanessa.lamb@unimelb.edu.au.

Carl Middleton, Director, Center of Excellence for Resource Politics in Social Development, Center for Social Development Studies, Faculty of Political Science, Chulalongkorn University, Bangkok, Thailand; Email: carl.chulalongkorn@gmail.com.

Saw John Bright, Water Governance Program Coordinator, Karen Environmental and Social Action Network (KESAN); Email: hserdohtoo@gmail.com.

Saw Tha Phoe, Campaign Coordinator, Karen Environmental and Social Action Network (KESAN), Email: phoehein@gmail.com.

Naw Aye Aye Myaing, staff member, Karen Environmental and Social Action Network (KESAN), Email: nawayeayemyaing@gmail.com.

Nang Hom Kham, member, Mong Pan Youth Association (MPYA) and ERS Alumni Fellow, EarthRights International; Email: aeliterann@gmail.com.

Sai Aum Khay, filmmaker and member, Mong Pan Youth Association (MPYA).

Nang Sam Paung Hom, member, Mong Pan Youth Association (MPYA).

Nang Aye Tin, member, Mong Pan Youth Association (MPYA).

Nang Shining, founder and Director of Mong Pan Youth Association (MPYA) and co-founder of Weaving Bonds Across Borders; Email: shining.cu@gmail.com.

Yu Xiaogang, Director, Green Watershed; Email: greenwatershed@163.com.

Chen Xiangxue, Project Officer, Green Watershed; Email: chenxx417@163.com.

Chayan Vaddhanaphuti, Professor and founding Director of the Regional Center for Social Science and Sustainable Development (RCSD), Faculty of Social Sciences, Chiang Mai University, Thailand; Email: ethnet@loxinfo.co.th.

C. Middleton and V. Lamb (eds.), *Knowing the Salween River: Resource Politics of a Contested Transboundary River*, The Anthropocene: Politik—Economics—Society—Science 27, https://doi.org/10.1007/978-3-319-77440-4_7

Myanmar—the river represents a multifaceted resource of cultural, ecological, historical and political values for residents who introduced those many names. This chapter highlights some of the many plans and ways of knowing the 'Salween River,' specifically highlighting a history of civil society research.[1]

As part of this review, we recognize both the values of and threats to this significant river system. At present, one of the biggest threats is the more than 20 large hydropower dam projects planned for the river across the basin countries of China, Myanmar, and Thailand. There are also a range of development and extraction activities ongoing and planned for the basin. Alongside this development attention, there is an identified need for information and baseline study, with the basin being referred to as "data-poor" (Salmivaara et al. 2013) and lacking baseline bio-physical information (Salween University Network Meeting 2016).

In this context, we argue that it is worth recognizing that much of what we do know of the values of and the threats to the Salween River Basin, its peoples, and ecologies, has been documented through civil society research, in the form of reports, films, and advocacy documents. Thus, here we review this extensive civil society scholarship alongside existing plans and policies for the development of the Salween basin, to provide an overview of the multiple knowledges of the basin, and to identify key knowledge gaps in support of a more inclusive, informed, and accountable water governance for the basin.[2] We also consider the different types of action linked to these efforts, and perhaps more importantly, how this research has actively engendered more inclusive and participatory ways of basin governance.

---

[1]The authors, the majority of whom identify as members of "civil society," recognize that civil society is not a homogenous entity and that there are many different actors and interests in this sphere. For instance, there are qualitative differences and distinct goals among different civil society actors, from locally embedded organizations, sometimes having no formal organizational structure, as well as locally embedded CSOs, as contrasted with typically larger and more formally organized and at times professionalized NGOs or international NGOs. We also recognize that there is a complicated trustee relationship between NGO and community which has been critiqued in academic work.

[2]One of the motivations for this "State of Knowledge" chapter stems from the persistence in academic work, particularly recent work on Myanmar, that appears dismissive of so-called 'civil society' or 'activist' work. While the authors do not believe that there is necessarily a clear, hard line to divide academic and activist scholarship—both rely in many instances on first-person accounts of environmental change and dispossession—we do hope that presenting this overview will provoke ideas for improved collaboration that includes researchers within the basin, rather than solely on consultant-led knowledge production that is many times instrumentalized for particular development interests.

7.2 Values and Existing Threats in the Salween River Basin

Many civil society groups have been observing for decades the state of the values, the basin's development, and the impacts of conflicts and poor governance of the Salween River (see Table 7.1: Civil Society knowledge production on the Salween River and the people and cultures within the basin). As we understand it, moving towards accountable and inclusive water governance of the Salween River starts with the way we understand its multifaceted values in addition to threats. While there is a recent increase in attention to the Salween, particularly in Myanmar since the 2010–11 political move toward democratic rule, much of what we know of the Salween has been documented prior to that by civil society organizations (CSOs) and non-governmental organizations (NGOs) in the region. This work was carried out at the personal and political risk of civil society groups and researchers.

Table 7.1 Civil society knowledge production on the Salween River and the people and cultures within the basin (English Language; list is illustrative, not exhaustive)

Author/organization	Date	Title of report	Location
Salween Watch	Ongoing	Current Status of Dams on the Salween River	Basin-wide
TERRA	1999	The Salween – My River, My Natural Belonging	Thailand
EarthRights International	2001	Fatally Flawed: The Tasang Dam on the Salween River	Myanmar
SEARIN and CSDS, et al.	2004	Salween Under Threat	Basin-wide
Karen Rivers Watch	2004	Damming at Gunpoint: Burma Army Atrocities pave the way for Salween Dams in Karen State	Myanmar
Committee of Researchers of the Salween Sgaw Karen	2005	Thai Baan Research at the Salween	Thailand-Myanmar border
Shan Sapawa Environmental Organisation	2006	Warning Signs: An Update on Plans to Dam the Salween in Burma's Shan State	Myanmar
TERRA, local villagers, and experts	2007	Salween: Source of Life and Livelihoods	Thailand-Myanmar border
Mon Youth Progressive Organization	2007	In the Balance: Salween Dams Threaten Downstream Communities in Burma	Myanmar
Kesan, local villagers and experts	2008	Khoe Kay: Biodiversity in Peril	Thai-Myanmar border
Karenni Development Research Group	2011	Report: Stop the Dam Offensive Against the Karenni	Myanmar

(continued)

Table 7.1 (continued)

Author/organization	Date	Title of report	Location
REAM, TERRA, MEENet	2013	Regional Conference on the Value of the Thanlwin/Salween River: Ecosystem Resources Conservation and Management Report	Myanmar
TERRA	2014	Proceedings: Regional Conference on the "Value of the Thanlwin/ Salween River: Ecosystem Resource Conservation and Management"	Myanmar
Paung Ku, REAM, TERRA, and MEENet Network Partners	2015	The Salween Basin Darebauk River Bank	Myanmar
KESAN	2017	Salween Peace Park	Myanmar

Source Created by the authors

Since at least the 1990s, groups in the country then more commonly referred to as 'Burma', provided continuing insights into critical human rights abuses that were not necessarily covered by academic documents, or at least, not from a local perspective. Many groups were not working in the national centers, but in the ethnic states and at the political boundary zones. They were mainly focused on issues of development, displacement and human rights. Thailand, despite its small share of only five per cent of the overall basin population, is active in civil society research where local NGOs are carrying out action research on the Salween River-border (Committee of Researchers of the Salween Sgaw Karen 2005; Chantavong/ Longcharoen 2007). Ongoing work in China has successfully mobilized broad attention to the river's challenges with a wide significance, not least because what happens in China "will have ripple effects far beyond national boundary lines" (Xiaogang et al. 2018: 2). This work in China, Myanmar, and Thailand, as well as further international research, has documented an ongoing exclusion of people's participation in decision-making processes and governance of the Salween, particularly related to marginalized groups, including ethnic minorities.

7.2.1 Livelihood Values

In assessing this body of Salween Civil Society Research, livelihood values are paramount. Mr. Nu Chamnankriripai, a village researcher in Thailand who participated in local research, explained, "*Engineers may see only water, rocks, and sand. But we see our fishing grounds, riverbank gardens, and our lives.*" In Thailand, village researchers like Mr. Nu are working with academics and NGOs and have identified 18 ecological systems specific to the Salween and a rich diversity of fish species that are a source of food and livelihoods income to those living along the river (Committee of Researchers of the Salween Sgaw Karen 2005).

In Myanmar, the river sustains fisheries, which provide a key source of dietary protein, and also supports floodplain agriculture and farmlands. Recent research by young people in Shan State has emphasized not only the significance of water access to remote villages along the Salween, but also the important gender dimension in the context of proposed water and natural resources developments in Shan State (Mong Pan Youth 2017).

In Karen State, local communities have documented effective local livelihoods and governance mechanisms for the river. They propose that a better representation of the Salween River values would be peace and conservation, as proposed in the Salween Peace Park, rather than conflict or exclusive development (KESAN 2017). In one case, a community living along the Salween, outside the city of Hpa An, conducted research around water access and governance. Their report on the situation shows the range of values documented within the research on water management of Daw Lar Lake, highlighting that the value of the lake is not measured solely by its economic productivity. It is also based on a holistic set of practices, "embracing religious, cultural, economic, and customary legal systems which are interconnected with the rich biodiversity of the lake's watershed" (KESAN 2018: 2).

Similarly, in Yunnan, research on local governance and values continues (Green Watershed 2017) and the Salween is now part of the China Nu River Gorge National Park. However, unlike the Salween Peace Park proposal, the role of local indigenous people in the creation and management this National Park is not clear. In China, while there is continuing research on the river from a technical or academic perspective, one point of critique is that there exists very little work outside of civil society reports, that has brought ethnic residents' concerns into formal and informal conversations about decision-making and development of the river.

7.2.2 *Biodiversity Values*

The work by civil society has also emphasized how the river serves important biodiversity functions across the basin. In China, the Nu is part of the Three Parallel Rivers World Heritage Site, which UNESCO calls "an epicentre of Chinese biodiversity" and one of the richest temperate regions of the world (UNESCO 2003). In Thailand, the river flows through the Salween National Park and Salween Wildlife Sanctuary, supporting wildlife biodiversity located along a 120 km long stretch of the border between Thailand and Myanmar. Downstream of this stretch there are two wildlife sanctuaries in Karen State and moreover, at the Khoe Kay river bend the river supports many endemic species, of which 42 are IUCN Red Listed species (KESAN 2008). Also significant is the unique Thousand Islands area in Shan State that not only supports rare species but is home to rare limestone formations (Action for Shan State Rivers 2016).

7.2.3 Values in Addition to Threats

The values of the river are diverse (and this summary is not exhaustive); they are also necessary to understand for effective and accountable water governance. As discussed at the 2018 Salween Research Workshop, in understanding a river system "the temptation is to start with threats" (Salween Studies 2018) rather than values. However, values are what define these threats and highlight what is at stake. It is mapping out these values by various civil society actors over past decades that elucidates the significance of the existing plans and threats to these cultural, livelihood, and biodiversity conservation values of the Salween.

Civil society actors have, then, been documenting the many values of the Salween across the basin since at least the 1990s. This work continues to focus on the exclusion of meaningful participation in decision-making processes and governance of the Salween, particularly concerning marginalized groups, including ethnic minorities.

7.2.4 Threats to the Salween

The threats at present stem from a range of competing development and governance pressures that tend to overlook or misunderstand the multifaceted values of the river. The threats include over 20 large hydroelectric dams proposed for the mainstream of the Salween River, with investments from the national governments of Myanmar, Thailand, and China alongside private firms in the pursuit of energy development in particular. Chapter 3, this volume, with details of all planned projects, presents the best available information of the proposed hydropower projects on the Salween mainstream, while also recognizing that the status of each project is not necessarily available to the public. How these projects would proceed is unclear, with anticipated effects to the lives and values of those who depend on the river and live in the basin. This concerns also the "potentially triggering disastrous earthquakes and dam breaks in this seismically active region" (International Rivers 2012). As such, this development is not proceeding without contestations over governance and decision-making.

While recent proposals are significant and represent an increase in the number of projects and electricity generation, it is also worth noting that a longer history of hydropower development planning in the river basin dates back to at least the 1970s (Paoletto/Uitto 1996; TERRA 2014a) where plans were developed for energy production and water transfer. In the 1980s, plans of establishing a Salween River Committee between Thailand and Myanmar were presented in the context of proposed Salween water diversion projects that would supply water to Thailand (TERRA 2014b). This history is important to note because many of the present-day plans rely on earlier planning schemes and technical reports.

7.3 Conflict and Peace

In Myanmar, while natural resource use is critically tied to the ongoing peace process, the Government of Myanmar has made plans to build eight projects on the Salween mainstream in Myanmar. Four of these eight dams—the Kunlong, Nongpa, Man Taung, and Mongton—are located in Shan State, while the Hatgyi and Ywathit dams are located in Karen[3] and Karenni States respectively (Salween Watch 2014, 2016). Two projects, Dagwin and Weigyi, were proposed along the stretch of the Salween that forms the political border between Thailand and Myanmar.[4] An important component of the projects proposed for the lower Salween basin is that much of the electricity generated would be sold abroad to China or Thailand; again, these would not be used to meet Myanmar's need for domestic power, despite electricity shortages in Myanmar itself.

The most significant recent development update in the Myanmar context is the pronouncement from Shan State. In July 2016, the State Minister for Finance and Planning announced that projects were suspended pending cost-benefit field analyses (Chan Mya Htwe 2016). This follows work by civil society organizations on the ground who have called attention to the lack of public participation in the decision-making and public consultations around the Mongton dam project (SHRF 2015; SWAN 2015). In a seeming reversal, however, in February 2018 the Union government allowed permission to Chinese surveyors from China Three Gorges Corporation to visit the dam site (China Three Gorges Letter to the Electricity and Energy Ministry, Feb 2018), underlining that even if some of the concerns civil society research shows are taken up, the decisions on these projects are very much in flux.

Of concern for these development plans in Myanmar are the lack of adequate processes and procedures for environmental assessments, and lack of available information about the plans and planning processes. For example, there are very few specific details in the public domain about the Mongton dam in Shan State, which if built would be the tallest dam in Asia.[5] While Myanmar's Environmental Conservation Law outlines Environmental Impact Assessment (EIA) procedures (Htoo 2016), the recent EIA experience with the Mongton dam has highlighted how hydropower development planning in Myanmar is discussed and decided behind closed doors, involving no or very little public participation. This holds despite the recent attempts at an EIA for the project and the significant media coverage attracted by the problematic EIA process. Here, it is worth noting that media has been a very important avenue for highlighting the findings from these various

[3]Karen State is also known as Kayin State. The latter name was designated by the Myanmar military Government in 1989.

[4]The 2017 SEA of the Myanmar Hydropower Sector Final Report, p. 37, fn 31, notes that these two projects are now cancelled.

[5]The *Asia Times* (11 Feb 2018) reports that the Myanmar government has asked that the project be "split in two".

research efforts. This is significant, compared to, for instance, documentation by the scientific community (journal articles) which are not necessarily formatted to reach public audiences or they lack links with or understanding of media.

In addition, in Myanmar subnational governance (and conflict) continues to be a pressing issue across the country, not only in terms of resource governance, but regarding broader political participation for the country's ethnic minorities as well (Burke et al. 2017). Forty-two of 50 large hydropower projects recently proposed in the country are located in ethnic states (Burke et al. 2017: 40). CSOs and NGOs have gone far to demonstrate the relationship between conflict and hydropower development and resource extraction, and as noted, work on displacement and human rights has been ongoing (Table 7.1). More recent work by 26 groups in Shan State has highlighted that the energy from Mongton dam will be sold to China, while local residents bear the costs, as seen in their letter to Daw Aung Sang Su Kyi in August 2016,

> We wish to remind you that the Salween river basin has been a conflict area for decades . . . Pushing ahead with these unpopular dams will inevitably lead to more Burma Army militarization, increased conflict, and ongoing atrocities. (Action for Shan State Rivers et al. 2016)

It is not only in Shan state, but in Karen State that there are serious concerns about ongoing conflict and development (KHRG/KRW 2018; see also, Middleton et al., Chap. 3, this volume).

7.4 Threats and Challenges of Governance

In Yunnan Province, China, different kinds of water governance challenges have arisen with the creation of the Nu River area national park. With the approval of the Nu River Gorge National Park, the Yunnan Provincial Government declared, in January 2016, that all small hydropower projects on the Nu River would be stopped (China News 2016). The final draft of China's 13th Five Year Plan (2016–2020), released in November 2016, excluded Nu River dam plans in hydropower development. Many take this as a sign that hydropower projects have been halted, particularly in the national park areas (Jing 2016). This is far from certain, however, as it comes after controversy in the early 2000s when the government halted all Nu-Salween projects due to lack of assessment and public outcry, but then re-opened the proposals.

In terms of Thailand's recent developments, while the Thai military government has not made any recent announcements on the construction of Salween dams, it has announced the revival of plans for a water diversion scheme to draw water from the Salween and Moei Rivers to central Thailand (Deetes 2016). Of note in Thailand are the recent administrative court cases brought about by Thai citizens against Thai companies investing abroad, such as the international arm of the Electricity Generating Authority of Thailand, EGATi, to hold them accountable to their 'extraterritorial obligations,' and recognize local community rights to rivers and resources (Deetes 2017; TERRA 2013). One notable outcome was the resolution, passed by the Thai Government in May 2016, to regulate Thailand's outbound investments in line with UN Guiding Principles on Business and Human Rights, and to ensure that the private companies respect the fundamental rights of communities.[6] These efforts to hold governments and companies accountable is important work that can have impacts beyond national borders.

In summary, a range of threats and opportunities face the Salween River. Civil society actors have been documenting local values of the Salween across the basin as well as cataloguing and following the development of the threats. Among them is the inconsistency between the need of developing hydropower in the rural parts of Myanmar for domestic needs and the proposals of selling hydropower to Thailand and China. Accountable and peaceful national governance of the Lower Salween in Myanmar remains a particular challenge with Thailand's evident plans of proceeding with investments abroad.

7.5 Linking Civil Society Actions and Improving Decision-Making for the Salween

There are important questions about the role of/for civil society research and action on the Salween in a changing context, where values, threats, and governance processes of the river are constantly in flux. In assessing past work there are multiple instances where we identify improved decision-making and governance processes in China, Thailand, and Myanmar, even if there is still room for increased accountability and participation of residents in the basin.

Local research and media advocacy in early 2000s in China resulted in serious reconsideration and temporary halting of the large hydropower projects planned for the upper Salween. The activity was led by civil society actors in collaboration with the broader public, responding to the fact that the planning process had been proceeding without impact assessment and consultation.

In Myanmar, recent actions and reporting from Shan State around the inadequacies of the Mongton dam consultations in 2017 has invoked a new set of governance processes at multiple levels. These actions in Shan State are reminiscent of research and action around the Hatgyi dam assessments and consultations in 2009–2011 in Thailand and Myanmar (ERI 2018).

In the Hatgyi case, civil society actors in Thailand and Myanmar not only documented local research in the lead up to the consultations and proposals, but also were in strong positions to work with the National Human Rights Commission of Thailand (NHRCT) to restructure the EIA research documentation and processes

[6]Work to see enforcement of this 2016 Cabinet Resolution is ongoing (Suk 2017).

of consultation. In 2012, for instance, "the NHRCT concluded that the dam would cause transboundary impacts to Thai communities on the Salween and those relying on the river for fisheries and other resources" (ERI 2018: 2). In addition, developers were required to consult communities not only in Thailand, as the initial process had done, but focus on both sides of the river.

In the Hatgyi case, we saw the work by civil society actors, including local residents, effectively raising the profile of a flawed EIA and consultation process, and show that such a move is one step towards more inclusive discussion of the river-border (Lamb 2014).

This raises an important point linked to the outcomes and processes of civil society work on the Salween to date. Much of this work by civil society actors over the past three decades has been about documenting impacts, and increasing the levels of meaningful participation by a range of actors in decision-making, including affected residents. Meaningful participation and improved accountability can shape outcomes and decisions about hydropower development. If concerns of local residents are included and taken seriously, this could shape outcomes in ways that are more supportive of the lives and livelihoods of people in the basin in the long term. This work in many ways is concerned beyond individual hydropower projects, to consider decisions about people's lives and right to self-determination more generally. This is underlined by comments at the 2018 International Day of Action for Rivers along the Salween, including those by Pati Saw Cher Tu Plor, head of an Internally Displaced Persons Camp:

> We need peace and freedom, as these are the only way to bring us equality and to ensure our community's rights to natural resources and land are respected. (as quoted in ERI 2018: 3)

Again, this work can be done and is evidenced, in a range of ways described above—from making change in the terms of a consultation (e.g., Mongton, Hatgyi consultations), to presenting policy advice (e.g., KESAN's proposals for Salween Peace Park), and sometimes seen in governments and companies reconsidering and halting large hydropower projects (e.g., halting Upper Salween dam development).

7.6 Coming Together: Bridging Epistemologies and Policies

While research on the Salween River Basin is proliferating, it is worth highlighting that much of what we do know of the Salween River Basin, its peoples, and its ecologies has been documented through civil society. With increased research attention to the Salween, the work of civil society groups and potential collaboration across NGOs, CSOs, academics, and governments is increasingly important for improved research and decision-making. Also essential is understanding this history of civil society work to document and re-frame development of the Salween, by putting its people and ecologies at the center of the analysis.

However, as we have illustrated here, much of the civil society research on the Salween in China, Myanmar and Thailand has been done on a case-by-case or community basis, or within a national framework for a national audience. There still remains a need and a space for cross-cutting and cross-border collaborative research and documentation.

In the service of expanding basin-wide collaboration, between and across activists, academics, and others, the idea of a "Salween Friendship Partnership" was floated at the January 2016 Salween University Network meeting. The "Salween Friendship Partnership" would bring together civil society actors, academics, community members, and other interested individuals across borders to emphasize transboundary cooperation and highlight economic, cultural, social, and political values of the Salween. In addition to the "Salween Friendship Partnership," academics and policy makers have discussed some ways how their work might support or complement one another to further highlight that there is a wealth of local research about the Salween. This book is one such collaborations that has emerged through these discussions. We hope for further, more dynamic collaborations moving forward.

References

Action for Shan State Rivers. (2016, September 29). *Drowning 1000 islands film* [Video file]. Retrieved from: https://www.youtube.com/watch?v=utwclwNAtVE.

Action for Shan State Rivers *et al*. (2016, August 17). *Open letter from 26 Shan community groups to Daw Aung San Suu Kyi to cancel Salween dams*. Retrieved from Shan Human Rights website: http://shanhumanrights.org/index.php/news-updates/251-open-letter-from-26-shan-community-groups-to-daw-aung-san-suu-kyi-to-cancel-salween-dams.

Asia Times. (2018, February 11). Myanmar power plans could spare the Salween. *Asia Times*. Retrieved from: http://www.atimes.com/article/myanmar-power-plans-spare-salween/.

Burke, A., Williams, N., Barron, P., Jolliffe, K. & Carr, T. (2017). *The contested areas of Myanmar: Subnational conflict, aid, and development*. The Asia Foundation. Retrieved from: https://asiafoundation.org/wp-content/uploads/2017/10/ContestedAreasMyanmarReport.pdf.

Chan, M. H. (2016, July 8). Hydropower dams, major development projects suspended in Shan State: minister. *Myanmar Times*. Retrieved from: https://www.mmtimes.com/national-news/21276-hydropower-dams-major-development-projects-suspended-in-shan-state-minister.html.

Chantavong, M. & Longcharoen, L. (Eds.). (2007). *Salween: River of life and livelihoods*. Bangkok, Thailand: Foundation for Ecological Recovery.

China News. (2016, January 25). 云南将停止怒江小水电开发 推进大峡谷国家公园申报 [Yunnan will promote the development of Nu River Grand Canyon National Park declaration, will stop small hydro]. *China News*. Retrieved from: http://www.chinanews.com/gn/2016/01-25/7732188.shtml.

China Three Gorges. (2018 February). *Letter to the Electricity and Energy Ministry*.

Committee of Researchers of the Salween Sgaw Karen. (2005). *Thai Baan research in the Salween River: Villager's research by the Thai-Karen Communities* (in Thai with English summary). [Documentary film]. Published and produced by SEARIN, Chiang Mai, Thailand. Retrieved from: http://www.livingriversiam.org/4river-tran/4sw/swd_vdo2-tb-research.html.

Deetes, P. (2016). *Salween water diversion project – recent updates.* Presented at Salween University Network Meeting, Chiang Mai, Thailand, January 29, 2016.

Deetes, P. (2017, May 22). Investments should recognise local rights. *Bangkok Post*. Retrieved from: https://www.bangkokpost.com/opinion/opinion/1253930/investments-should-recognise-local-rights.

EarthRights International (ERI). (May 2018). *ERI Briefer. The Hatgyi Dam: A case of Thai investment in Myanmar: Adverse impacts to Salween communities and key recommendations.* Retrieved from: https://earthrights.org/wp-content/uploads/HatgyiBriefer.pdf.

Green Watershed. (2017). *River Free Running, Community Flourishing* [Film]. Retrieved from Salween Stories website: https://www.salweenstories.org/nujiang/.

Htoo, T. (2016, January 15). New environmental impact rules released. *Myanmar Times*. Retrieved from: http://www.mmtimes.com/index.php/business/18490-new-environmental-impact-rules-released.html.

ICEM. (2017). *SEA of the hydropower sector in Myanmar. Baseline assessment report.* Hanoi, Vietnam: International Center for Environmental Management.

IR (International Rivers). (2012). *The Salween River Basin* [Fact Sheet]. Retrieved from International Rivers website: https://www.internationalrivers.org/resources/briefing-current-status-of-dam-projects-on-burma%E2%80%99s-salween-river-7868.

Jing, L. (2016, March 14). China's Grand Canyon: no more small hydropower plants for country's last wild river in scenic Yunnan. *South China Morning Post*. Retrieved from: https://www.scmp.com/news/china/policies-politics/article/1925053/chinas-grand-canyon-no-more-small-hydropower-plants.

Johnston, R., McCartney, M., Liu, S., Ketelsen, T., Taylker, L., Vinh, M.K., Ko Ko Gyi, M., Aung Khin, T., Ma Ma Gyi, Khin. (2017). *State of knowledge: River health in the Salween.* Vientiane, Lao PDR: CGIAR Research Program on Water, Land and Ecosystems. Retrieved from: http://hdl.handle.net/10568/82969.

Karen Rivers Watch. (2004). *Damming at gunpoint: Burma army atrocities pave the way for Salween dams in Karen State.* Retrieved from: http://www.burmariversnetwork.org/images/stories/publications/english/dammingatgunpointenglish.pdf.

KDRF (Karenni Development Research Group). (2006). *Dammed by Burma's generals: The Karenni experience with hydropower development from Lawpita to the Salween.* Retrieved from: https://www.internationalrivers.org/files/attached-files/proposed_salween_dams_revive_development_nightmare_for_karenni_in_burma.pdf.

KESAN (Karen Environmental and Social Action Network). (2008). *Khoe Kay: Biodiversity in peril.* Retrieved from: http://www.kesan.asia/index.php/reports/viewdownload/4-reports/13-khoe-kay-biodiversity-in-peril.

KESAN (Karen Environmental and Social Action Network). (2017). *Salween Peace Park: A vision for an indigenous Karen landscape of human-nature harmony in southeast Myanmar.* Retrieved from: http://kesan.asia/index.php/programs/salween-peace-park-initiative/93-salween-peace-park-a-vision-for-an-indigenous-karen-landscape-of-human-nature-harmony-in-southeast-myanmar.

KESAN (Karen Environmental and Social Action Network). (2018). *Community based water governance: A briefing report on Daw Lar Lake*. Yangon, Myanmar: KESAN. Forthcoming.

KHRG (Karen Human Rights Group) & KRW (Karen Rivers Watch). (June 2018). *Development or destruction? The human rights impact of hydropower development in villages in Southeast Myanmar.* Retrieved from: http://khrg.org/2018/07/18-1-cmt1/development-or-destruction-human-rights-impacts-hydropower-development-villagers.

Lamb, V. (2014). Making governance "good": The production of scale in the environmental impact assessment and governance of the Salween River. *Conservation and Society*, *12*(4), 386–397.

Mong Pan Youth Association. (2017). *Water and Gender* [A Film by Sai Aung Nawe Oo.] Retrieved from: https://www.salweenstories.org/mongpan/.

Paoletto, G., & Uitto, J.I. (1996). The Salween River: is international development possible? *Asia Pacific Viewpoint*, *37*(3), 269–282.

Salmivaara, A., Kummu, M., Keskinen, M., & Varis, O. (2013). Using global datasets to create environmental profiles for data-poor regions: A case from the Irrawaddy and Salween River Basins. *Environmental Management*, *51*(4), 897–911.

Salween Studies. (2018). Learning from other contexts for the Salween (Panel 8 Discussion and Presentation led by I. Rutherfurd). 2018 Salween Studies Research Workshop, University of Yangon, Myanmar, February 27, 2018. Retrieved from conference proceedings: http://www.csds-chula.org/publications/2018/4/30/conference-proceeding-salween-studies-research-workshop-the-role-of-research-for-a-sustainable-salween-river.

Salween University Network Meeting. (2016, January 29-31). *Shared Vision.* Retrieved from: https://www.academia.edu/29283801/Meeting_Report_TOWARDS_A_SHARED_VISION_FOR_THE_SALWEEN-THANLWIN-NU_.

Salween Watch. (2013, March 13). *Current Status of dam projects on Burma's Salween River.* Retrieved from: http://www.burmalibrary.org/docs15/salween_dams-2013-03-en-red.pdf.

Salween Watch. (2014). *Hydropower projects on the Salween River: An update.* March 14, 2014. Retrieved from: http://www.salweenwatch.org/images/PDF/Salween.ENG14.03.2014.pdf.

Salween Watch. (2016). *Current Status of Dams on the Salween River - February 2016.* Retrieved from: https://www.internationalrivers.org/resources/11286.

SHRF (Shan Human Rights Foundation). (2015). *Situation Update: Naypyidaw must immediately stop its attacks in central Shan State and let communities return home.* Retrieved from SHRF website: http://www.shanhumanrights.org/index.php/news-updates/233-naypyidaw-must-immediately-stop-its-attacks-in-central-shan-state-and-let-communities-return-home.

Shan Women's Action Network (SWAN). (2015, August 25). *Shans present 23,717 signatures against Salween dams to Australian consultants in Yangon.* Retrieved from: http://www.shanwomen.org/publications/press-releases/111-press-release-by-action-for-shan-state-rivers.

Suk, W. (2017). *Writing the rulebook: Will Thailand force its businesses to respect human rights?* Earth Rights International. Retrieved from: https://earthrights.org/blog/writing-the-rulebook-will-thailand-force-its-businesses-to-respect-human-rights/.

TERRA. (1999). The Salween – My river, my natural belonging. *Watershed*, *4*(2), 41–44. Retrieved from: http://www.terraper.org/web/en/key-issues/salween-river/.

TERRA. (2013). *Ethics of investment in Myanmar's power section.* Retrieved from: http://www.terraper.org/web/en/node/1547.

TERRA. (2014a). *Chronology of Hutgyi Dam From 1996-October 2014.* Retrieved from: http://www.terraper.org/web/sites/default/files/key-issues-content/1421399932_en.pdf.

TERRA. (2014b). *Proceedings of The Regional conference on the Value of the Thanlwin/Salween River: Ecosystem Resource Conservation and Management.* Yangon, Myanmar. May 28–29, 2013. Bangkok, Thailand: TERRA. Retrieved: http://www.terraper.org/web/en/key-issues/salween-river.

UNESCO. (2003). Three parallel rivers of Yunnan protected areas. Retrieved from: http://whc.unesco.org/en/list/1083.

Xiaogang, Y., Xiangxue, C., Middleton, C., & Lo, N. (2018). *Charting new pathways towards inclusive and sustainable development of The Nu River Valley.* Retrieved from: http://www.csds-chula.org/publications/2018/20/03/report-charting-new-pathways-towards-inclusive-and-sustainable-development-of-the-nu-river-valley.

Chapter 8
"We Need One Natural River for the Next Generation": Intersectional Feminism and the Nu Jiang Dams Campaign in China

Hannah El-Silimy

8.1 Introduction

Several hours into the two-day journey from Kunming to the Nu River Valley, I turned to the woman next to me, a well-known environmentalist, to discuss my research topic. "I'm writing about the civil society movement to stop the Nu Jiang dam," I explained. She looked surprised. "But there is no civil society. There is no social movement.[1] We don't have any…local people here" (Interview 16, March 2016).

Looking around the minivan full of journalists and prominent Nu Jiang anti-dam activists, I saw that nearly every person was from the urban centers of Beijing or Kunming, highly educated, of the Han ethnic group (the majority ethnic group in China), from elite or middle-class backgrounds, and employed as academics, journalists, and international non-government organization (NGO) workers. It was also striking that there was only one man in the van. Given that this group represents the main campaigners working to stop the dam project, this paper explores the dynamics of a campaign led not by affected people themselves, but predominantly by journalists and urban environmentalists.

[1]I use the term "civil society" to mean any groups or collections of individuals "which are independent from family, government or business, promote a public interest, and do not seek economic profit" (Matelski 2013: 154). This may include NGOs, CSOs, religious and interest groups, and social movements. Based on this definition, there is in fact a Nu Jiang anti-dams civil society even if it does not include local people.

Hannah El-Silimy, Ph.D. student, University of Hawai'i at Mānoa;
Email: hannahel@hawaii.edu.

C. Middleton and V. Lamb (eds.), *Knowing the Salween River: Resource Politics of a Contested Transboundary River*, The Anthropocene: Politik—Economics—Society—Science 27, https://doi.org/10.1007/978-3-319-77440-4_8

While research has been conducted on this significant case study of anti-dam activism, particularly as it represents one of the few success stories of anti-dam campaigns in the Mekong region, most existing research has either neglected or only briefly commented on the social identities of the campaigners, such as their gender, ethnicity or class.[2] This is not unique to analyses of the Nu Jiang campaign or Chinese environmental civil society; as Howell (2007: 416) notes, to date, "civil society theorists have paid scant attention to the gendered nature of civil society." Furthermore, the 'ethnic' nature of civil society has often gone unexamined in studies on civil society in China unless it is ethnic minority led; as Mullaney comments, in the academic world, Han identity "enjoys a powerful and hegemonic neutrality all its own" (2012: 2–3).[3]

Yet, looking around the van, it seemed clear to me that the ability to influence policy and lead environmental campaigns took place in the *context* of the social identity of the campaigners. To not acknowledge this would miss a key understanding of the dynamics in China of how civil society's influence, strategies and successes vary especially according to who is campaigning, but also when, where, and on what issue. While the campaign's end result was successful, the dynamics of the campaign also bring to light some of the gender, class, and ethnic inequalities and cleavages within Chinese society.

By and large, the local communities who would be impacted by the project and are mainly from rural and ethnic minority backgrounds, have a limited voice in the national and international debates about the project. In terms of gender, while women are leaders of the campaign, it is the predominantly male scientists and policy makers who are conferred with 'expert' status and with the most authority to influence decision makers. Therefore, the Nu Jiang case can also reveal the social and structural limitations as well as opportunities that actors in social movements need to navigate.

To present my research findings, this chapter will start with the project's methodology, followed by background and context on civil society and environmentalism in China as linked to the Nu Jiang dams project and campaign. I will then present an analysis of the research by examining how an intersectional feminist framework can help reveal a deeper understanding of China's social and political context, civil society, and the Nu Jiang campaign case. Finally, I discuss how gender, ethnicity, and class influenced the political opportunities available during the Nu Jiang campaign and its ultimate outcome both in making it 'successful' and in considering the potential to reinforce status-quo power relationships.

[2]For other Nu Jiang research, see Xie/Van der Heijin (2010) on political opportunities, see Yang/Calhoun (2007) for research on the role of media, and see Matsuzawa (2011) on the use of transnational ties and activism.

[3]For a more comprehensive examination of Han-ness, Tibetan-ness and ethnic identity in China, see Harrell (2012), Mullaney (2013), Zenz (2013).

8.2 Methodology

I used a primarily qualitative research approach, collecting both secondary and primary information through semi-structured in-depth interviews, informal interviews and participant observation. To conduct participant observation, I joined a group of environmentalists visiting the Nu Jiang River Valley and visited the offices of several environmental NGOs (ENGOs) to observe their activities. I also collected a small amount of quantitative data through surveys of 11 ENGOs in Yunnan and Beijing on the gender and ethnicity makeup of their staff, though considering the small sample size, I consider this to be more illustrative than definitive data.

I conducted a total of 21 semi-structured in-depth interviews and an additional 10 informal interviews. For each interview, I either recorded or took notes depending on the context, and then typed up the interviews afterwards using the qualitative software QDA Miner Lite. Interviews were conducted mainly with ENGO workers in Yunnan and Beijing, with the majority having a specific Nu Jiang focus, as well as several journalists, academics, and local residents of the Nu Jiang. Out of 21 in depth interviews, 17 were women and 4 were men; 13 were Han and 8 were Tibetan, Naxi, Bai, Lisu and Hani. To conduct the research, I visited Yunnan Province twice and Beijing once between November 2015 and April 2016.

This research was done with a research approach drawing on an understanding of intersection feminism. One issue I faced as a researcher was that of representation. Intersectional feminism "ponder(s) the question of who has the right to write culture for whom" (Behar 1995: 7). As a person of Asian descent from the West who works with NGOs in Asia, I am not an 'insider' but nor am I an absolute 'outsider'; this position made accessing interviews, some of which were conducted with friends and colleagues much easier, but there are still many limitations to my knowledge of social and ethnic dynamics in China. Throughout the research I have attempted to draw upon the lessons of my own experiences with gender, ethnicity and class as a Westerner and ethnic minority in my own context without assuming Western standards or values to be analogous with those in China. In this process, I have consulted peers and colleagues identifying from a diverse set of ethnicities from China to identify and reflect on my assumptions as a Western researcher; however, any errors or deficiencies of this chapter remain my own.

Finally, security and ethics were a significant concern, given the sensitivity of hydropower issues in China. As most interviewees preferred not to be publicly identified due to concerns about political repercussions, interviews are denoted by numbers and as a security precaution, exact dates or the location of interviews are not published. The only named interviewees are Dr. Yu and Wang Yongchen, two high profile figures who permitted the use of their real names.

8.3 Background: Civil Society and Environmentalism in Authoritarian China

Civil society groups and ENGOs in particular have proliferated to become a significant force in China since the early 1990s. As Sun/Zhao (2008: 144) note, ENGOs in China have "considerable mobilization capacity, international networks, and a history of several successful environmental campaigns". Despite restrictions which vary from bureaucratic obstacles in receiving registration and funding to police monitoring, intimidation and repression employed by the state, ENGOs are able to find strategic ways to operate and navigate such conditions, through strategies including "negotiation, evasion or feigned compliance" (Saich 2000: 125–6).

Moreover, civil society does not necessarily have to be oppositional to the state, and may even contribute to the stability of a regime when its activities and aims align with the regime's interests. Teets (2014: xi) demonstrates this point with the concept of "consultative authoritarianism", where civil society groups may be allowed and even encouraged by authoritarian regimes when their collaboration is found to be useful to the government and they are deemed sufficiently non-threatening—a condition which is frequently correlated in China with ethnic identity and the topic at issue. As one female Beijing based academic in her 40s explained,

> The government's attitude to NGOs is that *you can be my helper but you are not my master*. NGOs can have more political space compared to the past but only if you follow what is the government's interest…you can work as long as you help me with my agenda. You can't come in with your own agenda that could be more powerful than me, like women's rights groups….Environmental groups go in the same line with government policy, so they can be allowed and given more space. (Interview 18, April 2016, *emphasis added*)

As this quotation highlights, environmental groups have been able to grow as environmental protection becomes a top concern for China's authorities. This concern is due to multiple reasons, including apprehension among political leaders over the impact of environmental damage on both GDP and social unrest (Economy 2004). As a result, ENGOs in China have experienced political opportunities not available to groups working on other issues such as women's rights. In some cases, they have successfully been able to push their work beyond what Xue Ye from Friends of Nature labelled "bird watching, tree planting and garbage collection" (as cited in Büsgen 2006: 26) to work more on political issues such as hydropower or pollution, on the occasions that such work also allies with government interests.

While previous research on environmental civil society in China has rarely engaged with the issues of gender, class, and ethnicity, this research will demonstrate its importance. Without discussing identity in the context of civil society, we have less context to understand the tactics and strategies as well as successes or failures of groups in China and elsewhere.

First, groups in China which use activist tactics are often lauded and paid attention by Western researchers, donors, and stakeholders, while 'softer' tactics such as conservation and cultural preservation spark less enthusiasm. Yet, there is a lack of understanding that groups make strategic choices which are highly related to ethnicity and location, with Beijing, urban, and Han groups having more political opportunity to engage in direct advocacy on environmental issues. Meanwhile, ethnic minority groups are more likely to employ strategies such as cultural preservation to be able to conduct their work with less repression, as I discuss further below. Second, the success or failure of a group's activities should also be understood in their social context in order to avoid drawing wider conclusions based on incomplete information. It would be dangerous for example, to assume that the Nu Jiang campaign's strategies and successes could be replicated by other groups in China, without considering the diverse social and power structures that they must navigate.

8.4 Background: Nu Jiang Project and Campaign

The Nu Jiang, which means "Angry River" in Chinese, is known for being one of the only major rivers in China without mainstream hydropower development, as well as a site of spectacular beauty. Originating on the Tibetan Plateau, the Nu Jiang, or Nu River, flows through the Tibetan Autonomous Region before passing through the most western regions of Yunnan province in China, after which it crosses the border down to Myanmar and Thailand, where it is known as the Thanlwin and Salween respectively.

Hearing of the Yunnan provincial government's plan to build a cascade of 13 dams and two reservoirs on the mainstream of the Nu River in cooperation with the Huadian Power International company, environmentalists mobilized to take action to try to keep this 'last' free-flowing river undammed (Lin 2007: 169). The key leaders of the campaign were Wang Yongchen of Green Earth Volunteers and Dr. Yu Xiaogang of Green Watershed in partnership with other ENGOs, mainly International Rivers, Friends of Nature, Global Village Beijing and the now-defunct China Rivers Network, as well as independent NGO workers, volunteers, journalists, academics and scientists who played supporting roles. The campaign also worked closely together with individuals from the State Environmental Protection Authority (SEPA)[4] who provided key information and support to environmentalists (Sun/Zhao 2008: 151–156).

The ensuing campaign from 2003 to 4 resulted in the suspension of the dam in February 2004 by Premier Wen Jiabao and was "one of the most high advocacy campaigns....and controversial cases of NGO advocacy as [of] yet" (Büsgen 2006: 6).

[4]Now the Ministry of Environmental Protection (MEP); will be referred to throughout this paper as SEPA in reference to its name at the time of the events.

While the project was suspended in 2004, the campaign has continued until the present day, due to concerns that the project could be revived again. Since 2004, environmentalists have used a variety of strategies to keep pressure on the local authorities and central government, including an annual media trip for journalists and NGOs to the Nu Jiang, collecting scientific data and evidence about geological instability in the area, meeting with policy makers, making connections with downstream NGOs and international organizations, and public awareness activities such as photo exhibitions in Kunming and Beijing (Interview 13, March 2016; Interviews with Wang Yongchen, Dr. Yu Xiaogang, March 2016).

Since the Chinese government has not formally cancelled the projects, it is difficult to claim the campaign as an outright success if success is measured only by formal cancelation. However, campaigners believe the Nu Jiang dams are unlikely to be built because the Nu Jiang area has been established as a national park area. Two parks have already been approved as of May 2016, namely the Grand Canyon National Park and Dulong River National Park (Zhaohui 2016). Furthermore, the latest 5-year plan from the National Energy Administration released in late 2016 does not include mention of the dams on the Nu (Phillips 2016); the lead environmentalists believe this to mean that the dams are unlikely to be built in the future. While environmentalists also have concerns with the environmental impact of the large scale tourism development the national parks are likely to bring, they still unanimously expressed that this would be a superior result than the dam project in their opinion (Interview 13, March 2016; Interviews with Wang Yongchen and Dr. Yu Xiaogang, March 2016).

In the following sections, I bring an intersectional feminist analysis to these interviews and experiences of the Nu Jiang Campaign.

8.5 Intersectional Feminist Analysis of the Nu Jiang Campaign

Using an intersectional feminist lens, this section will show that the available political opportunities and outcome of the campaign were influenced by gender, ethnicity and class factors, starting by contextualizing this research within the field of existing intersectional feminist writings.

8.5.1 Intersectional Feminist Literature and Identity in China

Scholarship on intersectional feminism provides an important contextual setting for understanding the dynamics of civil society and women's experiences in the Global

South[5] such as the Nu Jiang campaign, for a number of reasons. Firstly, intersectional feminist and Third World feminist writers have successfully widened the scope of feminist writings to include the experiences and voices of women of color and women from the Global South, in particular emphasizing the importance of situating women's experiences within broader contexts such as colonialism and neoliberalism as well as patriarchy (Mohanty 1991; see also Crenshaw 1989; Hill Collins 2000).

Secondly, as a field, intersectional feminism has pushed feminist theorists to uproot what Anzaldúa terms "dualistic thinking" and go beyond essentialist categories of men versus women (1987: 80) towards a more nuanced understanding of identity and power. By bringing in analyses of factors that influence women's experiences beyond gender such as class, ethnicity, race, physical ability and sexual orientation, intersectional feminist writers explore what Mohanty (1991) terms the "relations of power" between multiple, cross-cutting, and sometimes fluid identities.

While including voices of women from the Global South is an important step towards a more inclusive feminism, it is still incomplete unless such writings also deal with the "relations of power" within Global South contexts such as China. Writings published under the framework of intersectional feminism frequently refer to "Chinese women" as one group without reference to ethnic or class divergences (see Ong 1995; Han 2000; Barlow 2004). Zheng notes this dynamic and the absence of class (although not ethnic) analysis among Chinese feminists, noting that "in sharp contrast to transnational feminist emphases on multiple systems of oppression and intersectionality of gender, class, race, ethnicity, sexuality, and so on, the absence of "class" in Chinese feminist articulation is glaring" (2010: 113).

In the context of China, 'woman' is an identity shared by more than half a billion people, and can mean a great range of human experiences, so to speak only in terms of gender is only one node among many intersections. 'Woman' could mean a Han migrant worker woman in Guangzhou; a Tibetan woman with a Ph.D. living in the city in Kunming, Yunnan; a Naxi woman farmer living in a remote village; it could also mean a highly educated Han woman from a politically connected family in Beijing. All of these women face different privileges and disadvantages related to their multiple identities; thus, an intersectional feminist perspective reveals why it is important to ask precisely which women we are talking about when we talk about women (Brooks/Hesse Biber 2007). For these reasons, this chapter aims to contribute towards a deeper analysis and interrogation of the "Global South" woman within feminist literature, not only in opposition to the West, but also within the Global South context, given that ethnic and class stratification and inequality also exists *within* China and other Global South contexts.

[5]While "Third World" is commonly used in intersectional feminist literature, in this chapter I prefer to use "Global South" as a term with less potentially pejorative connotations, particularly given that China would have been considered a "Second World" country according to the original meaning of the term.

8.5.2 *Nu Jiang Campaign Profile and Identities*

The Nu Jiang campaign profile shows a small group of highly educated, urban professional environmentalists, with limited grassroots participation. The campaign almost perfectly meets Jenkin's definition of a professional Social Movement Organization (SMO) with "outside leadership, full time paid staff, small or nonexistent membership, resources from conscience constituencies, and actions that 'speak for' rather than involve an aggrieved group" (1983: 533).

Han women from Beijing and Kunming make up the majority of the Nu Jiang environmentalists. Out of the three organizations still highly active on the Nu Jiang dam, all the staff and leaders are women except for Dr. Yu of Green Watershed, and all are Han or Western staff based in cities (Interviews 13, March 2016; Interview 4, November 2015; Interview with Wang Yonchgen, March 2016). On the other hand, the dam-affected area, the Nu Jiang Lisu Autonomous Prefecture is one of the more remote and least developed prefectures in China; over half of its residents live under the poverty line, and most are ethnic minorities (Hu/Li 2014). There are nine ethnic groups living in the Nu Jiang area, with the majority of the population being ethnic Nu, Lisu, Bai, or Tibetan (Grumbine 2010).

These dynamics are by no means limited to the Nu Jiang campaign, but I found to be common among Chinese ENGOs that I interviewed. Out of the eleven ENGOs I interviewed in Beijing and Yunnan province, eight had female leaders and almost half of the organizations had no male staff. Meanwhile, all eleven ENGOs were led by an individual of ethnic Han identity, and seven had no ethnic minority staff.

The reasons for these dynamics are multifaceted. I will go through three key issues in turn, gender, ethnicity, and then class. Firstly, in terms of gender, women are highly represented in civil society in China, for reasons both to do with women's leadership as well as gender inequality. While some interviewees felt positive about women's representation in civil society, the majority of women NGO workers interviewed saw this dynamic as a "gender burden" that women took on by working for low wages, facing risk of arrest and intimidation, and little social recognition in return. As one woman in her 30s working in an ENGO in Yunnan stated,

> In China we say that working in a NGO or CSO job is "outside the system". Working for government or official job "inside the system" is safer, more secure…Although we see many women working in civil society, it is also because civil society jobs are considered as women's work: low pay, risky, dangerous and undesirable. (Anonymous interview) (Interview 10, December 2015)

Despite acknowledging this inequality, many women interviewees did not view themselves as victims, pointing out that they also experienced freedom from the social expectations to be high wage earners that men face, and could instead do the work that they felt passionate about. This included, as another female ENGO worker in her 20s, explained,

> I think it's also positive: women want to do this work (in ENGOs), women also have more economic freedom to do this work. Men have no choice, they must get a high paying job, maybe they can only work in an NGO after they retire. (Interview 2, November 2015)

The female environmentalists working on the Nu Jiang are therefore both advantaged and disadvantaged by their gender; on the one hand, they have the freedom to work on issues they care about with less social pressure, but they also take on more risky and dangerous work. As civil society members, their work is considered "low status," so they work with 'experts' such as scientists, who are usually male, to bolster their credibility with the government.

Second, in terms of ethnicity, despite the Nu Jiang area being predominantly inhabited by ethnic minorities, all of the key environmentalists are of Han ethnicity and none are from the area itself. One reason for this is that while many people of ethnic minority backgrounds do work in civil society groups, ethnic minority staff or groups based in ethnic minority areas often avoid working on issues such as hydropower as doing so would result in being targeted for heightened surveillance (Interview 3, November 2015; Interview 8, December 2015). In order to avoid increased targeting by local authorities, their work tends to focus on less sensitive issues such as cultural preservation, conservation or environmental education (Interview 8, December 2015; Interview 9, December 2015).

For example, one male ethnic minority NGO worker in Yunnan responded, "I'm not allowed to talk about hydropower" when asked. Instead, he explained that their organization focuses on environmental education related to river issues, stating "we use a positive way. We make everyone like it (nature), if they like it, they will protect it. Then we are also safe" (Interview 9, December 2015). For the most part, only Han urban environmentalists had the political opportunity or space to openly address environmental justice issues such as dams, a point that will be further elaborated in this chapter.

As well as political repression, one female NGO worker who had previously been involved in the campaign shared her feelings about lack of trust between local communities and the NGOs preventing more ethnic minority and local participation by stating,

> We can't say that local people really trust us. Maybe it's not that they don't trust, but I think if you speak their language, it certainly would make a huge difference. Not only language, but also, I felt whenever I was in Nu Jiang, I never met anyone that I felt we can openly discuss anything, I'm just asking questions, questions, questions. They don't come to you, they don't really open up, and also I think the local people have an image of the Han people. (Interview 17, March 2016)

While surveillance may be the main reason for lack of local participation, lack of trust between local ethnic minority communities and Han-led ENGOs is also a factor to be considered, as the above quote illustrates. Of course, this challenge is likely exacerbated by political repression; as the ENGO workers are often followed by local police when visiting the area, this is also a factor which significantly hinders building relationships with local communities.

Thirdly, in terms of class, it is well known that most ENGOs in China come from the middle class. As Yang (2010: 123) notes, "Chinese environmentalists are well-educated urban professionals. On the spectrum of the burgeoning middle class, they represent the more intellectually oriented elements and are distinguished from business and political elites…environmentalists resemble intellectuals more than other social strata in Chinese society."

The intellectual status of Chinese environmentalists is a both an advantage and disadvantage in the Nu Jiang context. On the one hand, their status gives them what Yang terms "transnational competence" including an ability to understand international NGO culture such as proposal writing, as well as social capital, in particular media connections and access to some members of the political establishment. However, he also raises concerns that this limits their appeal beyond the middle class, arguing that "it may hinder the building of broad-based social alliances" (Yang 2010: 122–124), a dynamic which is reflected in the largely middle-class nature of the Nu Jiang campaign.

The next section of this chapter will discuss how these gender, ethnic and class dynamics influenced the strategy and outcome of the Nu Jiang campaign reading this work through the framework of political opportunities.

8.6 Political Opportunities and the Nu Jiang Campaign

This section explores how the Nu Jiang environmentalists, as well-connected and highly educated Han professionals, were able to operationalize the political opportunities and advantages necessary to influence policy change. I consider how these identities (the gender, ethnicity, and class of these 'professionals') matter to political opportunity, and how, together, they may provide particular kinds of political influence and opportunities, and restrain others. In fact, the Nu Jiang campaign is a classic example of how complex and dynamic political opportunity can be, and how an intersectional analysis can deepen our understanding of political opportunity. As O'Brien and Stern note when describing political opportunity and environmentalism in China, "there is not one unitary, national, opportunity structure, but multiple, crosscutting openings and obstacles to mobilization", further noting that "the most obvious way to unpack opportunity is by social group… opportunity also varies by region and issue" (2008: 14).

McAdam's (1996) framework of political opportunities identifies four aspects: increasing access to the political system; a divided elite; elite allies; and limited state repression as the most significant factors necessary for groups to be able to influence social change. In the case of the Nu Jiang campaign, I find that the most significant opportunities in this case were a divided elite, elite allies, and relatively low state repression on this particular issue and towards some individuals in the campaign. Through the lens of intersectional feminism, I will unpack how these opportunities were influenced by class, ethnicity and gender.

8.6.1 Elite Allies and a Divided Elite

Elite allies in the State Environmental Protection Agency played an influential role in furthering and legitimizing the Nu Jiang anti-dams movement. A personal friend of Wang Yongchen, Pan Yue joined SEPA in 2003 as a vice minister, and publicly described ENGOS as a 'governmentally' of SEPA. At that time, SEPA had limited resources – only 300 staff members – and struggled to wield significant authority (Economy 2004: 21). Wang Yongchen recounted this alliance,

> In 2003, I have a friend working in the environmental department (SEPA), he was thinking that in the whole of China, we have already set up large scale dams, except two rivers, one of which is the Nu River. So this high official in SEPA, he said *we need (to keep) one natural river for the next generation.* (Interview with Wang Yongchen, March 2016, *emphasis added*)

As recounted by Wang, individuals within SEPA who opposed the Nu Jiang dams saw ENGOs as allies who could help advance their environmental agenda, through influencing the general public, and using their ability to take more critical stances than a government agency could.

Alongside cultivating elite allies, the Nu Jiang campaigners also used the divide between the local authorities, SEPA and the Communist Party of China (CPC) and State leadership in order to push for their agenda. While the project was backed by the Nu Jiang authorities and the Yunnan provincial government as a source of potential revenue, key figures in SEPA opposed the project and the campaigners also believed the central government's top leadership could be persuaded to intervene against the project, given the central government's increasing interest and commitment to environmental issues. Considering these divided interests, Nu Jiang environmentalists focused much of their energies on advocating towards the CPC and central government leadership as potential allies who had the authority to stop the project from moving forward (Interview with Dr. Yu, November 2015).

These strategies of building alliances with elites and exploiting elite divisions were implicitly, rather than explicitly, based upon class, ethnic and gender identities. While Sun and Dingxin argue that "the top leadership tends to offer support to ENGOs when there is no or little opposition to an environmental campaign" (2008: 159), this support is very much dependent upon the identity of the group in question. As well-known and respected Han journalists and environmentalists, Wang Yongchen, Dr. Yu and other campaigners were able to cultivate and use their personal relationships or *guangxi* and their status to gain more support and legitimacy for the Nu Jiang dams cause.

Numerous experiences recounted demonstrate the difference in political opportunities available in China to urban Han groups versus ethnic minority and rural environmentalists. In an interview with an ethnic minority woman in her 30s working with a rural environmental organization in Yunnan province, she discussed the restrictions by the government that her organization faced and their inability to work on sensitive issues such as hydropower.

> Even as a company, the police are still watching us. They come to our activities, they see us as sensitive…After I book a flight ticket, every time, they call me, want to know where I'm going: I don't know exactly how they know when I book a ticket, of course they are monitoring us. But I think it's okay because everything we do is legal and good for people, just protecting culture, language…Working on dams, it's too difficult…And the country is too strong. Asking for compensation is possible, but stopping the dam is impossible. Policemen will put you in jail; they catch one or two people if they protest and make everyone afraid. (Interview 8, December 2015)

While rural and ethnic minority organizations face significant restrictions even for non-political activities, Han and urban organizations, particularly Beijing-based organizations, have been able to advocate on sensitive issues such as the Nu Jiang dams and gain the support of elite allies in SEPA and the central government, though they have also faced challenges and restrictions along the way. In Wang Yongchen's account on the work on the Nu Jiang, she describes using her network to build high level and elite support,

> After I know that the whole country is setting up hydropower, we found some scientists to explain why we need to preserve the Nu River…and in 2003 September we had a big conference in China, and in this conference I told more people about this issue. In this conference we had a lot of film stars. (Interview with Wang, March 2016)

In her account, Wang notes working with government officials, scientists, and film stars in 2003, later moving on to expand her network to high profile journalists, international NGOs, overseas governments and the IUCN. She identifies that she was able to successfully work on the hydropower issue despite the challenges involved,

> Before in 2003 we were thinking we couldn't do something, we are just journalists, we produce articles, we are NGOs, we couldn't influence policy, but 13 years after, so I think maybe we can do something. After we met each other in Chiang Mai, after seeing my film, a lot of people said congratulations, thank you for keeping the river. I think maybe it's not easy, but we can do. (Interview with Wang, March 2016)

As well as class and ethnicity, gender also played a role in the political opportunities available to the Nu Jiang campaigners. Whereas Wang Yongchen had a high profile role in the campaign and was able to make alliances with key elites, most of the other women involved played lower profile roles such as raising awareness and organizing activities; meanwhile men predominantly mainly played the role of 'experts' who were brought in to further bolster the campaign and increase its influence with allies in the government. One of the women activists who was highly involved in the campaign but did not take a high profile stated,

> The women's role in the campaign was to make it more popular and raise popular awareness. Wang Yongchen is a symbol of the campaign, she is emotional, she kept it going for 10 years….Men argue about the theoretical dimensions, while women did the practical work – writing articles, organizing events, trips…seventy percent of the "knowledge" work is done by men – on dams, science, etc. Men went to the EIA meetings, talked to the government. Women generated activities. Wang Yongchen invited men such as geologists to be experts on the trip to Salween. Men were more likely to do the

> negotiations also, such as former government officials, they could also provide knowledge on what they learned working. (Interview 11, December 2015)

While being a woman was not a hindrance to accessing political opportunities, according to those involved it was mainly men who dealt directly with elites and decision makers. One woman environmentalist who also worked closely with the campaign in its earlier years described her opinion of this gender dynamic,

> Wang Yongchen, she is…very emotional and I think her emotion really gets out and affects people…I think for outreach to the common people, to the general public she's very effective but on the opposite if we want to convince government or academics maybe we need people like Yu Xiaogang who can present arguments more rationally. (Interview 17, March 2016)

Therefore, we can say that political opportunities were accessed in this case through the use of gendered strategies. While women took on the 'emotional' and 'practical' roles of organizing events, building alliances and public sympathy for the campaign, men persuaded other male politicians and officials through 'rational' and 'knowledge' work. This division of labor was not consciously decided, or necessarily hierarchical, as the previous female campaigner described, "it [the division of labor] was collaborative though- more do what you can do" (Interview 11, December 2015). Rather, it was the result of wider social dynamics that result in NGO work being predominantly female, and political and scientific work being predominantly male spaces, as well as a rationalist worldview that deems scientific evidence more credible than emotional or personal testimony to some policy makers.

Therefore, while female campaigners such as Wang Yongchen as well-connected urban Han professionals could access political opportunities through alliances with elite decision makers, and take advantage of division among elites, men with the status of policy makers and scientists were also key to making the case convincing to government officials, thus demonstrating class, ethnic, and gender dynamics at play.

8.6.2 State Repression: How Identity and Location Matter

This section highlights how the activists involved in the Nu Jiang campaign were able to operate in a less-restricted space as linked to their position, and the state's propensity for repression *on this particular issue* and *for some of the individuals* involved in campaigning. In this case, the state's restrictions were weak enough to allow for the possibility of a successful campaign.

To explain this, it is necessary to separate the response of the local Nu Jiang authorities, the Yunnan government and the central Beijing government. Throughout the Nu Jiang campaign, a spectrum of repression existed where the degree of repression increased with the degree of 'local-ness' of the individuals involved.

While local, ethnic minority community members faced and still continue to face strong threats and intimidation for speaking about the project, Kunming-based Han environmentalists faced strong pressure but were still able to continue to be active on the issue, while Han Beijing environmentalists faced relatively low repression. Class and ethnicity, as well as location, strongly influenced political opportunity in this case, as demonstrated in interviews and field visits.

At the most local prefectural level, Nu Jiang community members, who are ethnic minorities, continue to be pressured by local authorities not to speak with outsiders about the project, as I saw and experienced while visiting the region. Once our group entered the Nu Jiang prefecture, we were followed by a group of police and local authorities at all times. During meetings with local people along the Nu Jiang, the police took videos and photographs of all our interactions with local people, and specifically instructed the local residents we met with not to talk about the project with us.

Without being able to ask the local authorities directly, we can only speculate why repression is most extreme at the local level; one environmental worker believed it was because of the local authorities' conviction that the dams would be the best solution for economic development and poverty alleviation in the region (Grumbine 2010). Many interviewees also noted that local police and local authorities are known in China to use more repressive tactics than national authorities.

Another reason is because of ethnic politics in China. One NGO staff noted that the majority of the local government officials were Han while the local people were ethnic minorities. Given the context of ethnic activism by Tibetan and Uighur groups and the state's response to such activism, it seems likely that any political activism involving minorities would be seen as more threatening by authorities. The interviewee further explained that the National Security Law passed in China in 2015 makes illegal any activity which could be labeled as divisive between Han and ethnic minority people, and is another tool used to prevent local ethnic minority activism as well as Han activists working together with ethnic minorities (Interview 16, March 2016).

ENGOs, particularly from Yunnan, are aware of the challenges for participation in activism for local communities, and frustrated by the situation. One female campaigner in Yunnan explained her experience trying to work with local communities along the Nu Jiang: "[We] did some workshops with local people, brought people to see the Manwan dam, did trainings for local people on dam impacts, law and so on but they (the local people) cannot actually act. They were threatened a lot, the local government is too strong" (Interview 11, December 2015).

Dr. Yu also detailed his experience trying to organize local communities on the Nu Jiang, explaining that the villagers he had worked with in the past were later resettled under the guise of the "New Rural Development Program" into much larger houses than they had previously lived in, and then strategically controlled, as the resettlement village has a police gate which monitors any visitors to the village. Following their relocation, it has become more difficult to continue communication with those villagers, although he continues to visit the area on a regular basis.

Furthermore, his attempts to bring local residents to national or international events in the past were prevented by local authorities refusing permission for the villagers to travel (Interview with Dr. Yu, March 2016).

Interviewees in Yunnan also expressed frustration at downstream ENGOs and funders who expected them to work with grassroots communities in the Nu Jiang, the Yunnan-based campaigner stating that "they need to understand, it's totally unlike downstream (countries), where people can claim their rights" (Interview 11, Dec 2015). Another Beijing-based environmentalist, comparing their work to downstream anti-dam groups in Thailand and Myanmar, stated with disappointment, "it is my dream to work together with local communities on the Nu Jiang, actually" (Interview 13, March 2016). At the same time, as discussed in the previous section, while local repression may be the main reason for the lack of local participation, the campaign has also faced challenges in building trust and relationships with local communities.

Moving to the provincial level, the most active ENGO in Yunnan Province on the Nu Jiang, Green Watershed faced repression from the Yunnan authorities, perhaps to a lesser degree than the local communities but still of a significant nature. Following the dam's suspension in 2004, Yunnan authorities retaliated by confiscating Dr. Yu's passport and seizing Green Watershed's computers (Büsgen 2006: 41); Dr. Yu also lost his position in the Academy of Social Sciences (Interview with Dr. Yu, March 2016). While some academics interpreted the higher level of repression faced by Green Watershed as the result of their more 'radical' or 'oppositional' tactics compared to Beijing groups (Economy 2005; Teets 2014), the Yunnan-based campaigner firmly rejected this explanation, arguing instead that "Green Watershed did get more punishment because they are a local group and the Yunnan government can control them" (Interview 11, December 2015). Teets herself also notes that the Yunnan and central government took diverging positions towards ENGOs, stating that while "Beijing allowed the formation of more autonomous groups, Yunnan developed more sophisticated tools of state control" (2014: 82). As a local Yunnan organization, Green Watershed played a key role in communicating with local communities and gathering evidence, but was also subject to heightened repression from Yunnan provincial authorities which has continually limited their work at the local level.

In contrast, Han environmentalists from Beijing were able to openly advocate on the Nu Jiang issue. They faced relatively less repression for their work, apart from being followed and monitored while visiting the Nu Jiang region (Interview with Wang Yongchen, March 2016). Of course, being followed while visiting local communities for 13 years is not trivial, thus the term "relatively less." Their limited state repression is a political opportunity which has had a significant outcome on the campaign's ability to sustain and continue at a national and international level over the years.

It should be noted that political opportunity is a dynamic issue, and even Beijing-based groups' ability to work on such issues is now challenged by political developments as of 2016 including the Overseas NGO Law, which severely limits the work of all domestic and international NGOs in China (Wong 2016).

Nevertheless, during the time of this campaign, the ethnic and class backgrounds of the Nu Jiang environmentalists as middle-class Han in Kunming and Beijing had a significant impact on their experience of limited repression.

8.7 Conclusion and Synthesis

As Whittier writes, "systemic inequalities of gender, race, class and sexuality shape both movements and the institutions they confront" (2002: 295). By analyzing how identity and social dynamics shape social movements, we are given greater context to understand the strategies and tactics, as well as successes and failures, of civil society.

While none of the interviewees, nor this chapter, argue that gender, class or ethnicity was the sole or primary reason for the success of the campaign, identity played an important part in the strategies used and ultimate outcome. Ethnicity and class influenced the political opportunities available, as the Nu Jiang environmentalists were able to mobilize popular awareness, and high level elite and political support for their agenda and experienced limited political repression to a degree that they could continue their work, in a way that would be far more difficult for ethnic minority, rural and grassroots environmentalists to do, due to the social and political constraints outlined in this chapter. Meanwhile, as mainly Han women, the Nu Jiang campaign was also shaped by gender dynamics, as accessing these opportunities was also done though collaborating with predominantly male policy makers, scientists and geologists with the status of 'experts.'

While this chapter argues that civil society is influenced by social dynamics, this also raises the point that civil society may in turn "reproduce as well as transform gender inequalities, structures and belief systems" (Kuumba 2001: 2–3), and I would add, class and ethnic inequalities as well. The Nu Jiang campaign, ultimately seems to both transform and reproduce social inequalities. On the one hand, the campaign can be read as an example of those with more opportunity using their power to advocate on behalf of those who are unable to speak out. It is fair to say that the ENGOs have successfully used their relatively privileged position to bring the concerns of local communities to a national and international audience. This viewpoint is evidenced by the ongoing repression of local communities in the Nu Jiang which does not allow them to speak openly about the dam projects without reprisal. At least, the campaign brings the voices of the affected communities to be heard by others, and in this way makes a contribution towards greater social equality in China.

On the other hand, without local people's participation, it is very difficult to know whether the campaign adequately took into consideration the voices of local people impacted by the proposed dam project. Without affected people's participation and leadership, social inequalities are intrinsically reproduced at some level through the campaign itself. In terms of gender, the campaign makes significant progress for women's leadership and women's representation, yet existing gender inequalities determine that scientists and policy makers who are predominantly male are ultimately those conferred with the status of 'experts' with the ability to persuade decision makers.

Due to the dynamics of power, privilege, and repression in China, there are no easy or simple solutions to any of these points raised. While the aim of those involved in social movements such as the Nu Jiang campaign may include an intent towards social justice and leadership of affected communities, this case illustrates why political and social contexts and challenges in China can make it ultimately difficult to fulfill such commitments. Nevertheless, the environmental groups in China must be lauded for finding ways to effect policy change and having ultimately won a significant victory to keep the Nu River flowing freely for future generations.

References

Anzaldúa, G. (1987). *Borderlands/La Frontera: The new mestiza*. San Francisco, CA: Spinsters/ Aunt Lute.

Barlow, T. (2004). Picture more at variance: Of desire and development in the People's Republic of China. In K.K. Bhavnani, J. Foran & P. Kurian (Eds.), *Feminist futures*. London and New York: Zed Books.

Behar, R. (1995). Introduction: Out of exile. In R. Behar & D.A. Gordon (Eds.), *Women writing culture* (pp. 1–33). Berkeley, CA: University of California Press.

Büsgen, M. (2006). *NGOs and the search for Chinese civil society: Environmental non-governmental organizations in the Nujiang campaign*. Working Papers Series No. 422. The Hague, The Netherlands: Institute of Social Studies.

Brooks, A., & Hesse-Biber, S.N. (2007). An invitation to feminist research. In S.N. Hesse-Biber & P. Leavy (Eds.), *Feminist research practice: A primer* (pp. 1–27). Thousand Oaks, CA: SAGE Publications.

Crenshaw, K. (1989). Demarginalizing the intersection of race and sex: A black feminist critique of antidiscrimination doctrine, feminist theory and antiracist politics. *University of Chicago Legal Forum, 140*, 139–167.

Economy, E. (2004). *The river runs black: The environmental challenge to China's future*. Ithaca, NY: Cornell University Press.

Grumbine, R. E. (2010). *Where the dragon meets the angry river: Nature and power in the People's Republic of China*. Washington, DC: Island Press.

Han, A. (2000). Holding up more than half the sky: Marketization and the status of women in China (pp. 392–409). In A.K Wing (Ed.), *Global critical race feminism: An international reader* (pp. 392-409). New York: New York University Press.

Harrell, S. (2012). *Studies on ethnic groups in China: Ways of being ethnic in Southwest China*. Seattle, WA: University of Washington Press.

Hill Collins, P. (2000). *Black feminist thought: Knowledge, consciousness, and the politics of empowerment*. New York, NY: Routledge.

Howell, J. (2007). Gender and civil society: Time for cross-border dialogue. *Social Politics: International Studies in Gender, State & Society*, *14*(4), 415–436.

Hu, Y., & Li, Y. (2014, January 28). Yunnan promises to raise incomes in poor areas. *China Daily*. Retrieved from: http://www.chinadaily.com.cn/china/2014-01/28/content_17262353.htm.

Jenkins, J. (1983). Resource mobilization theory and the study of social movements. *Annual Review of Sociology, 9*, 527–553.

Kuumba, M. B. (2001). *Gender and Social Movements*. Walnut Creek, CA: AltaMira Press.

Lin, T.C. (2007). Environmental NGOs and the anti-dam movements in China: A social movement with Chinese characteristics. *Issues & Studies, 43*(4), 149–184.

Matelski, M. (2013). Civil society and expectations of democratisation from below: The case of Myanmar. *Thammasat Review, 16*, 153–166.

Matsuzawa, S. (2011). Horizontal dynamics in transnational activism: The case of Nu River anti-dam activism in China. *Mobilization, 16*(3), 369–388.

McAdam, D., McCarthy, J.D., & Zald, M.N. (1996). *Comparative perspectives on social movements: Political opportunities, mobilizing structures, and cultural framings*. Cambridge: Cambridge University Press.

Mohanty, C.T. (1991). Cartographies of struggle: Third world women and the politics of feminism. In C.T Mohanty, A. Russo & L. Torres (Eds.), *Third world women and the politics of feminism* (pp. 1–51). Bloomington, IN: Indiana University Press.

Mullaney, T.S. (2012). Critical Han studies: Introduction and prolegomenon. In T.S Mullaney, J.P Leibold, S. Gros & E.A. Vanden-Bussche (Eds.), *Critical Han studies: The history, representation, and identity of China's majority*. Berkeley, CA: University of California Press.

O Brien, K.J., & Stern, R.E. (2008). Introduction: Studying contention in contemporary China. In K.J O'Brien (Ed.), *Popular protest in China* (pp. 11–26). Cambridge, MA: Harvard University Press.

Ong, A. (1995). Women out of China: Traveling tales and traveling theories in postcolonial feminism. In R. Behar & D.A Gordon (Eds.), *Women writing culture* (pp. 350–373). Berkeley, CA: University of California Press.

Phillips, T. (2016, December 2). Joy as China shelves plans to dam 'angry river'. *The Guardian*. Retrieved from: https://www.theguardian.com/world/2016/dec/02/joy-as-china-shelves-plans-to-dam-angry-river.

Saich, T. (2000). Negotiating the state: The development of social organizations in China. *The China Quarterly, 161*, 124–141.

Sun, Y., & Zhao, S. (2008). Environmental campaigns. In K.J O'Brien (Ed.), *Popular protest in China* (pp. 144–163). Cambridge, MA: Harvard University Press.

Teets, J.C. (2014). *Civil society under authoritarianism: The China model*. New York, NY: Cambridge University Press.

Whittier, N. (2002). Meaning and structure in social movements. In D.S. Meyer, N. Whittier, B. Robnett (Eds.), *Social movements: Identity, culture, and the state* (pp. 289–309). Oxford, United Kingdom: Oxford University Press.

Wong, E. (2016, April 28). Clampdown in China restricts 7,000 foreign organizations. *New York Times*. Retrieved from: http://www.nytimes.com/2016/04/29/world/asia/china-foreign-ngo-law.html.

Xie, L., & Van Der Heijden, H-A. (2010). Environmental movements and political opportunities: The case of China. *Social Movement Studies, 9*(1), 51–68.

Yang, G. (2010). Civic environmentalism. In Y-T. Hsing & C.K. Lee (Eds.), *Reclaiming Chinese society: The new social activism* (pp. 119–140). London, United Kingdom: Routledge.

Yang, G., & Calhoun, C. (2007). Media, civil society, and the rise of a green public sphere in China. *China Information, 21*(2), 211–236.

Zenz, A. (2013). *Tibetanness' under threat? Neo-integrationism, minority education and career strategies in Qinghai, P.R. China*. Leiden, The Netherlands: Global Oriental.

Zhaohui, J. (2016, May 3). Nujiang speed up the construction of national park. *China Environment News*. Retrieved from: http://www.chinaenvironment.info/EcoSystems/201605/t20160504_65554.html.

Zheng, W. (2010). Feminist networks. In Y-T. Hsing & C.K. Lee (Eds.), *Reclaiming Chinese society: The new social activism* (pp. 101–118). London, United Kingdom: Routledge.

Chapter 9
Local Context, National Law: The Rights of Karen People on the Salween River in Thailand

Laofang Bundidterdsakul

9.1 Introduction

Tension has existed for decades between the central Thai Government and the ethnic minority communities over access to and use of land and forest along the Salween River. Thailand's national law and policy have not acknowledged the rights to these resources for the ethnic minorities living along the Salween River, even as their livelihoods have long depended upon them (Vandergeest/Peluso 1995). Compounding these challenges is the Hatgyi Dam proposed to be built just across the border in Myanmar that the Electricity Generating Authority of Thailand (EGAT) has backed through a largely opaque decision-making (Magee/Kelley 2009; Middleton et al., Chap. 3, this volume).

This chapter focuses on three Karen villages along the Thai side of the Salween River in Sob Moei and Mae Sariang districts of Mae Hong Son province, and the implications for them of Thailand's national law and its ongoing evolution. I will show how state officials' activities toward these communities, backed by their claims of the legitimacy of the national law, have impacted the ethnic minority people living there. I also show how the communities have responded to these challenges from the state through their own assertions for legitimacy based upon their history, their local traditional practices, and their claims of indigeneity. Overall, I argue that the use of national law is not only about the state trying to regulate livelihood practices, but it is also a challenge to the definition of Karen identity in these villages.

Karen people have lived in this area along the Salween River for centuries. The rich forest is central to their way of living and livelihoods, including providing for food, medicine and building materials. However, with the establishment of the Salween National Park in 1993 in the area village lands were enclosed within it,

Laofang Bundidterdsakul, Director, Legal Advocacy Center for Indigenous Communities; Email: phaajxyaaj@yahoo.co.th.

C. Middleton and V. Lamb (eds.), *Knowing the Salween River: Resource Politics of a Contested Transboundary River*, The Anthropocene: Politik—Economics—Society—Science 27, https://doi.org/10.1007/978-3-319-77440-4_9

thus defining many of the livelihood activities of these communities illegal. Article 16 of the National Park Act (1961) prohibits community members from using or gathering any products from the land and forest within the National Park. This is further complicated by the fact that a significant portion of the population in these three villages do not have citizenship, and under the Land Act (1954) non-Thai citizens are prohibited from owning land. Within the communities themselves, Thai and non-Thai citizens have *de facto* ownership of land; however, this is not recognized by the Thai government.

This chapter argues that there remains an institutional bias in Thai laws and policies that serves to marginalize the interests of ethnic minority people. Government officials have racialized upland ethnic minority communities as backwards and as a national problem that contribute to deforestation, drug use and proliferation, and national security (Vandergeest 2003: 27). This is evident, for example, in the track-record of forest management planning of the Department of National Parks, Wildlife and Plant Conservation (DNP). It labels upland ethnic minority communities, like the Karen, as forest destroyers, accusing traditional practices like shifting cultivation as the major cause of deforestation within the country (Johnson/Forsyth 2002; Vandergeest 2003; Roth 2004; Forsyth/Walker 2008; Wittayapak 2008).

The structure of the chapter is as follows. First, I introduce the three case study villages and discuss about how the communities use the land, forest, and river. Then I discuss the national law and policies that affect these communities, specifically regarding citizenship, community rights, the land law, the forest law and policies related to hydropower projects, and the communities' response. In the chapter's conclusion, I analyze the tension between local practice and national law and how this is a challenge to the definition of Karen identity.

9.2 Local Context: Indigenous Karen and Livelihoods in Villages on the Salween

Community members along the Salween River in Thailand, particularly those that will be directly affected by the Hatgyi dam, are mostly Karen ethnicity (Karen Research Team et al. 2005). This chapter focuses on three communities: Sob Moei village and Mae Sam Lab village in Mae Sam Lab sub-district, Sob Moei district, and Tha Ta Fang village in Mae Youm sub-district, Mae Sariang district, Mae Hong Son province. Tha Ta Fang and Mae Sam Lab villages border the Salween National Park that was established in 1993, and Sob Moei borders the Salween Conservation Area established in 1963. All three villagers are predominantly Karen, but each have their own unique history, which is introduced briefly in the following sections (Fig. 9.1: Location of Salween National Park and Tha Ta Fang, Mae Sam Lab and Sob Moei villages).

Fig. 9.1 Location of Salween National Park and Tha Ta Fang, Mae Sam Lab and Sob Moei villages. *Source* Cartography by Chandra Jayasuriya, University of Melbourne, with permission

9.2.1 Sob Moei Village

Sob Moei village is located at the confluence of the Salween River and the Moei River before they flow into Myanmar. The village is named after this significant feature; Sob in Thai language means 'convergence,' and thus the village name reflects the convergence of the Moei River (with the Salween River). Before Sob Moei village was officially registered by the Thai government, Karen people had lived for centuries in more dispersed clusters in the general vicinity, on both the Myanmar and Thailand side of the Salween River, that centered around the location of their rotational agriculture (shifting cultivation).[1] Approximately 40–50 years ago, Mr. Pa-Kortoo's household became the first to settle at the village's current location. Surrounded by more mountainous terrain, the village's location has a relatively large flat area and is thus suitable for sedentary farming activity. Until 1973, there were only nine families in his village; however, around this time King Rama IX (King Bhumibol) came to visit the village. Before the visit, local government officials tried to gather those families who were still dispersed and living close to their rotational agriculture fields to move into Sob Moei village itself. The government officials led the community to prepare paddy fields and dig a simple canal for irrigation. When the King visited Sob Moei, he presided over a land ownership ceremony that assigned *de facto* rights for eight paddy fields to the village, although land documents were not issued at the time. After that, the majority of villagers, who had not yet relocated, moved to live in Sob Moei village (DCCN 2009).

As Sob Moei village grew, some important communal buildings were built in the village. In 1991, the (national and provincial) government built a healthcare center to be the central health care service not only for Sob Moei village, but also over ten villages nearby. Also in 1991, a school was built to service Sob Moei and five nearby communities, located near an older Buddhist temple that the community had built themselves with support from a monk from Myanmar. Although at the time the majority of villagers practiced Karen traditional Animist beliefs, in 1995 the Buddhist temple was moved beside the river near the convergence point, organized by Kor Tor, a well-known monk from Myanmar (DCCN 2009).

The majority of the population of this village are Karen ethnicity, and in 2005 there were 132 households with about 600 people (Montree et al. 2007: 103). Of those 600 people, about 30% of them do not have Thai citizenship; therefore they have a non-Thai citizen ID card (issued by the Thai Government). Non-Thai citizen ID cards restrict travel beyond the province of residence, whereby special permission is required to travel and work outside. There are two major kinds of

[1]Rotation farming, also called swidden agriculture or shifting cultivation, consists of partial forest clearance, multiple cropping, shallow cultivation, and field rotation to produce food and sometimes cash crops. It is a system, through use of a prolonged fallow phase, that allows woody vegetation to return to a site that had been cleared for annual crops, before it is once again cleared for cultivation.

non-Thai citizen ID card, the first kind permits for a permanent stay, while the second kind is for a temporary stay and needs to be renewed every ten years (DCCN 2009). In Sob Moei village, there is a mixture of both non-Thai citizen ID cards.

Sob Moei village is 72 km away from Mae Sariang city and it is very difficult to travel between them. During the dry season, it is possible to travel to Mae Sariang by motorbike; however, during the rainy season villagers must travel by boat on the Salween River to Mae Sam Lab and then by motorbike or car to the city. In Sob Moei village, livelihoods rely on agriculture, the river, and non-timber forest products (NTFPs). For agriculture, there are three kinds of farming, rice paddy fields in flat areas, rotation farming for rice and a variety of vegetables, and river bank gardens for vegetables. The Salween River, tributary, and streams provide fishing for local consumption and for income. The forest is for personal use and for income; people gather many NTFPs including forest vegetables and medicinal plants, and also hunt wild animals (Montree et al. 2007: 104).

9.2.2 *Mae Sam Lab Village*

Mae Sam Lab village is located alongside the Salween River and is roughly 45 kilometers from Mae Sariang city. They have a better road that can be used during the rainy and dry seasons to connect the community to the city. Mae Sam Lab was established in 1961 by three families who practiced agriculture in the area. As commercial trade increased between Thailand and Myanmar, goods would travel from Mae Sariang to Mae Sam Lab village by road, and at Mae Sam Lab they would be loaded on to boats bound for Myanmar. As this trade increased more people came to live in Mae Sam Lab. The population especially grew when the government of Thailand provided a logging concession to a private company to log along the Salween River. Logging occurred both on the Thai-side and the Myanmar-side, and has taken place for a long time. Because Mae Sam Lab is located in the center of the Salween forest and its role as a river port became important for transporting the logs, many people came to live in this village (DCCN 2009). After 1992, armed conflict between Burmese Military (the *Tatmadaw*) and the Karen National Union (KNU) escalated and many people fled from Kayin (Karen) State in Myanmar to Mae Sam Lab.

Now Mae Sam Lab is the largest village along Salween River in Thailand, and has had six generations of official leaders (DCCN 2009). The population of the community consists of three different ethnicities: Karen, Shan, and Karen-Muslim. This village is different from other villages in the same district because the number of people without Thai citizenship is greater than those with Thai citizenship. This is because many of the people in Mae Sam Lab village are first- or second-generation families who fled Myanmar (Montree et al. 2007). As of 2011, there were 280 households with 1,521 people, and approximately 20% have Thai citizenship and the remainder have non-Thai citizen ID cards (DCCN 2011).

Livelihoods in this village also differ from Sob Moei village and Tha Ta Fang village, due to the history of the village as a center of commercial trade among people in the Salween River Basin from both Thailand and Myanmar. Many people in the village work in transportation, as shopkeepers, in animal and agriculture product trade, in restaurants, or as labor at the river port. However, there are also some villagers, especially earlier settlers of the village, who have *de facto* ownership of land for rotation farming, river bank gardens and paddy fields. More land cannot be acquired for agriculture in this village because of the Salween National Park, which was official established in 1993. The acquisition of more land is further prohibited by the fact that most of the people in Mae Sam Lab do not have Thai citizenship (Montree et al. 2007: 96–99).

9.2.3 Tha Ta Fang Village

Tha Ta Fang village is one of the older and bigger Karen communities along the Salween River. Like Sob Moei, Tha Ta Fang has a long history. In the past, many people lived in more temporary settlements that were located near their rotational agriculture fields dispersed around the current location of the village. The village, as it is located today, was formed before 1941 by Mr. Jor-ou. At this time, the village was called Jor-ou village and was not registered by the Thai government. Before World War 2, commercial trade in the Salween River Basin was relatively prosperous, and travelling between Thailand and Myanmar for trade was common. Tha Ta Fang village was the place merchants would stay on the way, and there was also a Thailand Police check point in the village. In 1965, when there were more villagers, Thai solders built a school and in 1975 Jor-ou village was officially registered by the Thai government and changed its name to Tha Ta Fang village. After it was registered, the government appointed Mr. Jor-ou as the village leader (Montree et al. 2007: 84–89). The village's population is mostly Karen, and in 2007 had a population of 83 households with 743 people (Montree et al. 2007: 90).

This village is similar to Sob Moei village, in that it is far away from Mae Sariang city and the community's main livelihood is now agriculture, as trade now passes primarily through Mae Sam Lab. Forty families have *de facto* ownership of paddy fields, and at least 10 families have rotational farming fields. River bank gardens are also common and villagers rely on forest products and the river for fishing and transport (Montree et al. 2007: 90).

Tha Ta Fang village is also surrounded by the Salween National Park. Before the national park was established, the local administrative government negotiated with the DNP to allow Tha Ta Fang villagers to maintain access and use of their paddy and rotational fields. The village area was mapped and agreements were reached between the village and district officials, and the DNP officials. In these agreements, the DNP would not confiscate any more land and Tha Ta Fang villagers would not use the forest and land inside the national park boundaries. This has allowed Tha Ta

Fang villagers to hold more agricultural land than Mae Sam Lab. Yet, further growth of the village is limited because villagers are no longer able to build new houses, cultivate new land, or expand infrastructure like roads.

9.2.4 Community Use of Land

Most of the Karen people in the Salween River basin, including these three villages, practice farming for their livelihoods. The location of these communities, far away from large urban areas such as Mae Sariang city, limits commercial trade, meaning that subsistence farming is very important. Agricultural land along the Salween River can be divided into three forms: rotation farming, paddy field, and riverbank gardens, discussed in turn below.

The terrain along the Salween River in Mae Hong Son province is mountainous, with few flat areas. Therefore, most communities practice rotation farming, also called shifting cultivation. They cultivate rice and other vegetables such as cucumber, taro, and pumpkin on the slopes of the mountains. Rotation farming is a part of upland ethnic minority culture, and is a pattern of sustainable land use. It is also important to maintain food security among Karen people (Forsyth 1996). Of the three villages in this chapter, Sob Moei village relies on rotation farming more than the other two villages. On average, the families practicing rotation farming still own five to seven farming areas (Deetes 2005). Rotation farming takes place on customary land that is a form of common property; each year the land farmed will be rotated to a different area, and the community members share the land with each other. Community members distinguish between rotational farming land and forest. Rotational farming land has been used for a long time; community members will not expand the rotational farming into the remaining forest, which is for other livelihood purposes (NTFP collection, hunting etc). By local tradition, the land is reserved for people in the community only, and it is not permitted to sell the land or to be owned by anyone outside the community (Kanchanapan 2004).

Rotation farming is well known in Thai society and scholarship, and while it is still criticized as destructive to the environment by the DNP and Royal Forest Department (RFD) (Forsyth/Walker 2008; Roth 2004; Wittayapak 2008), it has also been recognized as indigenous peoples' way of living, which does not harm the forest (Kanchanapan 2004). In 2003, Thailand's Ministry of Culture, which is responsible to oversee culture, religion, and art, offered a resolution regarding the Karen way of life and issued statements about natural resource management, citizenship, traditional culture, and education. Regarding natural resource management, the Ministry supported Karen peoples' traditional practice of rotation farming as a culturally significant practice (Council of Ministry Resolution 2010).

Paddy rice growing is also a common means of agriculture for people along the Salween River. While there are not many flat areas, paddy fields can be modified and terraced to fit into the few narrow spaces along the river. In the paddy fields, rice is planted during the rainy season for household consumption (May–November) and soybean is cultivated during the dry season for sale (December–April). Ownership of paddy fields are *de facto* private and permanent ownership. Paddy fields are not officially registered and villagers don't have any kind of land title. However, possession of paddy field is acknowledged among the community. Normally, people will not sell them to each other, but instead pass the land on to the next generation. Paddy fields are important to people because they are significant sources of food security.

River banks of the Salween River and Moei River are available during the dry season to plant vegetables and other kinds of food, as well as tobacco. People prefer to plant vegetables on the river bank because it is easy to grow and does not require fertilizers, as the soils are rich with natural fertilizer from the river (Deetes 2005). River banks are also treated as common property amongst the community. People believe that land on the river bank is provided by the spirit or nature, which is why everyone in the community has the right to use it. Because of the seasonal changes made to the river bank year-to-year, the locations of river bank gardens correspondingly change annually. When the water level turns low, people will choose suitable land for themselves (Deetes 2005: 72).

9.2.5 Forest

Forest is also very important as a source of local food security and income. For household use, people collect NTFPs like bamboo shoots, herbs, other forest vegetables, and wood for construction. People collect NTFPs for sale in the market like dry leaves for roofs in January and February; konjac, a sort of tuber for food in August to November; honey in April; and mushroom in May and June. These wild products can generate important income for villagers (Deetes 2005).

The Karen's well-known proverb of "[if you] use forest [you] must maintain the forest" (Deetes 2005) shows common awareness in the community. When people collect any resource in the forest, they will always be careful and considerate not to destroy it, but maintain it for future collection (Montree et al. 2007). This customary belief is passed down through the generations and practiced in how people use and protect the forest. In Sob Moei village, this environmental ethos is reflected in how the community has tried to engage with local forest officials and the administrative government (see also, Hengsuwan, Chap. 11, this volume). Mr. Decha Sri-sawaidaoruang, a community leader said that the "community has tried to cooperate with the forest office and the local administrative government to maintain the forest, for example through wild fire prevention, monitoring the forest, and tree ordinations [Buddhist ceremony]" (Interview, 14 March 2017). Community members also designate sacred areas, such as the watershed, where people do not practice rotation farming or cut down the trees (Deetes 2005).

9.2.6 Water

Using the Salween River and its tributaries water resources for livelihood is common for people. The three main uses are fishing, agriculture, and transportation. On the Salween River itself, fishing is not only for subsistence, but also for selling the fish for income. Especially in the rainy season during fish migration, people can earn more from fishing. Importantly, fishers in communities alongside the Salween River, which include Sob Moei, Mae Sam Lab, and Tha Ta Fang villages, can earn daily income from fishing (Deetes 2005: 36). For now, the river is still largely a free-flowing system and is rich in fish species, with at least 70 known species (Deetes 2005). During the rainy season when people plant rice on their paddy fields, they commonly use water from the Salween River tributaries or streams to irrigate the paddy field. Moreover, when the water level of the Salween River increases in the rainy season, the river will bring natural fertilizer to the river bank, which benefits river bank farming. Additionally, as communities in this area have poor access to decent roads, many people prefer to travel by boat on the Salween River.

9.3 National Law and Challenges to the Salween Karen Community

9.3.1 Citizenship

In this area, the Salween River constitutes the border between Thailand and Myanmar. There are two main reasons for the movement of Karen people across the border. Firstly, some had little choice but to flee the fighting in Kayin (Karen) State, and secondly – mostly a long time ago – some migrated for farming and for commercial trade. Fighting was especially intense from the mid-1980s and during the 1990s, which led to major movements of people into Thailand that would not be possible nowadays. Many of those who fled the fighting are still not able to move back to their old community due to ongoing insecurity in their home villages (SCPP 2010).

Some migrants successfully gained Thai citizenship, depending on how long ago they moved to Thailand, as well as the season; in the rainy season transportation is more arduous making official processes such as registration more difficult to undertake. Thai citizenship is granted under the Nationality Act (1965). Section 7 states that:

> The following persons acquire Thai nationality by birth: (1) A person born of a father or a mother of Thai nationality, whether within or outside the Thai Kingdom; (2) A person born within the Thai Kingdom except the person under Section 7 bis paragraph one. "Father" in (1) means also a person having been proved, in conformity with the Ministerial Regulation, that he is a biological father of the person even though he did not register marriage with the mother of the person or did not do a registration of legitimate child.

According to this article, a person who is born in another country and escapes to Thailand or a person who is born from those people who escaped to Thailand are not eligible to get Thai nationality. However, the fourth amendment (2008) of the Nationality Act (1964) under article 23 offers an exception for those who were born in Thailand if they have official evidence to prove that they were born in Thailand between 1972 and 1992. It is also possible under Article 7.2 and 7 bis. that the Minister may grant Thai citizenship guided by other Ministerial Regulations and rules formulated by the Cabinet.

In summary, thousands of people entered Thailand without formal permission and without citizenship during the period of armed conflict. The Thai government created a policy to register them, providing them with non-Thai citizen ID cards, and allowing for a temporary (10 year) stay within a limited area. These ID cards allow for basic rights such as access to education and health care, and to legally work in the country, but they do not grant voter rights or land tenure rights. In 2009, in the three districts alongside the Salween River, namely Mae Sariang, Sob Moei, and Tah Songyang districts, 17,437 people held one of the two types of non-Thai citizenship ID card. In Sob Moei, Mae Sam Lab and Tha Ta Fang villages, 237 people, 1421 people and 359 people respectively held a non-Thai citizen ID card issued by the district registry office (SCPP 2010). People who are eligible to get Thai nationality have tried to submit a claim, but it is a slow and complex process that is difficult to complete.

9.3.2 Community rights

Rights to land and forest for indigenous communities are restricted by four forest laws: the Forest Act (1941), the National Park Act (1961), the National Forest Conservation Act (1964), and the Wildlife Sanctuary Act (1992). However, a high-profile campaign by civil society from 1990 to 1997 led to the recognition of "community rights" to natural resources (HRLA 2015). This first appeared in the Thai Constitution of 1997 article 46 as:

> Persons so assembling as to be a traditional community shall have the right to conserve or restore their customs, local knowledge, arts or good culture of their community and of the nation and participate in the management, maintenance, preservation and exploitation of natural resources and the environment in a balanced fashion and persistently as provided by law.

Such a provision remained largely unchanged in article 66 of the subsequent Thai Constitution of 2007. Recognition of community rights has remained in the current Thai Constitution of 2017 in article 43 as:

> A person and a community shall have the right to:
>
> (1) conserve, revive or promote wisdom, arts, culture, tradition and good customs at both local and national levels;

(2) manage, maintain and utilize natural resources, environment and biodiversity in a balanced and sustainable manner, in accordance with the procedures as provided by law;
(3) sign a joint petition to propose recommendations to a State agency to carry out any act which will be beneficial to the people or to the community, or refrain from any act which will affect the peaceful living of the people or community, and be notified expeditiously of the result of the consideration thereof, provided that the State agency, in considering such recommendations, shall also permit the people relevant thereto to participate in the consideration process in accordance with the procedures as provided by law;
(4) establish a community welfare system. The rights of a person and a community under paragraph one shall also include the right to collaborate with a local administrative organization or the State to carry out such act.

While these articles in the 1997, 2007 and 2017 Thai Constitutions do provide for the concept of protecting indigenous communities' right to natural resources, they do not define the precise nature of those rights. As a result, 20 years has passed since the original definition of "community right" to land, forest, and natural resources, and it is still unclear if those rights extend to tenure, access or use, or the ability to extract resources. Moreover, as many of the people from these indigenous communities hold non-Thai ID cards they are not afforded full rights under the constitution. According to precedents set by previous court judgments, when communities have claimed "community rights" under the constitution to protect the environment or for public interest purposes, such as in case of pollution, then the courts have upheld "community rights." However, in cases where communities claim "community rights" for the right to use or own land or natural resources, then those rights have been denied. Furthermore, the Administrative Court has demonstrated a tendency to recognize a wider interpretation for natural resource, environment, and public interest protection purposes, which means they are more likely to uphold "community rights," compared to the Civil Court and Criminal Court. For example, in case 278/2556 (2013) in the Administrative Supreme Court, a community sued the government to cancel a biomass power plant building license, citing the issue of pollution. The court verdict accepted that "community rights" must be protected by the government against pollution to air and water. In contrast, in case 660/2557 (2014) between Mr. Kor-ei Mimi and others versus the National Park Department the plaintiff sued for "community rights," claiming the community's history as a right to property on the land. In this case, the Administrative Court did not accept their claim (Prachatai 2016).

In the case of the Hatgyi dam, if it is built just across the border downstream in Myanmar, it will certainly result in some negative impacts on these upstream communities in Thailand. Yet, the precedent set by Administrative Court above suggests that if community members from Tha Ta Fang, Mae Sam Lab, and Sob Moei were to sue the government over damages from the dam claiming "community rights" they would likely be unsuccessful. The challenge is, furthermore, that as Tha Ta Fang and Mae Sam Lab are surrounded by the Salween National Park and Sob Moei borders the Salween Conservation Area, as the case with Mr. Kor-ei Mimi versus National Park Department demonstrated, their ability to sue for damage and their rights to property or land would also unlikely be upheld.

9.3.3 Land Law

The legal system for land in Thailand can be separated into two categories: land law and forest law. Almost all land with land titles falls under the Land Act (1954), except land for agriculture purpose, which falls under the Agricultural Land Reform Act (1975). The main idea of the Land Act is to grant land rights to private individuals and companies with Thai citizenship by providing a land title. Land rights under the Land Act (1954) allows for the transfer of land through sale or through inheritance. However, the Agricultural Land Reform Act (1975) prohibits the sale of land, although it does allow for inheritance. In practice, land documents are only provided to lowland flat areas and are not issued for forested areas.

Forests are governed by the four forest laws mentioned above, which all hold similar core principles. The objective of each is that of preservation. They also authorize the government to declare preservation areas without local people's consent. Each law regulates prohibitions and granting permissions and define sanctions for those who break the rules. Additionally, at the policy level, the Natural Resource and Environment Ministry released the National Forest Policy (1992) in which article 17 prohibits land titles for land with greater than a 35% slope. Thus, for the communities in Tha Ta Fang, Mae Sam Lab, and Sob Moei that in part live in mountainous areas with slopes greater than 35%, they are not permitted to be given land documents under the four forest laws and the National Forest Policy (1992).

Along the Salween River, many local people's rights to land have been deprived by the Land Act (1954), Agricultural Land Reform Act (1975), and the series of forest preservation laws. The government, via the Ministry of Natural Resource and Environment, has tried to maintain control over all forested areas in the country. Each year, actions enforcing these laws have resulted in the confiscation of land used by ethnic minority people for generations, and some people have been arrested and sued for criminal charges, and often judged to be guilty.

For example, the Supreme Criminal Court verdict of 10578/2559 (2015) Mae Sot Prosecutor versus Dih-Paepoe rejected the right to Karen rotation farming, favoring the forest law over local customary land rights. The case of Mr. Dih-Paepoe, better known as the "Mea Omki case" given that the defendant, Mr. Dih-Paepoe, was from Mae Omki village, occurred in April 2008 when Mr. Dih-Paepoe, a 79 years old man of Karen ethnicity was arrested by a government Forestry Officer while he was farming. The Royal Forest Department sued for a criminal charge of trespassing in the national preservation forest. The Human Rights Lawyer Association represented him and defended him by claiming community rights under the Constitution 2007 for rotation farming. The defense also argued he did not intentionally trespass in the forest. On March 22, 2017 the Supreme Criminal Court ruled that Mr. Dih-Paepoe was not guilty in his rotation farming because of his lack of intent to trespass in the forest, as the disputed land had been used in his family for over a generation. However, he was still forced to vacate the land because it was designated as preservation forest under the National

Preservation Forest Act, 1964. This Supreme Criminal Court ruling demonstrated that it would not recognize rotation farming as a constitutional "community right" of ethnic minority communities, even when the community's existence precedes the establishment of a preservation area (HRLA 2015).

This precedent also has been demonstrated in the Administrative Court verdict 660/2557 (2016) of Mr. Ko-eyh Mimi versus the Ministry of Natural Resource and Environment (2016). Known as the Grand Ko-eyh case, it addressed the case of a Karen community in Kanchanaburi Province who had lived in Keng Krachan National Park before it was established. In this case, it was ruled that being on ancestral land without land titles precluded the community from having legal rights to that land. Thus, the national park officer legally had the authority to burn the villager's house, as had occurred.

Access and use of land is very important to the life and livelihoods of upland ethnic minority people. Much of it is located in areas categorized by the Thai government as forest, making it a challenge for ethnic minority people. Ethnic communities have tried to negotiate the forest laws and policies with the local forest department for temporary permission. Sob Moei village and Tha Ta Fang village have provided land use evidence by participating in a land mapping project. Sob Moei village's mapping project is ongoing and supported by the Community Organizations Development Institute (CODI), and mapping is done in partnership with local forest department officers. Tha Ta Fang village's mapping project was supported by the local administrative department and completed roughly a decade ago. Even though these maps are not official land documents, because they are created with local administrative governments and local forest officers it gives community members a way to locally negotiate around the land and forest laws. Importantly, it grants them the ability to use customary land without fearing arrest from local forest officials.

9.3.4 Forest Law

As discussed above, Thai forests are subject to four forest laws: the Forest Act (1941), the National Park Act (1961), the National Forest Conservation Act (1964), and the Wildlife Sanctuary Act (1992). These laws grant ownership of forests to the central government, focus on the preservation of forests, and allows for permission to be given to private companies and individuals for specific purposes such as mining, forestry, and academic research. Under these laws, any use without official government permission is illegal. Therefore, wild resource collection by the community for their livelihoods is illegal, such as collecting wood to build, forest vegetables and herbs, hunting, and collecting leaves for roofs. However, many of these communities have dwelled in the forest for generations, well before the establishment of preservation forests (Kriyoonwong 2015).

According to the report of the Land Right Resolution Committee under the House of Representatives, in 2010, 635,916 people lived in areas classified as forest in Thailand (Kriyoonwong 2015), and between 2009 and 2015 an annual average of 2,652 cases per year were brought for forest trespassing (Forest Department 2017). The legal arguments in these forest cases do not acknowledge the context of indigenous communities nor their "community rights." 39 cases brought in April 2013 highlight the tension between legal deployment and community livelihoods. These cases occurred when an official field investigation joined by forest officers, police, and soldiers went to Tung Parka village with search warrants to investigate for illegal teak possession. The operation found 39 villagers possessed some teak in their house, which was considered to violate the Forest Act (1954) (Thai PBS 2015). At the province (lower) court all 39 defendants confessed to the allegation, and the judge decided to jail 21 of them and release 15 with a fine for illegal teak possession used for house building (iLaw 2015). A subsequent ruling in the Appeal Court emphasized that teak possession for house building of these defendants, without official permission, contributed to both global warming and flooding in lowland urban areas (Appeal Court verdict No. 425/2014).

As with Sob Moei, Tha Ta Fang, and Mae Sam Lab villages, Tung Parka village is a predominantly Karen village and the villagers practice their traditional livelihoods. However, the collection of wood for building houses is illegal in all cases. The villages discussed in this chapter are still not allowed to cut trees within the forest. This is regardless of their customary use of the land and even though they have negotiated some use of the forest with their local administration and forest officers. In these cases, the forest and land laws are upheld over "customary rights."

9.4 Tensions Between National Law and Local Practice: Implications for Karen Indigeneity and Culture

The discussion above on land and forest law reveals how the government's objective of preserving land and forests challenges ethnic minority people in their customary use of land. Yet, the history of the three communities detailed in this chapter reveals that they have practiced traditional livelihoods along the Salween River for a long time. The administrative part of the government has released some regulations to resolve this problem. Three of them are particularly important: the Cabinet Resolution of 30 June 1998 regarding land rights resolution for ethnic minorities; the Cabinet Resolution of 3 August 2010 regarding Karen traditional protection, which includes rotation farming; and the Office of the Prime Minister's regulation of 2010 regarding land right documentation for "community land titles." These three regulations offer alternative resolutions to protecting ethnic minority's rights to land that have been largely dominated by the four forest laws. The verdicts mentioned in this chapter have stressed that even though the communities have lived on and used the land before it was classified as preservation area, they are still vulnerable to being declared as illegal settlers (HRLA 2015).

The problem of using the law to limit, control, and deprive civil and natural resource rights to upland minority populations has a long history in Thailand (Vandergeest 2003; Wittayapak 2008). Since 1957, the government has created an image of ethnic minority groups in Northern Thailand as security risks and involved in forest destruction. The National Security Council has further stigmatized them by using the negative discourse of "hill tribe." Its meaning implies their diminished humanity as lower class and inferior people. This discourse becomes a way for the state to maintain control and deny rights to upland populations. Ethnic minority space in Thai society has been vigorously limited by law, policy, and field operations (Laungaramsri 1998). While indigenous community rights to land and forest are discriminated by the forest law, there is much evidence that the livelihoods of the ethnic minority and indigenous people do not harm forests or the environment. Research has clearly proven that Karen communities can use their traditional knowledge to protect the forest while they do rotation farming and collect resources from the forest for food and sale for earning an income (Kanchanapan 2004; Forsyth/Walker 2008).[2]

Although the country has changed and developed further since these discourses first emerged, the image of ethnic minorities still has not improved. Instead, arguably it has gotten worse. It continues to stigmatize ethnic minority and indigenous people as national security threats, and endangering the new government's drug policy and reforestation efforts. In practice, projects have been created to target these "hill tribe" communities and force their relocation from national parks, accompanied by policies for forest reclamation, or increasing the area of forest in the country. These policies have significantly affected many indigenous communities (Laungaramsri 1998). In Tha Ta Fang, Mae Sam Lab, and Sob Moei, over half of the population are still not able to obtain Thai citizen ID cards. Thus, they have little legal recourse against the government should the DNP or RFD decide they are violating forest policies or land laws. Moreover, policies, like the forest preservation policy, which prohibits land titling on mountainous areas with a greater than 35% slope, represent a systemic prejudice against upland minority populations within the legal system.

Although many upland minority people, like those in the villages discussed in this chapter, can claim customary use and rights, the law does not privilege those rights and instead the discourse of "hill tribe" labels them as offenders and forest destroyers. In 2011, the National Park, Wildlife, and Plant Species Department, under the Ministry of Natural Resource and Environment, released the "Economic Modeling of Some Environmental Impacts of Deforestation" that calculated the economic cost of forest destruction. The damage was determined in terms of the

[2]Although not the focus of this chapter, it is important to note that the Thai state has also sought to promote Thai culture with upland communities, including Thai language schools, pictures of the Thai monarchy, and temples with Thai monks to which local communities send young males to be ordained, all of which can be understood as a part of the process of nation-building, but also functions as a different means of influencing upland areas (Walker 2001; Laungaramsri 2001; Lamb 2018).

release of stored carbon and the cost of reducing the assumed increase in temperature based on the required electricity to reduce the temperature using air conditioning (Meetam 2016). Here, the same discourse that motivates claims about "hill tribe" as forest destroyers accuses the ethnic minority and indigenous community livelihoods as causing global warming by rotation farming and tree cutting.

9.5 Conclusion

Most of the community in Sob Moei, Mae Sam Lab, and Tha Ta Fang villages discussed in this chapter are Karen indigenous people. They have lived in these communities alongside the Salween River for more than a generation; however, many of the people are non-Thai citizens and thus are deprived of their civil rights. Their livelihoods rely: on land for rotation farming, growing paddy rice, and river bank gardens where they plant rice and vegetables; on forests for collecting natural resources for building materials and food, and for selling the products; and on the Salween River for fishing, farming, and transportation. Regarding these natural resources, they also have local governance arrangements to protect them and control their use through a combination of customary rules and unofficial agreements between the community and local officials.

However, the government's laws on forests, land, and citizenship have caused serious tensions between the government officers, especially forestry officers, and local community livelihoods. Meanwhile, the government has also strongly backed the Hatgyi dam project, which if built would create serious impacts on the livelihoods of these communities. The apparent purpose of these laws and policies is the central government's desire for 'pure' forest area and to support the business of water development projects, to the exclusion of human rights and natural resource protection. However, in practice the local forest officers cannot uphold these forest laws and policies fully, as there are also contradictions between the forest laws and government regulations that seek alternative mechanisms to solve the conflict between forest law and ethnic community livelihoods. Through these alternative mechanisms, the communities have tried to negotiate with the local forest officers on land and forest use rights.

This chapter reveals that the use of national law is not only about the state trying to regulate livelihood practices, but it is also a challenge to the definition of Karen identity in these villages. The conflict between law and policy with indigenous community rights has been a critical issue for more than three decades. The government has become increasingly aware of the problem, but still intends to retain these laws and the villagers' use of natural resources under its control. It reflects an imagination of the "hill tribe," which is influenced by a national security discourse that links ethnic minority people to drug selling, forest destruction, and being non-Thai citizens. Thus, even as many indigenous communities have lived in these areas for a long time, the national law and policy still deny their right to access natural resources and land.

Given the challenges described in this chapter, the community members in Sob Moei, Mae Sam Lab, and Tha Ta Fang villages request the right to maintain their livelihood. First, they ask that their right to land for traditional farming for food security be officially recognized by the Thai state, which would include both individual rights and collective rights depending upon the community's consensus. Second, they ask for the right to collect wild products be recognized to meet their daily needs, such as food, materials for building, as well as to collect products to sell for an income. The Hatgyi dam is also a common concern. Here, these communities want both the government and the private company to: ensure adequate community participation; recognize the challenges faced by the communities including how difficult it is to change their way of life, the limited options for resettlement available to them, and their relative disadvantage in negotiating given their current non-Thai citizen status; and acknowledge and incorporate community knowledge into decision-making.

References

Council of Ministry Resolution. (2010). *Council of Ministry Resolution: Policy for recovering the lives of Karen people*. Retrieved from: https://andamanproject.org/wp-content/uploads/2018/01/Karen-Cabinet-Resolution-in-English.pdf.

Deetes, P. (2005). *Thai Baan research at the Salween: Villagers' research by the Thai-Karen Communities*. Retrieved from: http://www.geozigzag.com/pdf/sw_tb_book.pdf.

DCCN. (2009). *Community history and documentation: Sob Moei village*. Development Center for Children and Community Network (DCCN).

DCCN. (2011). *Unpublished project proposal*. Development Center for Children and Community Network (DCCN).

Forest Department. (2017). *Case related to forest law abusing*. Retrieved from: http://forestinfo.forest.go.th/Content.aspx?id=88.

Forsyth, T. (1996). Science, myth and knowledge: Testing Himalayan environmental degradation in Thailand. *Geoforum, 27*(3), 375–392.

Forsyth, T., & Walker, A. (2008). *Forest guardians, forest destroyers: The politics of environmental knowledge in northern Thailand*. Seattle, WA: University of Washington Press.

HRLA. (2015). *Mae Omki: The life of the forest and ideals of the justice process*. Bangkok, Thailand: Human Right Lawyer Association (HRLA).

iLaw. (2015, May 15). *Lesson learned of Tung Parka to the justice system revolution of Thailand society*. Retrieved from: https://ilaw.or.th/node/3680.

Johnson, C., & Forsyth, T. (2002). In the eyes of the state: Negotiating a "rights-based approach" to forest conservation in Thailand. *World Development, 30*(9), 1591–1605.

Kanchanapan, A. (2004). *Shifting cultivation system: Status and changes*.

Karen Research Team, DCCN, & SEARIN. (2005). *Karen research: Way of life, way of forest for the Salween Karen People*.

Kriyoonwong, S. (2015). *Land and Natural Resource Management Bill: Reduce Inequality, Create Fairness 2015 (in Thai)*.

Lamb, V. (2018). Who knows the river? Gender, expertise, and the politics of local ecological knowledge production of the Salween River, Thai-Myanmar border. *Gender, Place & Culture*. https://doi.org/10.1080/0966369x.2018.1481018.

Laungaramsri, P. (1998). Discourse as hill tribe. *Social Science Journal,* 104.

Laungaramsri, P. (2001). *Redefining Nature: Karen Ecological Knowledge and the Challenge to the Modern Conservation Paradigm.* Chennai, India: Earthworm Books.
Magee, D., & Kelley, S. (2009). Damming the Salween River. In F. Molle, T. Foran, & M. Käkönen (Eds.), *Contested waterscapes in the Mekong Region: Hydropower, livelihoods and governance* (pp. 115–140). London, United Kingdom: Earthscan and USER (Chiang Mai University, Thailand).
Meetam, P. (2016). *Global warming case model.* Retrieved from: http://naksit.net/th/?p=276.
Montree, N., Chantawong, M., & Longcharern, L. (2007). *Salween River and live record under changing flow.* Bangkok, Thailand: Foundation for Ecological Recovery.
Phongpaichit, P., & Baker, C. (2002). *Thailand Economy and Politics (Second Edition).* Oxford and New York: Oxford University Press.
Prachatai. (2016, 8 September). *Verdict of "uncle Kor-ie" reflection the raising of official power retaining over natural resource.* Retrieved from: https://prachatai.com/journal/2016/09/67829.
Roth, R. (2004). On the colonial margins and in the global hotspot: Park-people conflicts in highland Thailand. *Asia Pacific Viewpoint*, *45*(1), 13–32.
SCPP. (2010). *Non-Governmental Organization's suggestion regard primarily solve stateless problem on the north of Thailand.* Stateless Children Protection Project (SCPP).
Thai PBS. (2015, 22 April). *Tung Parka decide not to file appeal the case of forest tress passing.* Retrieved from: http://news.thaipbs.or.th/content/567.
Vandergeest, P., & Peluso, N.L. (1995). Territorialization and state power in Thailand. *Theory and Society*, *24*(3), 385–426.
Vandergeest, P. (2003). Racialization and citizenship in Thai forest politics. *Society & Natural Resources*, *16*(1), 19–37.
Walker, A. (2001). The 'Karen consensus', ethnic politics and resource-use legitimacy in northern Thailand. *Asian Ethnicity*, *2*(2), 145–162.
Wittayapak, C. (2008). History and geography of identifications related to resource conflicts and ethnic violence in Northern Thailand. *Asia Pacific Viewpoint*, *49*(1), 111–127.

Chapter 10
An Ethnobotanical Survey in Shan State, Myanmar: Where Thanlwin Biodiversity, Health, and Deforestation Meet

Mar Mar Aye and Swe Swe Win

10.1 Introduction

The American botanist John Harshberger coined the term 'Ethnobotany' in 1896. Since that time, ethnobotany studies have covered a range of topics (Abbasi et al. 2011). The study of ethnobotany is of great importance in understanding the interrelations of material and intellectual culture. It can involve the study of plant species as well as the cultural practices of herbalists treating diseases. Ethnobotanical studies are also significant in communicating the medicinal values of locally important plant species to outside audiences and documenting them for future use, and can include traditional knowledge of plant diversity and use by indigenous communities. The documentation of traditional knowledge of plants has provided the basis for many modern drugs (Cox 2000; Flaster 1996). Ethnobotanists also explore how plants are used, not only for health, but also for necessities such as food, shelter, medicine, clothing, hunting, and religious ceremonies.

In Myanmar, wild plants have long been used as a source of medicine. This chapter presents research from an ethnobotany study to understand the traditional knowledge of herbalists and the uses of plants for healthcare in four communities in Myanmar's Shan State. We approach this work from a perspective that seeks to highlight the local knowledge of communities and the challenges they face. The research shows the ways in which local people in four villages in Lashio District along the Thanlwin (Salween) River, Shan State, rely on medicinal herbs for the treatment of disease. The research presented in this chapter is based on botanical surveys and interviews conducted from July 2015 to February 2016, when we collected and identified 21 medicinal plants from 14 plant families.

Mar Mar Aye, Professor and Head (retired), Department of Botany, Lashio University; Email: pro.marmaraye@gmail.com.

Swe Swe Win, Lecturer, Lashio University; Email: Drsweswewin.91@gmail.com.

C. Middleton and V. Lamb (eds.), *Knowing the Salween River: Resource Politics of a Contested Transboundary River*, The Anthropocene: Politik—Economics—Society—Science 27, https://doi.org/10.1007/978-3-319-77440-4_10

This research also shows how the practice of herbal medicine is threatened by a range of factors, including deforestation and agricultural expansion. While residents in all four villages rely on herbal medicines, residents of poorer, more isolated villages tend to rely solely on these practices while residents of better off villages also access hospitals and clinics. As a research site, Shan State is very much understudied, and through the research presented in this paper, we aim to contribute to the general knowledge of botanicals for health and the ways that local communities practice herbal medicine in Shan State and more broadly, across the Thanlwin River Basin.

This chapter also highlights the importance of medicinal herbs to the health of local communities, and evaluates the future availability of medicinal herbs in the four selected sites. The study finds that deforestation, the expansion of shifting cultivation, and an increasing trend towards permanent cultivation of cash crops have the greatest impacts on the medicinal herbs used by villagers in the study areas. It is the significance of medicinal plants and deforestation that we turn to first, as it is important to understanding the broader context and future of medicinal plants in Myanmar.

10.2 Significance and Context of Herbal Medicines in Myanmar

According to the World Health Organization (WHO 2013), approximately 80% of people worldwide use herbal remedies as part of their overall health care. Herbalists across many different cultures throughout the world use plants and other substances to improve health, promote healing, and prevent and treat illness (Kamboj 2000). In Myanmar, while precise data on the number of people who use herbal medicines is not available, herbal medicine, which relies on plants and plant products, topically or orally, to treat illnesses, has widespread usage. However, what is perhaps unique to Myanmar is that in the face of rapid deforestation, local communities are maintaining their medicinal plant strategies.

A Global Forest Resources Assessment by the Food and Agriculture Organization of the United Nations (FAO 2010: 21) shows that Myanmar has one of the highest annual rates of deforestation in the world.[1] The 2015 report shows that Myanmar has continued to lose forest cover (FAO 2015: 1). Although Shan State has experienced less deforestation in the last twenty years than the rest of country, it is now considered a critical deforestation front (EIA 2015). The conversion of forested land into agricultural land for paddy rice, corn, rubber, and other exportable cash crops (WWF 2015) threatens areas where local people cultivate and collect medicinal plants. While it is hard to find precise data on the amount of land

[1]The report lists Myanmar as one of the "Ten countries with the largest annual net loss of forest area, 1990–2010."

already cleared to make way for cultivation, it is clear that agricultural expansion is a factor in the decline of medicinal plants (Woods 2013) and that agricultural expansion will continue as the demand to produce more crops for export increases (Phochan 2015).

This broader context of deforestation is of particular concern in this study as it is linked not only to herbal medicines, but also local people's livelihoods and local economies more generally.

10.3 Methods

To understand the importance of ethnobotany in Shan State, this study employed a mix of qualitative and quantitative data collection techniques. The authors and the research team conducted a botanical survey, interviews with villagers, and site visits in the selected study areas in Kunlong township. Field visits were conducted from July 2015 to February 2016 and the authors were assisted by local translators and research assistants at the four villages described below.

As part of this research, a botanical survey was conducted in each of the villages. Specimens were collected with the help of local informants and then processed using herbarium stand methods. Fresh specimens of both vegetative and reproductive parts of plants were identified using available literature (Dassanayake/Fosber 1988; Hong Kong Herbarium and South China Botanical Garden 2007, 2008, 2011, 2012; Hundley/Chit Ko Ko 1961; Hutchinson 1959; Kress et al. 2003; Lawrence 1966; De Padua et al. 1999). Both the vegetative and reproductive parts of the specimens were pressed, dried, preserved, and mounted on herbarium sheets in the Botany Department of Lashio University. Both the vegetative and reproductive parts of the fresh specimens were used for morphological studies. In the study, 21 medicinal plants were collected and identified as part of 14 plant families. This was done with the help of taxonomists in the Department of Botany of Lashio University.

A total of 80 interviews were conducted across four villages (20 per village), in addition to interviews with each village's herbalist or traditional healer (see Table 10.1: Interviews in the Study Area Across Four Villages). Table 10.1 summarizes the number of interviews in each village. Interview guides were used (see below for the list of open ended questions). Topics covered during the interviews included: the use of herbal medicine versus modern medicine; uses of plants and importance to the community; which species are cultivated locally, which are collected wild, areas where plants are collected; changes in the availability of wild plants, and reasons for these changes.

All interviews were conducted in Myanmar, Shan, or Kokang languages with the help of local translators. In general, most of the herbalists are men; however, their wives assist them in their work, including in the treatment of diseases. Most of the local people who use herbal medicine, however, are women, because the men are

Table 10.1 Interviews in the study area across four villages

Village	No. of interviewees		Herbalist
	Male	Female	
Yae Lei Kyun	8	12	1 (male)
Tone Kyat	4	16	1 (female)
Wa Soke	5	15	1 (female)
Ohn Tone	8	12	1 (male)

Source The authors

frequently away doing agricultural work in the forest. The majority of interviewees were between 48 and 75 years old.

The interviews were guided by five key questions:

1. How do you treat your illnesses, by herbal or modern medicines?
2. What species are used for various alimentary diseases?
3. How do you collect herbal plants, cultivated or wild?
4. Where do you collect medicinal herbs?
5. What are the local names of the plants collected?

The study sites for this research were located in Lashio District of Northern Shan State. Lashio District is positioned along the Thanlwin (Salween) River, which originates on the Tibetan Plateau and flows southeast through Yunnan Province of China to arrive in Myanmar, where it cuts through the Shan Plateau (Deetes 2012). This area is ethnically very diverse, and includes Shan, Kokang, Wa, Lahu and Kachin residents. A history of conflict has limited studies in this area. The study site of this research was Kunlong township, and the four villages selected were Yae Lei Kyun, Tone Kyat, Ohn Tone and Wa Soke (see Fig. 10.1: Map of Study Area: Four Selected Villages).

10.3.1 Site Rationale: Geography and Livelihoods of the Four Villages

The four villages of Yae Lei Kyun, Tone Kyat, Wa Soke, and Ohn Tone were selected for this research because they represent communities at different elevations and with a range of different livelihoods, adding to the potential diversity of herbal plants. The four villages are also ethnically and culturally diverse. Yae Lei Kyun village is a majority Shan community, while both Kokang and Shan people reside in Tone Kyat village. Wa Soke is majority Kachin, and the fourth village, Ohn Tone, includes Kachin, Shan, and Lahu residents.

Two of the four villages, Tone Kyat and Yae Lei Kyun, are situated on the banks of the Thanlwin River. Wa Soke and Ohn Tone villages are located in a mountainous area. Each village in the study area is located at a different elevation, and the authors relied on a GPS device to measure elevation.

Fig. 10.1 Map of study area: Four selected villages. *Source* Cartography by Chandra Jayasuriya, University of Melbourne, with permission

The research showed that herbal medicine played an important role in all four villages, even though livelihoods, economic status, and access to services (hospitals, markets, roads) varied, particularly between the mountainous and lowland villages. For instance, the lowland villages of Tone Kyat and Yae Lei Kyun are generally economically better off, and enjoy better access to hospitals in nearby towns. While they still rely on herbal medicine, they can more easily combine it with western medicine. Interviewees in all four villages noted concerns around threats to traditional medicine.

10.3.1.1 Village 1: Yae Lei Kyun

Located on the bank of the Thanlwin River, Yae Lei Kyun Village is home to a total population of about 200 people of Shan ethnicity and is made up of 36 households. In the summer and winter, the villagers cultivate seasonal vegetables and medicinal herbs on the riverbank and sell the products to local consumers. In the rainy season, Yae Lei Kyun is only accessible by boat travel across the Thanlwin River from Kunlong town. While the local people also use modern medicine, the majority of the villagers still rely on medicinal plants given by herbalists for their health problems. For serious health problems, residents visit the public hospital for treatment with modern medicine by health providers in Kunlong. This village was selected as a community with access to a hospital to analyze the people's reliance on herbal medicine, their economic situation, and their ways of addressing health issues.

10.3.1.2 Village 2: Tone Kyat

Tone Kyat village is also located on the bank of the Thanlwin River. With about 170 households and a total population of 1288 individuals, it is home to Kokang and Shan ethnic groups. Most of the villagers depend on agriculture. The local people cultivate on rubber plantations and grow corn for export, mostly to China. According to interviews and observation, most of the youth no longer live in the village; they have migrated for either study or work. Because of their export crops, the socioeconomic status of households in Tone Kyat is higher than the other riverbank village of Yae Lei Kyun, and higher than the two mountainous villages in this study (Wa Soke and Ohn Tone). There is a government-supported primary school for basic education in the village, with instruction in Myanmar language. There is also a hospital nearby in Kunlong; Tone Kyat is about 1 mile away.

10.3.1.3 Village 3: Wa Soke

Located at the top of a mountain, Wa Soke village is east of Kunlong city and about nine miles away from the mainstem of the Thanlwin River. The village is the largest of the four, with a population of about 214 people and 41 households. It is home to residents who identify as members of the Kachin ethnic group. Villagers usually travel by car or motorbike to access the town or other villages. Due to difficult road access during the rainy season, villagers must rely heavily on medicinal herbs for their health. Some villagers earn a living by cultivating sugarcane and corn as cash crops; however, they still earn less money than the residents of Yae Lei Kyun village. The majority of the population are poor and rely on locally collected and cultivated food resources. Men usually work in agriculture, while women stay at home to care for the children. Only one certified midwife serves pregnant women and those needing newborn care in the village. The midwife plays an important role in the village, as the nearest hospital in Kunlong city is quite far away and transportation is difficult.

10.3.1.4 Village 4: Ohn Tone

Ohn Tone is located at the top of the same mountain, approximately two miles away from Wa Soke. There are about 200 people, with a total of 35 households in this village. It is home to Kachin and Lahu ethnic groups. Some villagers cultivate rice and corn on the slope of the mountain to earn income. However, the majority of villagers are very poor. There is a primary school supported by the government with Burman teaching staff from the lowland areas of Myanmar. Because this village is located in an upland area far from public health facilities, they also rely heavily on medicinal herbs to treat their ailments. Similar to Wa Soke, there is no hospital nearby.

10.4 Research Challenges: Translators and Transportation

The challenges of conducting this research are related to access, language barriers and transportation. In terms of transportation challenges, the study sites of mountain areas such as Ohn Tone and Wa Soke are difficult to reach. In the summer, they can only be reached by car or motorbike. Throughout the rainy season, the roads are slippery and dangerous.

In regard to language barriers, the majority of villagers speak a range of local languages and dialects, such as Wa, Shan, Kokang, Lahu and Kachin. To address this challenge, the authors worked with translators to translate from local languages to Myanmar (Burmese) language. Most of the translators assisting in this research were herbalists and leaders of the villages. The position of the translators within the villages was important, because the villagers interviewed trusted and respected them, which aided in their willingness to be interviewed. We also worked with motorbike taxi drivers, car taxi drivers, and local botany students who attended Lashio University. All these individuals lived in the selected areas, had the skills to help negotiate access to the research area, were able to help introduce the authors to herbalists and village leaders, and were available to translate.

10.5 Collected Botanical Species and Local People's Lives and Livelihoods

10.5.1 Summary of Collected Plants

Twenty-one medicinal plants were collected from the four villages and identified as belonging to 14 plant families. Specifically, the plant families *Acanthaceae, Lamiaceae, Asparagaceae, Euphobiaceae, Anarcadiaceae, Polygonaceae, Dioscoriaceae, Solanaceae, Nytaginaceae, Caprifloliaceae, Taccacaceae, Chloranthaceae, Menispermaceae,* and *Laganiaceae* were cited. Identification was done with the help of taxonomists in the Department of Botany of Lashio University. Out of the medicinal plants collected, 40% are wild plants and 60% are both wild and cultivated. The most widely useful plant parts in the preparation of remedies (see Fig. 10.2: Percentage of plant parts used for herbal medicine) are the whole plants (28.6%), leaves (47.6%), stems (9.5%) and tuberous roots or bulbs (14.2%).

The plants collected can be found growing in the wild in forests, on the slopes of the mountains and within or nearby the villages. These plants can be found in a wide range of habitats including woodlands, rocky surfaces, forests, grazing areas and farmlands, home gardens, road and riversides, farm borders and live fences (boundaries made of living herbs or shrubs).

Some local medicinal plant users get the plants from herbalists who sell them in the 'five day' markets as fresh or dried plants, discussed further below. In addition, some villagers, especially herbalists, cultivate medicinal plants in their home gardens for treatment of ailments. Among the 21 collected plants, 12 plants are both cultivated and wild plants. The remaining nine plants are wild plants (see Table 10.2).

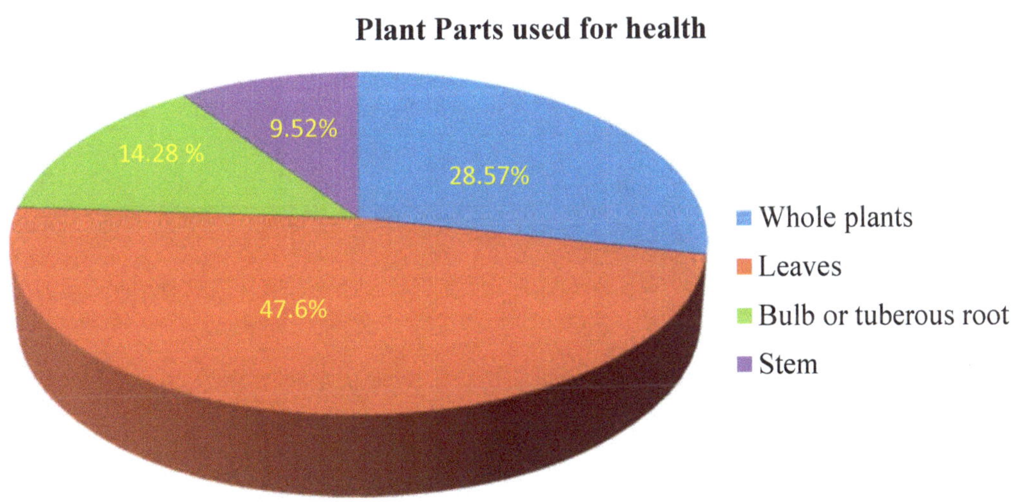

Fig. 10.2 Percentage of plant parts used for herbal medicine. *Source* The authors

Table 10.2 Collected plants for health

Photo	Nomenclature	Medicinal uses
	Scientific name *Datura metel* L. Common name Angel's trumpet Local name Ba-daing (Burmese) Family *Solanaceae*	Asthma and pneumonitis (as inhaler); toothache
	Scientific name *Dioscorea bulbifera* L. Common name Air potato Local name Lay-arr-lu, Putsa-o (Burmese) Family *Dioscoreaceae*	Hypertension
	Scientific name *Asparagus spp.* Common name Asparagus Fern Local name Shint-ma-tet, Kanyut-gala (Burmese) Family *Asparagaceae*	Tonic for general health
	Scientific name *Bryophyllum pinnatum* (Lam.) Oken. Common name Bryophyllum, Air Plant, Life Plant Local name Ywet-kya-pin-paunt (Burmese) Family *Crassulaceae*	Bone fractures; muscle swelling reduction

(continued)

Table 10.2 (continued)

Photo	Nomcnclature	Medicinal uses
	Scientific name *Buddleia acutiflolia* Common name Butterfly bush Local name Pone-ma-gyi (Burmese) Yar-punn-pan (Shan) Family *Loganiaceae*	Women's general health; reduce swollen gastrointestinal tract
	Scientific name *Chloranthus spicatus* (Thunb.) Makino Common name Charan Local name Tha-nat-khar-pan (Burmese) Family *Chloranthaceae*	Reduce swelling in different parts of the body (as bandage)
	Scientific name *Sambucus javanica* Bl. Common name Elderberry Local name Pale-pan (Burmese) Family *Caprifoliaceae*	Laxative; dysentery
	Scientific name *Cleodendrum japonicum* (Thunb.) Sweet. Common name Glory bower Local name Phet-kha ni (Burmese) Family *Lamiaceae*	Reducc swclling in general (as a dressing); reduce gastrointestinal tract illness (taken orally); women's general health (as bath)

(continued)

Table 10.2 (continued)

Photo	Nomenclature	Medicinal uses
	Scientific name *Tinospora crispa* (L.) Hook. F & Thornson Common name Heart-leaved moonseed Local name Sin-don-ma-new (Burmese) Wu-kinn-htin (Kokant) Family *Menispermaceae*	Jaundice; heart ailments; relieving flatulence
	Scientific name *Clerodendrum infortunatum* Common name Hill glory Local name Phet-kha-phu (Burmese) Mai-lu-hpawng (Shan) Family *Lamiaceae*	As Glory bower (above)
	Scientific name *Mirabilis jalapa* L. Common name Four o'clock flower Local name Lay-nar-ye-pan, Mye-su-pan (Burmese) Family *Nyctaginaceae*	Laxative
	Scientific name *Pedilanthus tithymaloides* Peit. Common name Zigzag plant Local name Gongaman (Burmese) Family *Euphorbiaceae*	Skin infection (as protective barrier); bandage for damaged skin tissue

(continued)

Table 10.2 (continued)

Photo	Nomenclature	Medicinal uses
	Scientific name *Dendrobium aphyllum* Common name Leafless dendrobium Local name Thit-khwa (Burmese) Family *Orchidaceae*	Tonic for general health
	Scientific name *Pseuderanthemum latifolium* Common name Malabar false eranthemum Local name Not known Family *Acanthaceae*	Women's general health (as bath); reduce gastrointestinal tract illness (orally)
	Scientific name *Leonurus sibiricus* L. Common name Motherwort Local name Pingu-hteik-peik (Burmese) Family *Lamiaceae*	Protective barrier for treatment of injury; diarrhea and treatment of dysentery
	Scientific name *Rhus semialata* Murs. Common name Nutgall tree Local name Ma-phwet (Shan), Chin pyut (Burmese) Family *Anacadiaceae*	Dysentery

(continued)

Table 10.2 (continued)

Photo	Nomenclature	Medicinal uses
	Scientific name *Solanum indicum* L.	Malaria (Tone Kyat Village)
	Common name Poison berry, Indian nightshade	
	Local name Khayan-kazawt (Burmese)	
	Family *Solanaceae*	
	Scientific name *Cryptolepis buchanani*	Treatment of bone fractures (protective barrier); skin disease (as bath)
	Common name Wax leaved climber	
	Local name Nasha-gyi (Burmese)	
	Family *Asclepiadaceae*	
	Scientific name *Plantago major* L.	Heal wounds; diuretic; malaria; edema; and burning sensation of the body
	Common name Way bread	
	Local name Pharr-gyaw (Burmese)	
	Family *Plantaginaceae*	
	Scientific name *Fagopyrum cymosum* Meissn.	Reduce high blood pressure; bone fracture (used with Life plant and Lemongrass)
	Common name Wild Buckwheat	
	Local name Buckwheat, Shari-mum, Phet-mu (Shan)	
	Family *Polygonaceae*	

(continued)

Table 10.2 (continued)

Photo	Nomcnclature	Medicinal uses
	Scientific name *Tacca integrifolia* Ker.Gawl Common name White batflower Local name Zawgyi-moke-seike (Burmese) Family *Dioscoreaceae*	Reduce swollen tissues (paste of rhizomes); malaria (decoction of rhizomes)

Source The authors

The collected medicinal plants are described by interviewees as beneficial to local people for addressing health problems. In this research, most plants are commonly used by local people in the study areas for stomach disorders and as protective barriers for edema (tissue swelling) or injury. According to research findings, there were some commonly used species across the four villages. For example, "Air Plant" or Bryophyllum (*Bryophyllum pinnatum*) is used as a remedy for bone fractures across all villages and all ethnic groups in the selected villages. It is collected wild by herbalists and is cultivated (see also, Table 10.2: Collected Plants for Health). The fresh crushed leaves of this species are used to apply on bone fractures and to reduce muscle swelling. This species can be easily reproduced in its natural habitat, and is therefore easily accessible for collection and home garden cultivation.

In addition, species such as asparagus fern (*Asparagus densiflorus*) are widely used both by local people in Tone Kyat and across Myanmar. According to the herbalist in Tone Kyat, a tonic is prepared by cleaning the fresh or dried rhizome and boiling it in water. Both herbalists and villagers, male and female, prepare this tonic, but it is mainly consumed by men. They believe that the tonic will keep them healthy and strong. The asparagus fern is also collected by herbalists for sale, either dried or as a tonic, and may be mixed with other plants. There is no shortage of this species in the forest near the village.

Additionally, wild buckwheat or Sharimum (*Fagopyrum cymosum Meissn.*), known locally in Shan as 'Phetmu,' is very popular in Northern Shan State for reducing high blood pressure. It can be eaten fresh, or cooked. The cleaned and crushed leaves are combined with lemon grass and Bryophyllum to make a paste and are used as a bandage for bone fractures. There is no shortage of this plant reported.

Generally, most of the collected plants were applied in the same way to treat the same ailment across different villages, locations, and ethnic groups. However, the research did show that some herbalists had distinctive herbal remedies and applications. According to the botanical survey, Tone Kyat residents used multiple plant species that were only recorded in that village. This included several plants, such as:

Charan (*Chloranthus spicatus*), Elderberry (*Sambucus javanica Bl.*), Four o'clock flower (*Mirabilis jalapa L.*), Zigzag plant (*Pedilanthus tithymaloides Peit.*), and Poison berry (*Solanum indicum L.*), which were used only by herbalists in Tone Kyat, located along the Thanlwin River. In both Tone Kyat and Ohn Tone, Glory bower (*Clerodendrum japonicum*), Hill glory (*Cleodendrum infotunatum*), and Malabar false eranthemum (*Pseuderanthemum latifolium*) were turned into a paste and applied directly to reduce swollen tissue and were used orally to address gastrointestinal tract illnesses.

Only the herbalist at Ohn Tone used Nutgall tree (*Rhus semialata*), Wax leaved climber (*Cryptolepis buchanani*), and White batflower (*Tacca integrifolia Ker. Gawl*). Nutgall tree was used not only for the treatment of dysentery but also in cooking to add a sour flavor to the food. Wax leaved climber was used as a protective barrier for the treatment of bone fractures and as a bath for the treatment of skin diseases. White batflower was also used to reduce swollen tissue, and a decoction of rhizomes was applied for the treatment of malaria. The leafless dendrobium (*Dendrobium aphyllum*) was used by Kokang villagers in Tone Kyat as a tonic for general good health and longevity. To make the tonic, they use leaflets boiled in water. It is also cultivated for export to China.

In sum, a wide range of plants are used to treat a number of ailments; some are only grown in a particular context, like the species mentioned in Tone Kyat along the Thanlwin River. For more details and a full listing, please see Table 10.2.

10.5.2 The Role of Herbalists

Across these four villages, herbalists are highly respected for the important role they play in the collection and application of medical herbs. The physical and economic limitations of the villagers make it difficult to buy modern medicines and access public healthcare services, which are only located in Kunlong city. This gap in healthcare is covered by village herbalists, who are also generally poor and rely on agriculture for their livelihoods.

Most villagers in Yae Lei Kyun highly respect U Sai Aung Myat, a Shan herbalist with a long experience in the treatment of various ailments in this village. He cultivates medicinal plants in a home garden, which he tends with his spouse near the bank of the Thanlwin River. The plants grown here are used either for medical care for patients or are sold.

In Wa Soke village, local people likewise show respect to herbalists who live in their village, and rely on them for the treatment of ailments, not least because of transportation difficulties and their lower economic status (Herbalist interview, 20 December 2015). In Wa Soke village, the herbalist used a combination of more than one species. Herbal plants can be collected around the village easily, and are soaked in a traditional rice brew or decoction of water to reduce the swelling of internal organs. Although villagers said that they sometimes used Myanmar medicine (specifically mentioned Ywet-hlay brand) for health problems, the majority reported relying on herbal medicines.

Most herbalists interviewed have no formal education, but work from generational experience and knowledge that has been passed down. Among the interviewed herbalists, only Mr. U Sai Than Sein of Ohn Tone village became an herbalist through more formalised herbal medicine training after his work as a primary school teacher. He received his training in medicinal herbs from the Nationalities Youth Resource Development Training School. This is in contrast to the other herbalists in Ohn Tone and Wa Soke who said they received medicinal knowledge from their ancestors in order to understand how to use medicinal herbs to treat aliments.

Mr. U Sai Than Sein told us that in Ohn Tone, villagers usually opt to take herbal medicines to treat ailments, but sometimes also use modern medicine when they can afford it. During an interview, his mother showed us the earthen pot used to boil medicinal plants daily. As U Sai Than Sein described it:

> My mother boils a mixture of dried collected plant species and takes it as medicine daily, and she is healthy and over 80 years old. Applying a combination of species [rather than one single species] is more effective for curing disease (11 November 2015).

This was a key point made by herbalists, whether they were formally educated or not; in many cases it was more effective to use multiple species for treatments.

In addition to collecting medicinal herbs for treating ailments at home, herbalists and some villagers, particularly in Tone Kyat, collect plants to sell in the five-day market. The five-day market is very important for herbalists and local communities. It is held in rotation over a five-day period among five different sites around Kunlong township, namely Kunlong, Holi, Nartee, Karmine and Hopan. Local upland people, especially herbalists, sell both fresh and dried medicinal herbs in this market. As an herbalist who sells at the Kunlong five-day market described: "I collect herbal plants from the slopes of mountains which are located behind my village for selling and treatment" (Interview, 20 September 2015).

The five-day market provides an important economic link between rural and urban communities, supporting the local economy of rural upland villages like Ohn Tone and Wa Soke, and allow greater access to medicinal herbs. This arena for economic activities links rural and urban communities together through medicinal herbs. Herbal medicines thus play an important role in providing income for remote communities in addition to treating illness.

10.5.3 Local Medicinal Plants for Trade

Beyond the five-day market, some medicinal plants are also sold regionally and internationally. For example, the leafless dendrobium orchid is not only used locally, but is also sold commercially. This species can be found wild in the forest and can also be cultivated in home gardens. In the five-day market, there were only a few of these orchids available for sale; most of the time, this plant is exported to neighbouring countries such as China. Thus, while it is clear that across the four villages local people rely on medicinal plants, there is also interest and pressure from outside actors and markets. Dr. Saw Lwin, an orchid expert who runs Myanmar Flora and Biotech, suggests that people, particularly in China, believe that the stem of some orchids can prevent and cure cancer and help people look younger (Phyu 2014).

Locally, the dried stem of leafless dendrobium is used by members of the Kokang and Wa ethnic groups as a tonic for general health. In the study sites, the dried stems are exported as medicinal plants to China for income for local people. According to an interview with local villagers, fresh orchids are sold for 200,000 Myanmar Kyat per viss (1 viss is equal to 3.6 lbs or 1632 g) in the local market (Interview, 20 December 2015). Dr. Saw Lwin, however, warns that the "over-collection [of *dendrobium* orchids] has resulted in the near-extinction of some species" (Phyu 2014).

There are questions surrounding the legality of such trade. Wild orchids are protected in Myanmar under a 1994 law, and the Convention on International Trade in Endangered Species of Wild Fauna and Flora also places restrictions on the international orchid trade. A range of orchid experts have discussed these challenges. In a recent *Myanmar Times* article, Mr. U Win Naing Thaw, director from the Ministry of Environmental Conservation and Forestry's Nature and Wildlife Conservation Division, explained that only one company, Myanmar Flora and Biotech, had applied for a permit to legally trade orchids. He also reported that most orchid growers and sellers did not know they are in breach of the protection of Wild Animals, Wild Plants and Conservation of Natural Areas Law, which also prohibits the sale of cut orchid flowers without a permit (Lwin 2011). In sum, the commercial or even local trade of this plant can impact local availability for medicinal uses and may be putting those who harvest it in breach of national laws.

10.6 Threats to Herbal Medicine: Deforestation, Cash Crops, Trade and Local Knowledge Transfer

In addition to the sale of orchids described above, other threats to medicinal plants and the practice of herbal medicine include the expansion of cash crops, overcutting, and logging for fuel wood (Sai Thein Sein Interview, 20 February 2016). Although current forest laws are being rewritten (Lwin 2017), according to a Forest Department officer, these laws are difficult to enforce in Kunlong township and the current fines are too small to discourage logging and overcutting (Interview, 24 February 2016).

The main threat to wild medicinal plants is deforestation and agricultural expansion, particularly cash crops and rubber plantations in the areas surrounding Kunlong. For instance, in Ohn Tone and Wa Soke, villagers traditionally grew rice for subsistence through shifting cultivation. Now, swidden farming techniques are being used to clear portions of the forest to cultivate cash crops such as ricc, corn, and sugarcane. According to interviewees, this is a more permanent form of cultivation and is occurring on the mountain slopes. Villagers cultivate cash crops both for local consumption and for export to neighboring countries such as China. Local people reported that they do not make enough money selling cash crops because the inputs like seeds and fertilizers cost more than they can make from selling their crops. Villagers also rely on synthetic chemical fertilizers to cultivate cash crops. These factors, combined with the natural vulnerability of such arid and semi-arid lands, may lead to further reduction in the natural habitats of medicinal plants.

In addition to being collected in the wild, traditional medicinal plants are also cultivated in home gardens, and harvested along the roadside and live fences. One reason that traditional healers are planting in their home gardens and sourcing the plants from distant places along the Thanlwin River is because medicinal plants are becoming harder to find in the Kunlong area. With medicinal plants becoming harder to find, an herbalist from Yae Lei Kyun village explained:

> *Medicinal plants are important to us, so I cultivate them in my garden for sale and treatment* (20 December 2015).

Not only is the direct overharvesting of medicinal herbs a challenge, but the overharvesting of other non-timber forest products also disturbs the habitat of these herbs. Near Wa Soke and Ohn Tone, some villagers live temporarily in the forests at certain times of the year, building huts and cooking fires while they collect bamboo shoots to sell in the local markets. Personal observations and conversations with the herbalist Mr. Sai Thein Sein showed that these activities also threaten medicinal plants.

Lastly, the continuing practice of herbal medicinal knowledge is also at risk due to the absence of intergenerational knowledge transfer, as young pcoplc tcnd to bc more interested in other livelihoods such as growing cash crops and working in the city or abroad (Personal communication, 30 August 2015). Knowledge sharing is crucial not only to herbal medicine practitioners, but also to local policy makers in their efforts to conserve and sustain valuable medicinal knowledge in Kunlong and throughout Myanmar, where large numbers of people depend on herbal medicines for health.

10.7 Conclusion

Overall, this study documented plants from 14 families with medicinal value as indicated by local residents and herbalists. These plants can be found in a wide range of habitats, both wild and domestic, including forests, grazing areas, farmlands, home gardens, roadsides and riversides, farm borders and live fences. The majority of herbalists used one or more plants in their treatments, particularly in cases of bone fractures and stomach disorders, which is probably an indication of the frequency of these ailments in the area. The most widely sought after plant parts in the preparation of remedies are the roots, leaves and stems in this order.

Also, while villagers from both lowland and highland villages relied on medicinal plants and herbalists to treat illness, this reliance was more acute in highland villages with limited access to health care in the form of medical doctors, hospitals, and clinics. In drawing a link between access to transportation, socioeconomic status and public health across the selected villages, a pattern emerges. For instance, among the selected villages, the highland areas (Wa Soke and Ohn Tone) are more difficult to access, and residents report less income and poor access to public healthcare, particularly when compared to lowland villages in the same area.

Threats to herbal medicines in Shan State are a cause for concern for communities who rely on these plants to treat illness. Agricultural expansion for cash crops like rice, corn and sugarcane, rubber plantations, and logging for fuel wood is threatening medicinal plants and contributing to declining availability. The continuing practice of traditional healing is also threatened in the absence of intergenerational knowledge transfer in villages, as young people desire to go abroad to work. Knowledge sharing is crucial not only to local practitioners but also to local policy makers so that efforts are initiated to conserve and sustain the valuable medicinal knowledge of ethnic groups in Shan State and throughout Myanmar and the Thanlwin River Basin.

As researchers, we encourage further research on the effects of the synthetic chemical fertilizers used on cash crops, as this may affect soil fertility and damage native medicinal herbs. As some interviewees reported a decline in returns from cash crops, this presents an opportunity for additional research to understand the drivers for local people to cultivate the cash crops, which requires the use of chemicals and results in forest loss, impacting local herbs directly. Research into possible livelihood alternatives could lead to recommendations for more sustainable income.

By highlighting local knowledge on plants and medicinal treatments and the challenges facing the survival of this knowledge, this chapter intends to also inspire a broader outlook regarding the cultural and social importance of traditional healing practices among ethnic communities in Myanmar, and particularly Shan State. The chapter also seeks to raise awareness of the uses of medicinal plants commonly available in villages, and encourage a practical and sustainable approach to the conservation of these plant species in Kunlong Township, Lashio district.

References

Abbasi, A.M., Khan, M.A., Ahmad, M., & Zafar, M. (2011). *Medicinal Plant Biodiversity of Lesser Himalayas-Pakistan.* New York: Springer Science and Business Media.

Cox, P.A. (2000). Will tribal knowledge survive the millennium? *Science*, *287*(5450), 44–45.

Dassanayake, M.D., & Fosber, F.R. (1988). *A revised handbook of the flora of Ceylon* (Vol. 6). CRC Press.

Deetes, P. (2012, March 24). *The Salween River Basin* [Fact Sheet]. International Rivers. Retrieved from: https://www.internationalrivers.org/resources/the-salween-river-basin-fact-sheet-7481.

De Padua, L.S., Bunyapraphatsara, N., & Lemmens, R.H.M.J. (Eds.). (1999). *Plant resources of South-East Asia Medicinal and Poisonous Plants* (Vol. 12, No. 1). Bogor, Indonesia: Backhuys Publishers.

EIA (Environmental Investigation Agency). (2015). *Organized chaos: The illicit overland timber trade between Myanmar and China*. London, United Kingdom. Retrieved from: https://eia-international.org/report/organised-chaos-the-illicit-overland-timber-trade-between-myanmar-and-china/.

FAO (Food and Agriculture Organization). (2010). *Global forest resources assessment 2010: Main report*. FAO forestry paper (Vol. 163). Rome, Italy: FAO. Retrieved from: http://www.fao.org/docrep/013/i1757e/i1757e.pdf.

FAO (Food and Agriculture Organization). (2015). *Global forest resources assessment 2015: Desk Reference*. Rome, Italy. www.fao.org/3/a-i4808e.pdf.

Flaster, T. (1996). Ethnobotanical approaches to the discovery of bioactive compounds. In J. Janick (Ed.), *Progress in new crops* (pp. 561–565). Arlington, VA: ASHS Press.

Hong Kong Herbarium and South China Botanical Garden (Eds.). (2007). *Flora and Fauna of Hong Kong* (Vol. 1). Hong Kong: Agriculture, Fisheries & Conservation Dept.

Hong Kong Herbarium and South China Botanical Garden (Eds.). (2008). *Flora and Fauna of Hong Kong* (Vol. 2). Hong Kong: Agriculture, Fisheries & Conservation Dept.

Hong Kong Herbarium and South China Botanical Garden (Eds.). (2011). *Flora and Fauna of Hong Kong* (Vol. 4). Hong Kong: Agriculture, Fisheries & Conservation Dept.

Hong Kong Herbarium and South China Botanical Garden (Eds.). (2012). *Flora and Fauna of Hong Kong* (Vol. 5). Hong Kong: Agriculture, Fisheries & Conservation Dept.

Hutchinson, J. (1959). *The families of flowering plants* (2nd ed.): *Dicotyledons* (Vol. 1–2). Oxford, United Kingdom: Oxford University Press.

Hundley, H.G., & Chit Ko Ko, U. (1961). *List of trees, shrubs, herbs and principal climbers, etc.: Recorded from Burma with vernacular names*. Rangoon: Government Printing and Stationary, Union of Burma.

Kamboj, V. (2000). Herbal medicine. *Current Science, 78*(1), 35–39.

Kress, W.J., DeFilipps, R.A., Farr, E., & Daw Yin Yin Kyi. (2003). *A checklist of the trees, shrubs, herbs, and climbers of Myanmar* (revised from the original works by J.H. Lace, R. Rodger, H.G. Hundley, and U Chit Ko Ko on the "List of trees, shrubs, herbs and principal climbers, etc., recorded from Burma"). *Contributions from the United States National Herbarium, 45*.

Lawrence, G.H.M. (1966). *Taxonomy of vascular plants*. New York, NY: Macmillan.

Lwin, E.E.T. (2011, November 28). Government targets illegal orchid trade. *Myanmar Times*. Retrieved from: http://www.mmtimes.com/index.php/national-news/1783-government-targets-illegal-orchid-trade.html.

Lwin, S.T. (2017, February 20). U-standard forestry agreement to be set up. *Myanmar Times*. Retrieved from: http://www.mmtimes.com/index.php/national-news/mandalay-upper-myanmar/25008-eu-standard-forestry-agreement-to-be-set-up.html.

Phochan. (2015, September 30). Shan State, Sagaing region in danger of deforestation. *Eleven*. Retrieved from: http://elevenmyanmar.com/local/shan-state-sagaing-region-danger-deforestation.

Phyu, A.S. (2014, March 10). Orchid smuggling putting rare species at risk, warn experts. *Myanmar Times*. Retrieved from: http://www.mmtimes.com/index.php/national-news/9796-orchid-smuggling-putting-rare-species-at-risk-warn-experts.html.

Woods, K. (2013). The politics of the emerging agro-industrial complex in Asia's final frontier: The war on food sovereignty in Burma. In *Proceedings of Food Sovereignty—A Critical Dialogue*, New Haven, CT, September 14–15, 2013.

WHO (World Health Organization). (2013, November 30). *Main messages from World health report 2013: Research for universal health coverage.* Retrieved from: http://www.who.int/whr/2013/main_messages/en/.

WWF (World Wide Fund for Nature). (2015). *Living forests report, chapter 5: Saving forests at risk*. Gland, Switzerland: WWF. Retrieved from: https://c402277.ssl.cf1.rackcdn.com/publications/793/files/original/Report.pdf?1430147305.

Chapter 11
Not only Anti-dam: Simplistic Rendering of Complex Salween Communities in Their Negotiation for Development in Thailand

Paiboon Hengsuwan

11.1 Introduction

Since the 1970s, hydropower dam projects in the lower Salween River Basin have been planned as joint undertakings of the Myanmar, Thai and Chinese governments, state-owned enterprises, and multinational corporations (Lamb et al., Chap. 1, this volume). In response, civil society groups at various levels have mobilized against the proposed Salween dam projects, including Thai and international environmental organizations and human rights groups, academics, and journalists, with considerable support from locals who oppose these controversial dams (see Middleton et al., Chap. 3, this volume).

In 2007, the author began observing residents along the Salween, finding that, generally, local ethnic groups have continuously opposed these controversial dam projects. Hence, these villages have been represented as anti-dam communities by project developers (Modinator 2014). At the same time, local communities are situated in poor conditions of "development" (in Thai language, kanphatthana), or lacking development; lacking access to resources, roads, government services, land rights and citizenship (Hengsuwan 2012).

Although community protests may be interpreted as 'anti-development,' I argue that their reaction to development is actually more complex. Community members in fact seek development, but on their own terms (Keyes 2014: 135, 161). Furthermore, given the cultural and political complexity, this engagement cannot be reduced to a simplified struggle between state domination and community resistance (Orther 2006, cited in Walker 2012: 150). This is because local villagers have played a part in the extension of state power through development programs by pursuing, engaging in, and reframing state-sponsored development initiatives (Walker 2012: 150–151).

Paiboon Hengsuwan, Lecturer, Department of Women's Studies, Chiang Mai University; Email: paiboon1998@yahoo.com.

C. Middleton and V. Lamb (eds.), *Knowing the Salween River: Resource Politics of a Contested Transboundary River*, The Anthropocene: Politik—Economics—Society—Science 27, https://doi.org/10.1007/978-3-319-77440-4_11

The purpose of this chapter is to first analyze how Salween dams have been framed through state-project developer collaboration as hegemonic discourses of development. Underlining these discourses are that dams are development (Bakker 1999), and thus, that resistance to dams is anti-development. For villagers at the Salween who are anti-dam, they find themselves therefore positioned as anti-development by project developers (Modinator 2014). At the same time, local communities are situated in poor conditions of 'development' (in Thai language, *kanphatthana*), or lacking development; lacking access to resources, roads, government. In this chapter I analyze how individuals and communities have engaged in state development by proposing local development initiatives. The chapter thus shows how the villagers' anti-dam movement, in working with civil society, is one of the arenas through which communities are spurring local development. In this sense, it is overly simplistic for dam project proponents to claim that these communities are merely anti-dam. Actually, the Salween dam projects are just one of the local communities' concerns when contemplating future livelihoods. They are also advocating for land rights, community and citizenship rights, and access to basic infrastructure and government services. This chapter therefore proposes that we reposition and reframe the anti-dam movement as not a singular issue, but rather part of a wider array of movements surrounding local development. While both project developers and civil society campaign groups may focus on a single project or issue, both local communities and the government require a holistic approach to development planning.

11.2 Methodology

This chapter is an analysis of information from dissertation research conducted on environmental politics and resource management in Thailand, along the Salween borderlands, during 2007–2010, along with fieldwork conducted in the same communities along the border from 2016 to 2017. Rather than avoiding communities exposed to political advocacy in development conflicts where NGOs and academics have influenced villager's perceptions (Sangkhamanee 2013: 12), three villages central to anti-dam campaigning along the Salween River in Thailand were chosen: Saw Myin Dong, Bon Bea Luang, and Muang Mean (See Fig. 11.1: Map of study sites (3 villages) along the Thai-Myanmar border). In-depth interviews with 25 key informants from these three villages were conducted, along with interviews with 11 local and national activists. Participant observation, along with participation in meetings, activities, and NGO field visits also shed light on the complexity of villagers' techniques to deal with 'development' while at the same time sustaining their lives, communities, and cultures.

The three villages are located in Mae Hong Son Province on the Thai side of the Salween River where it forms the Thai-Myanmar border. Diverse ethnic communities have settled along the Salween borderland, including Karen, Myanmar, Shan, and Khon Muang (Northern Thai). Most of the villagers in Saw Myin Dong and Muang Mean are ethnic Karen. Saw Myin Dong is comprised of 128 households,

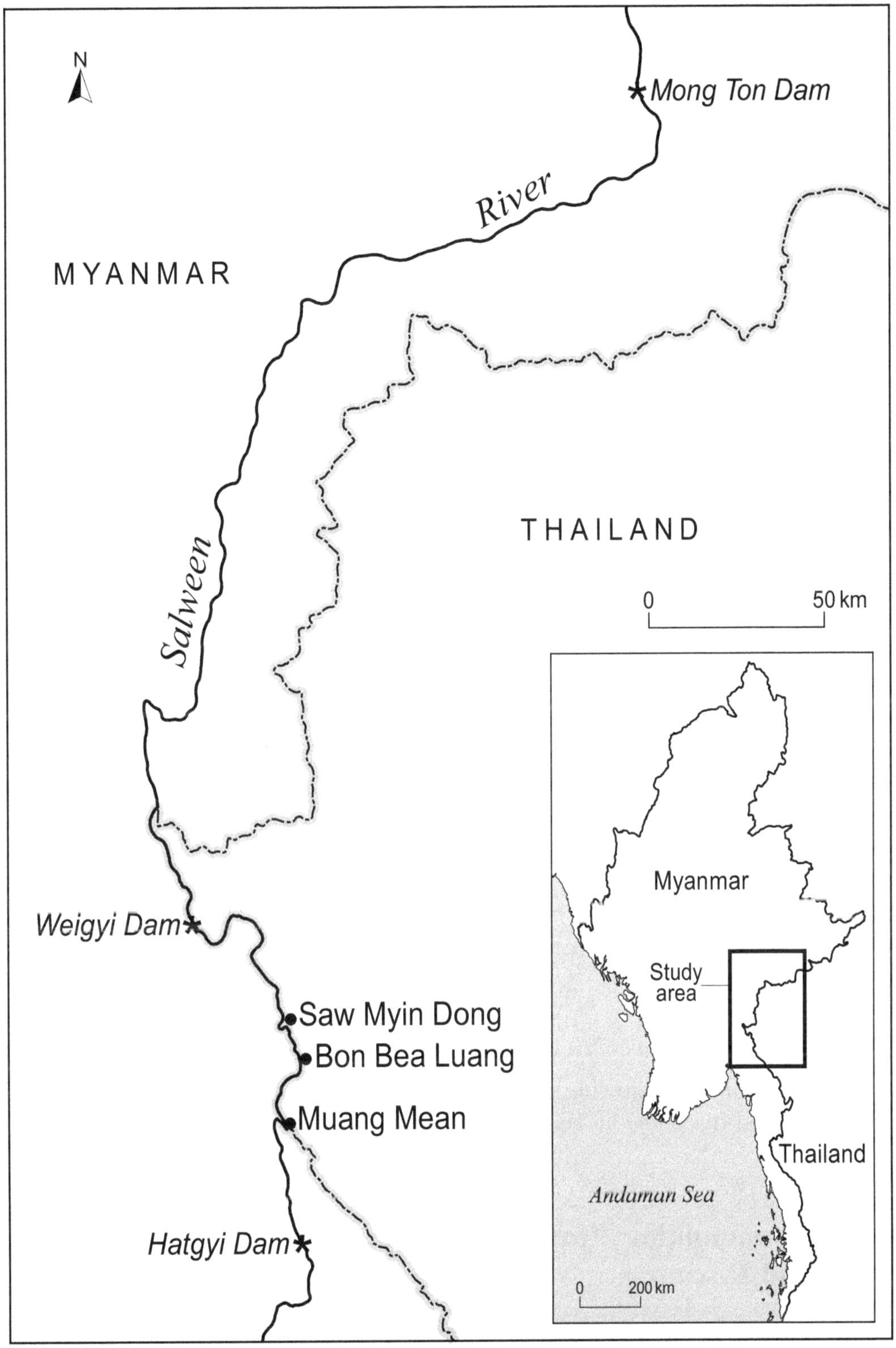

Fig. 11.1 Map of study sites (3 villages) along the Thai-Myanmar border. *Source* Cartography by Chandra Jayasuriya, University of Melbourne, with permission

with a population of 926. Of this number, 661 people possess Thai identity cards. Muang Mean has 89 households (only one hamlet), with a population of 884. In Muang Maen, 631 people possess Thai identity cards. The residents of Bon Bea Luang mainly identify as Karen, Karen Muslim, and Shan. Bon Bea Luang has 279 households, with a population of 1,542. Only 603 residents possess Thai identity cards. On the Myanmar side of the Salween opposite the towns of Saw Myin Dong and Bon Bea Luang are Myanmar military camps, but opposite Muang Mean is a Karen National Union-held area. The proposed Wei Gyi and Dagwin Dams would be located upstream of these villages, and the proposed Hatgyi Dam would be located downstream.

Various discourses rationalizing plans for dams on the Salween River have emerged since the 1980s. The production of the meaning of, for instance, 'development' or 'nature' is a discursive practice, and such meaning emerges not only because of language or speech, but also because of actual practice to define the truth (Foucault 1977; Yoon 2001). Therefore, instead of taking these meanings for granted, one should analyze 'truth' and 'knowledge' as the practice of 'power' (Foucault 1980; Lorenzini 2015).

Hydropower dams as 'development' is one of these meanings. This 'truth' is co-constructed by states/capitalists as a dominant hegemonic discourse, serving as a guide for how to use a river to generate economic growth and national progress, and expressing the production of a Salween resource frontier by means of regulation. Project developers frame the river, via scientific discourse, as prime for hydro-electricity production, which in turn generates development (Bakker 1999). Meanwhile, the Myanmar and Thai states also use 'development discourse' (Escobar 1995; Gupta 1998) and 'civilizing the margins' (Duncan 2004) as a means to accelerate territorialization (Vandergeest 1996) and to enhance capitalist expansion into the borderlands. It is important to investigate how development is framed, and how development discourse may be changed to community development.

However, it is not only the state that can create 'truth'. The villagers and organizations I spoke to also make claims. For instance, community groups claim community development as a sub-discourse of development, albeit one which is set against the state/capitalist regime of dams as development 'truth'. I discuss the details further in the following sections as linked to production of environmental knowledge and discourse in Thailand.

11.3 Communities' Production of Knowledge and Discourse

Since the 1980s, Thai communities and social movements have evolved new approaches in their engagement with the state, including people's movements, mobility, and counter-knowledge production. This section briefly introduces these approaches in the broader context of Thailand, before we turn more specifically to the Salween borderlands. Firstly, the late 1980s saw the emergence of several new organizations representing rural demands, such as the *Assembly of the Poor*, a peasant movement for rights to land, water, and forests. They have used both the street and the media to open up space in national politics for rural demands. The strategies of these movements reflect a sophisticated appreciation of the wider national political economy, and this sophistication is attributed to the role of urban NGOs and leaders who possess both rural origins and urban experience (Baker 2000: 5–11).[1]

A second factor contributing to increased engagement with the state has been the mobility of rural villagers in response to a growing labor market. Since the 1950s, northeastern villagers began to migrate to Bangkok and later overseas in search of wage employment. Migration indicates that most northeastern villagers have made 'development' of their rural worlds a personal goal. They are "cosmopolitan villagers" who have gained a sophisticated understanding of the world and have become considerably less compliant than in the past. They found they could now turn to NGOs as well as to certain academics, newspapers, and even a few elected representatives for support of their concerns (Keyes 2014: 149–173). We can say that they have a "cosmopolitan vision" (Petras 1997: 19–20).

Thirdly, we can look at knowledge production found in the case of *Thai Baan* research. *Thai Baan*, or 'Villager' Research, is kind of local knowledge research that is positioned as a "counter-hegemonic approach" to conventional research and another means that local people have employed to contest development (Vaddhanaphuti 2004: v). In doing so, local knowledge in environmental management is used as a political force for dam-affected villagers to express their views and engage with government institutions to influence development planning and decision-making processes. In other words, it can be used to facilitate greater participation and empowerment of local resource users in the long term (Scurrah 2013: 45–47).

Along with these approaches in engagement with the state, the researcher will focus on social movements and communities' counter-knowledge production along the Salween borderlands. It is proposed that these strategies create a site of negotiation for the purpose of initiating dialogue with state power. As shown in the case of rural communities in Northeastern Thailand, the *Isan* people have embraced the concept of *khwamcharoen:* committing oneself to the pursuit of progress—improvement in living conditions in the village (Keyes 2014: 156). Even though they have been strongly influenced by 'development' discourse originating from the

[1]Baker (2000: 11) mentions that the Assembly of the Poor is a direct result of the development of a new rural political economy that reflected in the background and life-experience of its leaders. The improvement of the road network expands the cheap communications that mean rural people can access to Bangkok by overnight bus. There are also affordable motorcycles. The rural TV broadcast network was rapidly expanded that rural people increasingly owned a TV. Even sons of peasants who had trained as teachers or lawyers returned and supplanted city-based NGO workers and organized the peasantry movement. Many of them had worked outside the village, especially in Bangkok and as overseas migrant labor in the Middle East. Compared to the peasant movement in the Latin America, those who are leading the struggle, travel to the cities, participate in seminars and leadership training schools, and engage in political debates, even as they are rooted in the rural struggle, and engage in agricultural cultivation (Petras 1997: 19–20).

Thai government, rural Northeastern communities have reworked this discourse with reference to their own cultural traditions. With a cosmopolitan vision, they have not only opposed government-sponsored development programs, but also pursued development with the intent of effecting improvements in their standard of living (Keyes 2014: 168–170).

Similarly, Jarernwong (1999) studied the development of the meaning of "rural development" in Northern Thailand. He found that rather than either passively accepting or totally resisting the government-defined meaning of development, villagers possess their own vision of development. Since the 1980s, the meaning of development as defined by the state has shifted from state-sponsored infrastructure and construction to the distribution of economic benefits with people's participation. As a result, villagers fully accepted projects that responded to their needs, and only passively accepted projects with only partial benefits to them. Through various measures, both interpretive and otherwise, they adjusted unprofitable projects in a way that was more gainful to communities, both in the short and the long term (Jarernwong 1999, 2001).

11.4 Targeting the Salween Borderlands: Civilizing the Margins by the States

The Salween borderlands, which are marginal spaces of both the Thai and Myanmar states, have been modernized, or 'civilized,' by both governments (Bryant 1997). Civilizing the margins (Duncan 2004) induces dramatic modernization in the Salween borderlands. I take development as referring to the raising of people's standard of living, but development projects also reinforce the beliefs that those defined as in need of development are primitives, further reinforcing them as subjects in need of development. This is also embedded in the state policies from which the drive for modern development stems. In the case of the Salween dams, the main actors in this process are powerful transnational companies and state-owned enterprises, which are increasing the commodification of nature (MacLean 2008; Nevins/Peluso 2008). This section discusses the process of commodification of the Salween River at the borderlands as a means of civilizing the margins, embedded in the wider political context of neo-liberalism at the Thai-Myanmar border.

In 1988 the Thai Prime Minister, General Chatichai Choonhawan, initiated an economic vision to transform Indochina from a "battlefield into a marketplace." This turning point brought capitalist development to the Thai-Myanmar border regions (Battersby 1999: 479). By 2003, the Thai government was engaged in bilateral cooperation with Myanmar on four mega-projects: the Tasang Dam on the Salween River, to be constructed by Thailand's MDX company; a coal mine in a Myanmar town opposite Prachuap Khiri Khan, Thailand; a port project in Dawei; and a Mae Sot-Yangon road project (Fawthrop 2003). Despite the fact that these development projects would affect the livelihoods of local people, the villagers were not informed about the projects by the Thai or the Myanmar governments. Most of the people interviewed for this study, for example, did not possess clear information about the Salween dam projects. The information they did receive was from NGOs rather than state actors.

Since the 1970s, dam developers have proposed to build large dams on the Salween River. These projects are currently being developed by companies from Thailand, China, and Myanmar. The Electricity Generating Authority of Thailand International (EGATi), Chinese corporations and Myanmar investors have joined to develop plans and proceed with the development of a number of hydroelectric dams on Myanmar's stretch of the Salween River (see Middleton et al., Chap. 3, this volume). This includes several projects, such as the Weigyi, Dagwin, and Hatgyi dams.

Two of the projects proposed along the border, Weigyi and Dagwin, have been delayed and potentially cancelled. Hatgyi, however, is still under consideration. The proposed Hatgyi Dam is a large dam at 1360 MW. The project investors are EGATi, Sinohydro Corporation, Myanmar's Department of Hydropower Planning (DHPP), and the Myanmar private investor International Group of Entrepreneur Company (IGOEC) (EGATi 2010; Salween Watch 2014). Hatgyi would be located on the Salween River, in Karen State, Myanmar not far downstream from the Thai border where this research was conducted.

The project has been met with strong opposition from civil society, with both local, regional and international groups joining in protests against Hatgyi and other Salween dams (Nang Shining 2011: 26–30; TERRA 2007; Salween Watch 2014). Opponents assert that the project should not move ahead based solely on the decisions of the state and private sector without the participation of local people. The project has been characterized by a lack of information disclosure (Chantawong 2011; TERRA 2007). Local people on both sides of the border are concerned about cross border impacts on local ecology and fisheries, and the inundation of residential areas and farmlands along the Salween River. The Myanmar government is supporting the Hatgyi Dam project as part of a strategy to remove ethnic armed groups from the dam site, as the project is to be located in an area of conflict between Karen insurgents, known as the Karen National Union (KNU) and the *Tatmadaw*, the Myanmar military (Nang Shining 2011; Hengsuwan 2013; Simpson 2013).

The Thai state-owned EGAT has issued numerous claims its activities will provide optimum benefits for communities in the long run (EGAT 2009a, cited in Nang Shining 2011: 2). EGAT has conducted a public relations campaign to promote the dam in communities on both sides of the border, and commissioned Chulalongkorn University's Environmental Research Institute to conduct an EIA of the Hatgyi Dam, which was completed in 2008. Although the dam developer EGATi promises to support affected communities, the Thai National Human Rights Commission proposed that the Thai government order EGATi to abandon plans to build the Hatgyi Dam (NGO-CORD North 2007, 2008). However, in 2009, Abhisit's government recommended further studies, and refused to stop the Hatgyi Dam project (Salween Watch 2010). In the recent years, EGAT has continued to push the project forward by asking KNU leaders to concede to EGAT's demands to conduct surveys at the dam site. Most recently, EGAT and a Chinese team have conducted new research in the area. Meanwhile, there has been increased militarization by the Myanmar army around the dam sites (Lambrecht 2000, 2004). EGAT has also approached local people in an effort to convince them to agree to the project (Salween Watch Coalition 2016).

The Salween dam projects – both inside Myanmar and on the Thai-Myanmar border, are being developed by Chinese-backed companies, state-owned Thai enterprises, and Myanmar companies, with backup from the Myanmar military in sensitive border regions. Development discourse is being used by states and energy investors as an entry point to turn the river into hydro-electricity for trade.

On the Thai side, the government has run community development projects, such as infrastructure development and construction of a health center and a school, and is framing the projects as an expansion of regional power development along the Salween borderlands. However, these projects are not simply good or bad. As Duncan (2004: 18) points out, "development projects are not the benign empowerment schemes that they claim to be, nor are they the all-powerful machines of ethnocide and cultural destruction that many activists claim." The minorities' choice between incorporation into government projects or resistance to government schemes depends on the particular context. In this case, local people in Thailand have chosen to participate in community development projects, but are resisting the Salween dam projects. This implies that the Thai state is not able to exert total control over populations on the border. Salween residents are prepared to confront the Thai government if they perceive that the Salween dams will be disruptive to their livelihoods and security.

International financial institutions such as the Asian Development Bank (ADB) and transnational energy investors also use regional energy security as a rationale to justify the Salween dam projects through power development projects such as the GMS Power Grid (ADB 2006). GMS countries also follow this theme: every state should think about its national power security and make plans for national power development (Middleton et al. 2009). These states-plus-capitalists have legitimized power over the Salween. They have identified the function of the Salween River as a source of energy, one used to serve the nations' and the region's prosperity. For this reason, they have attempted to impede local resource management and take control of the Salween River through the discourse of development and regional energy security. The way that states and international financial institutions envision power sharing and regional power trade is contested, however. Civil society groups continuously challenge international financial institutions, particularly the ADB, at the regional level, while at the same time challenging states to improve their national power development plans and energy projections.

Whether or not the Salween dams will be built depends neither upon the intention of local people nor that of individual states. Decision-making power is, rather, held by a combination of various parties. Transnational dam investors must therefore seek alternative ways to convince local people to acquiesce to dam construction. Several kinds of development are being introduced into the Salween borderlands in an effort to convince local people to agree to the Salween dams. Therefore, alternative development as conducted by states/capitalists has become a depoliticized program operating as a development apparatus to expand bureaucratic state power (Ferguson 2003). However, the Salween dam projects are still widely opposed by residents along the Thai-Myanmar border.

11.5 Life and Development in Local Communities

Remote Salween communities do not receive much attention from the central Thai government. Infrastructure development projects are slowly being introduced into villages. The main road from Mae Sariang and Sob Moei District is of poor quality, and poor infrastructure makes it difficult for villagers to travel during the rainy season. Local water supplies are limited. Bon Bea Luang village enjoys higher infrastructure development than Saw Myin Dong and Muang Mean villages due to its central location, and several government offices have been set up there, including the National Park Unit and the main check point of the Border Patrol Soldier Unit. However, local people face difficulties due to limited residential and agricultural land, low incomes, and minimal health services.

The lives of border people are threatened by Thailand's natural resource conservation policies. Both a wildlife sanctuary and a national park have been established at the Thai-Myanmar border, where many small communities have settled down in forest areas along Salween tributaries. The forestlands around these villages are now part of the Salween National Park and Forest Reserve, posing conflicts with government authorities such as national park officers and foresters over resources use (Bundidterdsakul 2013, Chap. 9, this volume, and Lamb/Roth 2018). These authorities tend to perceive of local people as destroyers of the forest. Local residents must therefore struggle for their right to continue living in protected forest areas. The villagers see this as their most urgent problem, with no immediate solution pending.

Villagers in two communities I conducted research in, Saw Myin Dong and Muang Mean, expressed that they are proud of their lives because they have plenty of rice and fish. Most of them do rotational shifting cultivation and a few do paddy cultivation to produce rice for consumption. They also catch a lot of fish in the Salween and Moei rivers and gather wild plants and animals from the forest. Their livelihoods rely on the abundance of aquatic and forest resources. However, they are worried about the proposed Salween dam projects, particularly the environmental and social impacts of the Hatgyi Dam downstream. I observed that local people are particularly concerned that Hatgyi Dam would cause flooding and inadequate/unstable water levels in the river, which would prevent them from cultivating crops along the riverbanks. The river ecosystem would change, potentially causing the extinction of a number of local fish species. Additionally, affected people from many villages inside Myanmar would have to resettle, causing further displacement. Due to these concerns, the villagers have resisted the Salween dam projects through engagement in the anti-Salween dams movement, involvement in public participation processes under the Abhisit Government, and by undertaking Thai Baan research as a means of counter-knowledge production.

11.6 Civil Society and Communities Resisting the Hatgyi Dam Project

Local communities and the state have had a long-term engagement over broader issues of development of the borderlands. The issue of the Salween dams is a recent one, and one of many ways that the state and community have interacted. In resisting the Salween dams, local residents regularly interact with national and international civil society groups that are concerned about the environmental and human rights impacts of these projects. These civil society groups have mobilized a broader anti-Salween dam movement, and brought increased awareness of the potential environmental and social impacts of the dams to affected communities.

Coalitions of NGOs, including Towards Ecological Recovery and Regional Alliance (TERRA), NGO-CORD North, Salween Watch and Living Rivers Siam (SEARIN), have organized several forums and debates on the Salween hydropower projects in Thailand, communicating with state parties, particularly the National Human Rights Commission of Thailand (NHRCT), to urge the government to order EGAT to stop the Hatgyi Dam and other Salween dam projects (NGO-CORD North 2011; TERRA 2007). This coalition of NGOs persuaded the NHRCT to scrutinize the Hatgyi Dam project in light of potential human rights and environment impacts. The issue of increased militarization at the upper dam construction site was noted as potentially exacerbating fighting between Myanmar soldiers and the KNU, which could have implications for the Thai government as refugees flee from the Myanmar side into Thailand (Nang Shining 2011: 30).

By collaborating with local, national, and international NGOs, local communities have organized their own Salween River Basin Network, consisting of 21 river basin management organizations across seven districts in Mae Hong Son Province, Tha Song Yang District and Tak Province (Nang Shining 2011: 30). They have conducted their campaigns in many ways, through rallies and protests, by collecting significant information, and submitting petition letters to relevant organizations and authorities. They submitted petition letters to the NHRCT and the Thai government, requesting the cancellation of the Hatgyi Hydropower Dam project (NGOs-CORD North 2008).

In 2009, the NHRCT submitted letters to the PM office to urge the government to halt the Hatgyi dam project for 90 days and a committee was formed to investigate potential human rights and environmental violations. Afterwards, the Information Disclosure Subcommittee on the Hatgyi Hydropower Project on the Salween River was set up to collect information and make recommendations to the Prime Minister. A coalition of organizations representing communities living along the Salween River in Myanmar and Thailand submitted a letter to Prime Minister Abhisit Vejjajiva calling for the Hatgyi Dam to be stopped immediately (Salween Watch Coalition 2011; Nang Shining 2011: 28–32).

Moreover, local representatives made recommendations to the Information Disclosure Subcommittee of the Hatgyi Hydropower Project on the Salween River and submitted a letter to the Prime Minister and Chairman of the Committee for the Investigation of Human Rights Violations regarding information disclosure and public hearings for the Hatgyi Dam (Electricity Generating Authority of Thailand International 2010; Salween Watch 2010). In 2011, EGATi held a public information disclosure meeting following the recommendations of the NHRCT. Despite this intervention by the sub-committee, the planning of the Hatgyi Dam has continued after Prime Minister Yinluck Shinawatra came to power in 2011. However, the subsequent military coup saw the military take power in May 2014 and the sub-committee was no more (Salween Watch Coalition 2016; Thepgumpanat/Tanakasempipat 2017).

There is still interest from the Thai government, however. The Thai government has expressed that it wants to continue energy cooperation with the Myanmar government and Aung San Suu Kyi insisted that the Thai government continue cooperation on the dam project during her visit in June 2016 (Prachachat 2016; Public Statement 2016). EGATi is willing to invest more than 100 billion baht in Hatgyi project. In addition to the dam project, the Thai government also expects to build a pumping station to divert water to the Bhumibol dam to alleviate drought in Central Thailand. Mr. Watchara Hemratchatanan, the president of EGATi, is on the record as expressing support, and that "we need to wait and see the Myanmar government and its negotiation with ethnic groups" (Prachachat 2016).

11.6.1 Villagers Conduct Thai Baan Research

As mentioned above, Thai Baan research carried out by villagers was an alternative to conventional research. In collaboration with SEARIN and a local NGO, Salween communities conducted Thai Baan Research with the initial findings released as a book in 2005 (Committee of Researchers of the Salween Sgaw Karen 2005). They attempted to involve local government in the research, and made the results of the study available to the Mae Hong Son provincial government, especially findings around various fish species found in the Salween and its tributaries. An additional Salween Study was conducted in 2005 through collaboration with the Foundation for Ecological Recovery based in Bangkok, the Mae Yuam Civic Group in Mae Sariang District, and representatives of Salween communities. The Salween Study compiled information on the socio-cultural and economic significance of the Salween River (Chantawong/Longcharoen 2007: 23).

The Thai Baan Research found 70 fish species in the Salween River and its tributaries (Committee of Researchers of the Salween Sgaw Karen 2005) and the

Salween Study identified up to 83 species of fish and aquatic animals. Some are endemic and rare species (Chantawong/Longcharoen 2007). Moreover, previous studies conducted by fishery experts identified at least 170 fish species in the Salween River Basin, many of them endemic. According to fishery experts, there could be between 200 and 500 species of fish in the Salween Basin (Chantawong/Longcharoen 2007: 26). The biodiversity and ecology of the Salween River Basin would be severely affected if the Salween dams were built, with drastic consequences for local people's livelihoods.

In line with other findings on local knowledge (i.e., Agrawal 1995, 2002; Laungaramsri 2001) this local knowledge has also been used in a range of ways. It is used by villagers to express their views and engage with government institutions to influence development planning and decision-making processes. Or, as Lamb (2018: 1–2) explains, "local knowledge is an agentive tool for communities to critique and find alternatives to status quo development schemes, which overlook resources users and their knowledge, but that in some ways the focus on the 'local' in local knowledge elides or limits that same struggle for representation." In my analysis, having voiced strong concern around the potential loss of their livelihoods, Salween communities are through these publications represented as 'anti-dam.' Although these communities are or may be anti-dam, however, they are not against development. The next section shows how villagers have tactically negotiated development with state agencies and NGOs, defining 'development' on their own terms and expressing their own demands.

11.6.2 The Villagers and **khwamcharoen**

The development of hydropower dams on the Salween River is one of many broader development issues that border communities have engaged in with the state. It is difficult to judge to what extent local people are being governed by the state.[2] Although they express determination and have proposed their own models for future livelihoods and community development, as seen in some of the local knowledge publications, this does not mean that villagers in Thailand reject government assistance or intervention.

The Salween borderlands are marginal zones, where both the central Thai and Myanmar governments try to exert control over their respective citizens and territories. As this chapter focuses on the Thai side, it is important to ask: how do local

[2]In modern Thai usage, the word can be made into a noun by adding the prefix 'khwam' [–ness], and the word *charoen* can be found in the 14th century and probably earlier. In the old sense, it means cultivating, growing, increasing, building up or expanding until complete in a positive sense, which applied mostly to nonmaterial matters. However, its meaning is changed in 19th century to imply secular or worldly development, material progress, and technological advance. Therefore, the emphasis of *charoen* is on material conditions, such as paved roads, electricity, machines, and modern buildings (Winichakul 2000: 531).

people perceive of the Thai state and its activities, and why are they unable to reject the state? For the local people, the Thai state plays an important role in the improvement of their wellbeing, even as it takes control of their territory.

Local people want to improve their difficult living conditions, such as poor roads, lack of agricultural and residential land, limited access to forestland and lack of citizenship status; they need the state and local government's help to improve these conditions. This section examines how local people articulate their own concept of development as self-proclaimed subjects. Of particular consideration are improvements in local infrastructure, access to basic services, citizenship, historical linkages with the Thai King, and disaster response.

11.6.3 Local Infrastructure and Basic Services

Transportation is a common concern among local people, particularly in Saw Myin Dong and Muang Mean. Both these villages are located far from paved roads, and boat travel can be limited and difficult, depending on season, time of day (no travel allowed after dark), and depending on security issues. Ai Srithong, a resident of Saw Myin Dong, stated that the quality of the roads in Muang Mean is low: "The concrete road in the village was built ten years ago. Riding a motorcycle to Bon Bea Luang village takes four hours, or three hours by car, but only one hour by boat. Unfortunately, there is no public transportation here."

Road construction is linked to local politics of development, as discussed between Ai Srithong and Sowan, a research assistant, in February 2010:

> Sowan: "Did people attempt to find money to improve the road? Was there any proposal submitted to the government?"
>
> Ai Srithong: "Yes, we submitted a proposal to the Sub-district Administrative Organisation (TAO). Only a small caterpillar machine came. I asked them to expand the road along the sharp curves, but they could not do it because there was no machine big enough… They didn't bring a good one. So, the road is still not good. It was okay when they first repaired it, but it collapsed during a day of heavy rain in the rainy season. People travelling here have to bring digging tools along with them. They sometimes have to push and pull motorcycles stuck in the mud."

For these villagers, paving the road is a good project, but they are suspicious about whether it will actually begin, even when they see construction workers come to survey in nearby villages.

11.6.4 Citizenship

The main concern of the state is security, and the military takes the role of national protector, as evidenced at military checkpoints along the border on both the Thai and Myanmar sides of the Salween River. One of the political tools used by the state to control its subjects is the granting or denial of citizenship. Internal borders are manipulated by new alliances of border polities; district officials, sub-district and village heads (Sakboon 2011). Villagers perceive citizenship as a means to access basic social services, such as health care and education, as well the right to live in Thailand (Grundy-Warr/Wong Siew Yin 2002; Sakboon 2011: 229; Hengsuwan 2017: 92).

Nevertheless, advocating for citizenship rights is an uneasy process for border residents due to registration and procedures that are tactically exploited by state agencies at a local level, including district officials, sub-district and village heads. These state officials have the power and authority to enforce policy and approve or deny villagers' applications, and manipulate citizenship registration laws and processes as an apparatus to control and abuse local residents. As a result, villagers must find alliances and networks to gain more bargaining power in negotiations for citizenship, and a provocative ally are the local NGOs (Sakboon 2011: 235–240).

Villagers in the Salween borderlands have long struggled to obtain citizenship. Data collected by a local NGO in 2016 shows that nearly half of the residents of three villages are stateless people (about 1,452 out of 3,352 people). In July 2016, the researcher talked to Lung Jai, a Muslim resident of Bon Bea Luang village, regarding these matters. His main concern is citizenship. He said:

> I often see villagers with no ID cards, and they cannot use Thai language to communicate with outsiders. They are questioned by the authorities regarding whether or not they are Thai, and are often arrested. So, I think that Thai language is an important skill to make officials understand us, and I've been learning to speak and write Thai language for two years. I've continued to improve my knowledge, reading books about law and citizenship, and attaining NGO trainings. After I was appointed a leader of the Muslim community in 2008, I worked with local NGOs to help stateless villagers attain legal status and to register for Thai citizenship. Now I am retired and a new leader is working on this issue, since a lot of us still don't have Thai citizenship.

As Lung Jai mentioned, local NGOs play a crucial role in building cooperation and connecting villagers to district officials. This has led to advances in the citizenship process. Noh La, a Karen woman NGO activist, stated that in 2016, 476 villagers from three villages had received Thai ID cards, but 976 villagers were still stateless. Noh La herself is a local resident who has accessed education and transformed into an NGO activist. Although the citizenship registration process is going better than in the past, by 2016 only one-third of surveyed villagers had received Thai ID cards.

11.6.5 Village Histories and Relationship with the King of Thailand

Another means of gaining recognition by the Thai state is by identifying yourself or your village as a state subject through village history (see also, Lamb 2014a). The late King Bhumibol visited the Salween border area three times between 1960 and 1970 in relation to national security concerns. Pa Te Sae Yi, a resident of Muang Mean village, recalled the first time the late King of Thailand visited:

> I don't remember the exact year, as I was a young child, but my mother and father remember clearly. When the King visited us, this was still a small village of only 5-7 houses. The only way to make a living at that time was by doing rotational farming in the mountains. So the King granted a royal patronage, meaning government money, to dig two irrigation canals so we could do lowland rice farming. Those two canals are still in use today.

He was touched by the late king's mercy, saying that the late King saw the difficulty the villagers faced and wanted to support them to do lowland rice farming. After the King's visit, the government brought new development projects to the Salween communities, building public health clinics and schools, as well as support from the police and the border patrol. Villagers believe that their quality of life improved after the King's visits, as the government came to recognize them as Thai subjects. Hence, this is applying a politics of belonging in relation to the late King and the state.

11.6.6 Disaster Response

Bon Bea Luang village is located in the foothills on a tiny piece of flat land along the Salween, as the villagers have not been allowed to settle further inland. The houses are built in a row, close to one other, along a narrow road. Due to the lack of land, the front rooms of the houses protrude onto the edge of the road. A view at the back of these houses shows that they are perched precariously on high wooden poles above a deep creek.

This is a dangerous place to live, particularly in the rainy season when the villagers face erosion and flooding. In 2009, Nongnut, a Shan resident of Bon Bea Luang, said that in the rainy season, a mass of water ran down the mountainsides, causing severe erosion and sweeping away the houses located near the stream. More than ten houses were ruined. Until now, the Karen and Shan families who lost their homes have not received any land to build new houses. The local government came to help them later, by offering some financial support.

"The affected families had to move to stay with their relatives. No official came to provide them land. They didn't know where to build new homes. My mom also moved to stay at my relative's house. TAO staff eventually came to help them, giving each family amount 30,000 baht. But my mom received only 21,500 baht, and no land," said Nongnut.

In 2011 and 2015, many villages located in mountainous zones in Mae Hong Son Province, including Bon Bea Luang village, were severely affected by heavy rains, floods, and landslides. As the case of Bon Bea Luang shows, affected villagers had no choice but to move in with relatives. They want to expand their communities and build homes in safer areas, but there is no space available. The forestlands surrounding the village are protected areas where the state does not allow them to settle. In my observations, this has meant that the state has pushed them onto increasingly marginal land. For this reason, villagers have joined the land rights movement, advocating for community rights and engaging in the on-going process of land field surveys and map digitization. They hope it will give them the legal right to settle in better, less marginal land, in the future.

All of these efforts demonstrate that the villagers are actually pro-development, however, development (in Thai: *khwamcharoen*) must take place on their own terms. Local people have long negotiated development that they identify as *khwamcharoen* based on their experiences, and serving their own vision, which is different from that of the state and the dam developers. The next section shows how communities reject development when it comes from a source they do not trust, and when they worry that agreeing to this development will be perceived as sanctioning the dam projects.

11.7 Negotiating *khwamcharoen*: EGAT and the State

This section argues that the state and dam developers use 'modern-ness' as a discourse of development to convince local communities to accept the Salween dam projects. EGAT tries to persuade villagers to acquiesce to their vision of development (*khwamcharoen)* through donations and other incentives. Local communities' relationship with EGAT shows that these communities are anti-dam but not anti-development, as villagers want development on their own terms. First we will examine how EGAT has sought to work in Thai communities via its Corporate Social Responsibility (CSR) Programme, when faced with strong civil society resistance to usher in development of the Hatgyi and Weigyi projects.

Local communities, however, widely conclude that this is not a sincere effort to bring development to their areas, but a trick to persuade them to support the dam projects. We will then examine how communities have sought to negotiate with the state over a longer period to bring 'development' to their communities on their own terms.

11.7.1 EGAT and khwamcharoen

The state-owned EGAT has played a major role in feeding Thailand's energy demand for the past five decades. Its privilege is embedded in its national duty to drive the country to the *khwamcharoen*. The nation's progress and EGAT's profit are intertwined. Hence, EGAT has begun to invest in neighboring countries, Laos in particular. Thailand receives advantages from this outbound investment, obtaining less expensive electric power and electricity system security and avoiding indebtedness from international loans for dam construction that would affect Thai natural resources (Prasityuseel 2001: ix).

EGAT publicly claims that they promote CSR to support society, communities, and the environment. "EGAT has strictly observed all applicable laws and regulations in all processes of its operations and activities, both before and during the project development and throughout the operating life of its power facilities. Particular emphasis has been placed on the implementation of all environmental and social impact prevention and mitigation measures as well as environmental monitoring programs" (Electricity Generating Authority of Thailand 2009a: 76–77). CSR appears to be good in theory, but it has not been successfully applied in practice. EGATi engaged in very limited CSR during the preparation process of the Hatgyi Dam project (Nang Shining 2011: 92). EGAT donated money to environmental conservation projects and tried to show that they care for the environment, but they still exploit natural resources through their large-scale projects. EGAT ramped up CSR activities in local communities to create an image of a socially conscious and ethical business (Nang Shining 2011; Vathanavisuth 2009).

EGAT also tries to work indirectly in communities. As Mr. Pornchai Rujiprapha, secretary-general of the Ministry of Energy and the governor of EGAT stated, "EGAT has designated 120 million baht (US$3.6 million) to assist in public health, education and employment for local villagers in the Salween dam area" (Sai Silp 2007). In reality, this is part of EGAT's mass propaganda effort (*karn tam mounchon*). As Nang Shining, a civil society activist, put it, "research findings reveal that EGAT international has exercised a philanthropic approach to community investment as its CSR activity, which villagers felt was undertaken to sway those opposed to the dam project" (Nang Shining 2011: 92).

At the local level, as part of its CSR strategy, EGAT has tried in many ways to gain access to communities in order to persuade them to support the Salween dam projects. Where the villagers are poor, EGAT uses *khwamcharoen* and the King's sufficiency economy theory as ideological tools to approach Salween communities. EGAT persuaded local people to join their sufficiency agriculture program, which included such activities as producing organic fertilizer and dishwashing liquid. Local leader Ai Chamnan said,

> EGAT (*karnfaifah*) is not involved with making *khwamcharoen*, but it intervenes to support the TAO with money to run activities, particularly agricultural activities such as catfish farms (*liangpladuk*), and making biological fermented liquid (*nammakchewapap*). This allows EGAT to promote its name in helping communities.

In 2008, EGAT provided books and other materials to the school in Saw Myin Dong village. This resulted in disagreements among villagers over supporting EGAT because the villagers were skeptical of EGAT's long-term tactics. Ai Chamnan said:

> There were public health mobile services in Saw Myin Dong village two or three years ago. The first time a medical team came here we accepted it. Villagers came to use the service. The second time, they came again. We wondered where the doctors come from. A doctor said that EGAT hired them to travel from Mae Sariang Hospital to start the mobile service.

> All of the funding came from EGAT. So the villagers rejected it. They didn't come to use the services once they understood that it was EGAT's trick.

EGAT also establishes good relationships with government officials. They were also suspected of having connections to high-level military personnel in order to use soldiers to coerce villagers to give in to EGAT's plans. Many residents in the community, including a villager named Ai Sanong, alleged that EGAT was providing funding to official departments that work closely with the community, including the education department and the military.

Lung Sanit, a Karen elder of Saw Myin Dong, also confirmed that soldiers based in the village tried to coerce the villagers to accept the Salween dam projects in 2008. Lung Sanit remembers well what the soldiers often said to villagers: the government will bring development to you. You should not resist.

In July 2009, I met Ai Sanan, a local Thai Karen activist, and Ai Chamnan, involved in supporting local people to protect their rights and natural resources. They have engaged in a number of anti-Salween dam campaign activities. I discussed the matter of EGAT's public relations strategies in local areas with them. Ai Sanan said:

> EGAT is trying to gather mass support (*duengmounchon*) for the Hatgyi dam in the Thai-Burmese [Myanmar] border zones. Last year, they gave people on the Burma side mosquito nets. Once people realized that the mosquito nets were from EGAT, they sent hundreds of them back to show that they rejected EGAT.

Ai Chamnan then told me this story:

> Making *khwamcharoen* through organizations for villagers happened in the past. Villagers knew that EGAT helped to develop the village, so if EGAT wanted to construct a dam, they would have to agree with EGAT's plans. That was EGAT's aim. They are very clever… Most villagers and government officials thought that the dam was good because it would produce electricity for us. Villagers said, 'the dam brings *khwamcharoen* to our villages. Why are you against that?' I argued 'what does *khwamcharoen* mean? It means money, right? People in Mae Sariang, for instance, used to receive money from logging the Salween forests. Where did the money go? They no longer have money. The meaning of *khwamcharoen* is sustainable development. When we explained that to the villagers, they understood better. The dam project will be approved if we don't oppose it. However, it is difficult to oppose it because the decision-making process is not transparent. If we do not struggle, they will build it.

In brief, EGAT has tried to persuade local people to agree to dam construction on the Salween by exercising a philanthropic strategy within local communities at the Thai-Myanmar border and swaying them away from opposing dam construction. Due to widespread resistance to the dam projects, EGAT has publicly promoted their CSR strategy to support good work for society, communities, and the environment. At the local level, EGAT has tried many ways to gain access to communities in order to persuade villagers to support the Salween dam projects. EGAT has also used *khwamcharoen* and the King's 'sufficiency economy' theory to approach communities by persuading villagers to join their development initiatives.

However, local communities are very concerned about the Salween dams. They are suspicious of the notion of dams as a means to achieve national development, and thus understand EGAT's tricks and tend to not support EGAT's CSR program. Villagers continue resisting large dam projects because of concerns over losing access to local resources, and they are also negotiating *khwamcharoen* to initiate community development on their own terms. The next section will analyze how development is negotiated, examining how villager-led *khwamcharoen* and EGAT-led *khwamcharoen* have been articulated.

11.8 Analysis: Villagers, the Thai State and Power Relations

The uneven power relations are evident when considering where does this electricity come from and who bears the costs. Lung Kaew's view is that dams and *khwamcharoen* are related. Large dams generate electricity, and then electricity makes *khwamcharoen* possible. Because of this, conflicts occur not only at the village level, but also at higher levels.

In addition, electricity projects can cause or exacerbate conflict in the village, ethnic conflicts in particular. The villagers are aware that they live in poor and risky conditions, and are denied rights and opportunities such as citizenship, access to state services, and legal land tenure. Villagers view the Thai state as a development-centric state that provides them with development based on the notion of material progress. They think that the state should be involved in improving the quality of their lives. Hence, they focus on material progress, which causes them to put their faith in the state. However, they still question the consequences of development projects, in which the idea of *khwamcharoen* is embedded. They worry that development projects might disrupt their traditional livelihoods and quality of life. For this reason, villagers request specific kinds of development that they feel will benefit them. They selectively accept or reject development programs introduced by the government and supported by local NGOs. As a result, knowledge production is a contested terrain that both pro-dam and anti-dam sides use to claim legitimacy and power over Salween resources (Vaddhanaphuti 2004; Scurrah 2013; Lamb 2014a).

11.9 Conclusions

Salween villagers are frequently represented as anti-development due to their widespread opposition to the Salween dam projects, which threaten their communities and livelihoods. Labeling Salween communities as anti-development is an oversimplification, however. Of course, they oppose the Salween dams due to the potential environmental and social impacts of these projects. However, a more sophisticated tool used by local communities is to negotiate with state agencies to assert their legitimacy as Thai subjects. They do this by narrating their communities' history of belonging to the Thai state, and by working with local NGOs to gain Thai citizenship. This is not simply participation in rural improvement projects initiated by the government agencies as Walker (2012) and Jakkrit Sangkhamanee (2013) mention, but is rather strategic (Lamb 2014a). The more the villagers engage with anti-Salween dam and land rights movements run by NGOs, the more support they get from those NGOs. They also negotiate with the local government to get support and gain benefits from community development. This chapter shows how villages have had a far longer engagement with the state, which can also be understood as linking the 'state' to the borderlands (Lamb 2014b). Therefore, they have been involved in long-term negotiations over development, both local and national. Salween communities are not anti-development, in this view, but rather seek development on their own terms.

Nevertheless, Salween villagers with connections to NGOs tend to be simplistically framed as anti-dam and thus anti-development. As the villagers oppose the Salween dam projects, the Thai and Myanmar states and the transnational companies, particularly EGAT, have attempted to reduce and/or remove this resistance. The Thai state and EGAT have also introduced alternative development as a discourse to expand their power over the Salween borderlands, making the local people submissive and swaying them from opposition to dam construction activities along the Salween. Some individuals, and organizations, use *khwamcharoen* and the Thai King's 'sufficiency economy' theory to approach border communities and persuade them to join their development programs. Corporate Social Responsibility is publicly promoted by EGAT to develop a good image for the company, and EGAT exercises a philanthropic strategy within local communities. These discursive practices bely the high level of exclusion inherent in the Salween dam projects, which would turn the Salween River into a source of hydro-electricity and exclude border people from accessing the Salween's resources.

When dams come to signify development, anti-dam communities are simplistically framed as anti-development. Salween communities are in fact much more sophisticated than frequently represented, and participation in the anti-dam movement is one of a number of arenas through which communities are seeking local development. Local communities have attempted to engage in development production by proposing community development initiatives based on their own desires and in accordance with their traditional livelihoods, and by participating in a number of movements to improve their lives. Individuals and communities have engaged with land rights movements and networks, and in struggles for citizenship rights.

As Keyes (2014: 172) points out, villagers in Thailand have reworked the discourse on 'development' to be meaningful with reference to their own experience, and challenge government policies that tend to obstruct their interests, or policies imposed on them without consultation. Local communities see their participation in the anti-dam movement as a strategy to improve local development. Participation in the anti-Salween dam movement is one of the villagers' many strategies to articulate their own meaning of local development. Therefore, it is necessary that we reposition the anti-dam movement as not a singular movement, but as a wider network of movements surrounding local development. While the anti-dam movement is often positioned as issue-based, this chapter proposes that we instead position the movement as one that is directly related to other issues of interest to local communities. Local villagers are flexible in their thinking, taking into account a wide array of influences on their livelihoods at the same time. Their anti-dam campaign activities must thus be understood in relation to other community initiatives. Communities are not simply anti-dam; they selectively accept or reject development on their own terms.

References

Agrawal, A. (1995). Dismantling the divide between indigenous and scientific knowledge. *Development and Change, 26*(3), 413–439.

Agrawal, A. (2002). Indigenous knowledge and the politics of classification. *International Social Science Journal*, *54*(173), 287–297.

Asian Development Bank (ADB). (2006, November 30). Developing the Greater Mekong subregion energy strategy. *Asian Development Bank.*

Baker, C. (2000). Assembly of the poor: Background, drama, reaction. *South East Asia Research*, *8*(1), 5–29.

Bakker, K. (1999). The politics of hydropower: Developing the Mekong. *Political Geography*, *18* (2), 209–232.

Battersby, P. (1999). Border politics and the broader politics of Thailand's international relations in the 1990s: From communism to capitalism. *Pacific Affairs*, *71*(4), 473–488.

Bryant, R.L. (1997). *The political ecology of forestry in Burma 1824–1994*. London, United Kingdom: Hurst & Company.

Bundidterdsakul, L. (Ed.). (2013). แม่อมกิ: ชีวิตแห่งผืนป่ากับมายาคติในกระบวนการยุติธรรม [Mae Omki: Life of forest and myth in justice process]. Bangkok: Human Rights Lawyers Association.

Chantawong, M. (2011, February 7). *The statement of NGO coordinating committee on rural development: Information disclosure and public hearing for the Hatgyi Dam*. NGO-CORD North.

Chantawong, M., & Longcharoen, L. (Eds.). (2007). สาละวิน: บันทึกแม่น้ำและชีวิตในกระแสการเปลี่ยนแปลง. [Salween: Note of life and river in transition]. Chiang Mai, Thailand: Foundation for Ecological Recovery.

Committee of Researchers of the Salween Sgaw Karen. (2005). วิถีแม่น้ำวิถีป่าของปกาเกอญอสาละวิน: งานวิจัยปกาเกอญอ [The *Thai baan* research at the Salween: The way of the river and forest]. Chiang Mai, Thailand: Development Center for Child and Community Network and SEARIN.

Duncan, C.R. (2004). Legislating modernity among the marginalized. In C.R. Duncan (Ed.), *Civilizing the margins: Southeast Asian government policies for the development of minorities* (pp. 1–23). Ithaca, NY: Cornell University Press.

Electricity Generating Authority of Thailand (EGAT). (2009a). *40th Anniversary growing together with Thai society: 2009 Annual report electricity generating authority of Thailand.* Nonthaburi: Electricity Generating Authority of Thailand.

Electricity Generating Authority of Thailand. (2009b). *Thailand power development plan 2008–2021 (PDP 2007: Revision 2)*. Bangkok: System Planning Division, Electricity Generating Authority of Thailand.

Electricity Generating Authority of Thailand International. (2010, August). *Information disclosure and public hearing of Hatgyi Dam on Salween River, Myanmar* [Booklet]. Sob Moi village, Mae Sam Reap, Mae Hong Son Province.

Escobar, A. (1995). *Encountering development: The making and unmaking of the third world.* New Jersey, NJ: Princeton University Press.

Fawthrop, T. (2003, January 17). Thai-Myanmar ties: Drug lords cash in. *Asia Times Online.* Retrieved from: http://www.burmalibrary.org/TinKyi/archives/2003-01/msg00032.html.

Ferguson, J. (2003). *The anti-politics machine: 'Development', depoliticization, and bureaucratic power in Lesotho.* Minneapolis, MN: University of Minnesota Press.

Foucault, M. (1977). *Language, counter-memory, practice: Selected essays and interviews Michel Foucault* (D.F. Bouchard & S. Simon, Trans.). Ithaca, NY: Cornell University Press.

Foucault, M. (1980). *Power and knowledge: Selected interviews and other writings 1972–1977* (C. Gordon, L. Marshall, J. Mepham & K. Sopers, Trans.). New York: Pantheon Books.

Grundy-Warr, C., & Wong Siew Yin, E., (2002). Geographies of Displacement: The Karenni and The Shan Across the Myanmar-Thailand Border. *Singapore Journal of Tropical Geography,* 23(1), 93–122.

Gupta, A. (1998). *Postcolonial development: Agriculture in the making of modern India.* London, United Kingdom: Duke University Press.

Hengsuwan, P. (2012). *In-between lives: Negotiating bordered terrains of development and resource management along the Salween River.* (Unpublished doctoral dissertation), Chiang Mai University, Thailand.

Hengsuwan, P. (2013). Explosive border: Dwelling, fear and violence on the Thai-Burmese border along the Salween River. *Asia Pacific Viewpoint, 54*(1), 109–122.

Hengsuwan, P. (2017). Living with threats and silent violence on the Salween borderlands: An interpretive and critical feminist perspective. *Regional Journal of Southeast Asian Studies, 2* (1), 69–121.

Jarernwong, S. (1999). *The development of meaning of "rural development" in Thai society: A case study of the construction of meaning in "A model village for development" in northern Thailand.* (Unpublished Masters thesis) (in Thai), Chiang Mai University, Thailand.

Jarernwong, S. (2001). ถอดรหัสการพัฒนา *[Decoding development].* Bangkok, Thailand: Rongphim Duantula.

Keyes, C. (2014). *Finding their voice: Northeastern villagers and the Thai state.* Chiang Mai, Thailand: Silkworm Books.

Lamb, V. (2014a). *Ecologies of rule and resistance: Making knowledge, borders and environmental governance at the Salween River, Thailand.* (Unpublished doctoral dissertation), York University, Canada.

Lamb, V. (2014b). "Where is the border?" Villagers, environmental consultants and the 'work'of the Thai–Burma border. *Political Geography, 40,* 1–12.

Lamb, V. (2018). Who knows the river? Gender, expertise, and the politics of local ecological knowledge production of the Salween River, Thai-Myanmar border. *Gender, Place & Culture.* https://doi.org/10.1080/0966369x.2018.1481018.

Lamb, V., & Roth, R. (2018). Science as friend and foe. Chapter 10 in S. Mollett & T. Kepe (Eds.), *Land Rights, Biodiversity Conservation and Justice: Rethinking Parks and People.* New York, NY: Routledge.

Lambrecht, C.W. (2000). Destruction and violation: Burma's border development policies. *Watershed, 5*(2): 28–33.

Lambrecht, C.W. (2004). Oxymoronic development: The military as benefactor in the border regions of Burma. In C.R. Duncan (Ed.), *Civilizing the margins: Southeast Asian government policies for the development of minorities* (pp. 150–181). Ithaca, NY: Cornell University Press.

Laungaramsri, P. (2001). *Redefining Nature: Karen Ecological Knowledge and the Challenge to the Modern Conservation Paradigm.* Chennai, India: Earthworm Books.

Lorenzini, D. (2015). What is a "regime of truth"? *Le Foucaldien, 1*(1), 1–5. Retrieved from: http://www.fsw.uzh.ch/foucaultblog/featured/28/what-is-a-regime-of-truth.

MacLean, K. (2008). Sovereignty in Burma after the entrepreneurial turn: Mosaics of control, commodified spaces, and regulated violence in contemporary Burma. In J. Nevins & N.L. Peluso (Eds.), *Taking Southeast Asia to market: Commodities, nature, and people in the neoliberal age* (pp. 140–157). Ithaca, NY: Cornell University Press.

Middleton, C., Garcia, J., & Foran, T. (2009). Old and new hydropower players in the Mekong region: Agendas and strategies. In F. Molle, T. Foran, & M. Käkönen (Eds.), *Contested waterscape in the Mekong region: Hydropower, livelihoods and governance* (pp. 23–54). London, United Kingdom: Earthscan and USER (Chiang Mai University, Thailand).

Modinator. (2014, March 18). Local activists protest Salween River dam construction with prayer ceremony. Retrieved from *rehmonnya* website: http://rehmonnya.org/archives/3109.

Nang Shining. (2011). *Evaluating the implementation of EGAT International's corporate social responsibility policy for the Hatgyi Dam Project on the Salween River, Myanmar.* (Unpublished Masters thesis), Chulalongkorn University, Thailand.

Nevins, J., & Peluso, N.L. (2008). Introduction: Commoditization in Southeast Asia. In J. Nevins & N.L. Peluso (Eds.), *Taking Southeast Asia to market: Commodities, nature, and people in the neoliberal age* (pp. 1–24). Ithaca, NY: Cornell University Press.

NGO-CORD-North. (2007, February 28). *The statement of NGO-CORD-NORTH: Request withdrawal from cooperation with Burmese military regime for the construction of hydropower dams on the Salween River.*

NGOs-CORD North. (2008, November 23). *The statement of NGO-CORD North: Request to cancel Hatgyi Hydropower Dam Project.*

Orther, S.B. (2006). *Anthropology and social theory: Culture, power, and the acting subject.* Durham, NC: Duke University Press.

Petras, J. (1997). Latin America: The resurgence of the left. *New Left Review I*, 223, 17–47.

Prachachat. (2016, September 14). Thailand plans to dust off Hatgyi Dam; EGATi awaits Myanmar Govt to negotiate with ethnic group. (Grittayanee Tamee and Areeya Tivasuradej, Trans). *Prachachat Online Business*. Retrieved from: https://www.mekongeye.com/2016/09/14/thailand-plans-to-dust-off-hatgyi-dam-egati-awaits-myanmar-govt-to-negotiate-with-ethnic-group/.

Prasityuseel, K. (2001). รายงานการวิจัยฉบับสมบูรณ์ความสัมพันธ์ระหว่างไทย-ลาว: ศึกษากรณีโครงการพลังงานไฟฟ้าเขื่อนน้ำงึม [The relationship between Thailand and Lao PDR: A case study of Nam Ngum Hydro-electric project]. Thailand: Faculty of Engineering, Chiang Mai University.

Public Statement. (2016, June 21). Statement on the Visit of Daw Aung San Suu Kyi to Thailand – June 2016. *International Rivers*. Retrieved from: https://www.internationalrivers.org/resources/statement-on-the-visit-of-daw-aung-san-suu-kyi-to-thailand-june-2016-11501.

Sai Silp. (2007, October 31). Salween River Dam Project essential for Thai energy security, says minister. *The Irrawaddy*. Retrieved from: http://www2.irrawaddy.org/article.php?art_id=9156.

Sakboon, M. (2011). The border within: The Akha at the frontiers of national integration. In C. Vaddhanaphuti & A. Jirattikorn (Eds.), *Transcending contesting development: Social suffering and negotiation* (pp. 205–243). Thailand: Regional Center for Social Science and Sustainable Development, Faculty of Social Sciences, Chiang Mai University.

Salween Watch. (2010). Hatgyi Dam: Abhisit appointed committee recommends more study, but refuses to stop EGAT's destructive project. *Salween Watch Newsletter, 2*, 1–2.

Salween Watch. (2011, April 28). *Salween Watch coalition statement: Escalation of civil war in Burma necessitates immediate halt to Salween dam plans.*

Salween Watch. (2014, March 14). Hydropower projects on the Salween River: An update. *International Rivers*. Retrieved from https://www.internationalrivers.org/resources/hydropower-projects-on-the-salween-river-an-update-8258.

Salween Watch Coalition. (2016). *Current status of dam projects on the Salween River* [Fact sheet]. Retrieved from: https://www.internationalrivers.org/sites/default/files/attached-files/salween_factsheet_2016.pdf.

Sangkhamanee, J. (2013). Representing community: A water project proposal and tactical knowledge. In R. Daniel, L. Lebel, & K. Manorom (Eds.), *Governing the Mekong: Engaging in the politics of knowledge* (pp. 11–25). Petaling Jaya, Malaysia: Strategic Information and Research Development Centre.

Scurrah, N. (2013). Countering hegemony and institutional integration: Two approaches to using *tai baan* research for local knowledge advocacy. In R. Daniel, L. Lebel, & K. Manorom (Eds.), *Governing the Mekong: Engaging in the politics of knowledge* (pp. 27–48). Petaling Jaya, Malaysia: Strategic Information and Research Development Centre.

Simpson, A. (2013). An 'activist diaspora' as a response to authoritarianism in Myanmar: The role of transnational activism in promoting political reform. In F. Cavatorta (Ed.), *Civil society activism under authoritarian rule: A comparative perspective* (pp. 181–218). Abingdon, United Kingdom: Routledge.

TERRA. (2007). *Chronology of Salween dam plans (Thailand and Burma).*

Thepgumpanat, P., & Tanakasempipat, P. (2017, May 21). Three years after coup, junta is deeply embedded in Thai life. *Reuters*. Retrieved from: https://www.reuters.com/article/us-thailand-military/three-years-after-coup-junta-is-deeply-embedded-in-thai-life-idUSKCN18G0ZJ.

Vaddhanaphuti, C. (2004). Preface. In C. Cretthachau & P. Deetes (Eds.), *The return of fish, river ecology and local livelihoods of the Mun River: A Thai baan (villager's) research* (p. 5). Chiang Mai, Thailand: SEARIN.

Vandergeest, P. (1996). Mapping nature: Territorialization of forest rights in Thailand. *Society and Natural Resources*, *9*(2), 159–175.

Vathanavisuth, S. (2009, October 8). EGAT chief values social engagement. *Bangkok Post.*

Walker, A. (2012). *Thailand's political peasants: Power in the modern rural economy*. Madison, WI: University of Wisconsin Press.

Winichakul, T. (2000). The quest for "*Siwilai*": A geographical discourse of civilizational thinking in the late nineteenth and early twentieth-century Siam. *The Journal of Asian Studies*, *59*(3), 528–549.

Yoon, P. (2001). The political philosophy of intersubjectivity and the logic of discourse. *Human Studies*, *24*(1–2), 57–68.

Chapter 12
Powers of Access: Impacts on Resource Users and Researchers in Myanmar's Shan State

K. B. Roberts

12.1 Introduction

Research on resource use has at its core focused on what shapes people's access to natural resources, highlighting key issues ranging from conflict, state territorializion, market forces to social difference (Rocheleau/Edmunds 1997; Ribot 1998; Peluso 1996; Schroeder 1999). In Southeast Asia, scholars have contributed to broader understandings of the links between violence and access (Le Billon 2001; Hengsuwan 2013; Koubi et al. 2014; Peluso/Vandergeest 2011), and recent work in Myanmar has shown how an increase in capital flow into regions as a result of ceasefire agreements has altered access to natural resource for people in these areas (Hunsberger et al. 2017; Woods 2011). Access rights, however, are not only property rights, but also include an ability to derive benefits, and this includes more than just material benefits like livelihoods, but also knowledge (Elmhirst 2011; Peluso/Lund 2011; Ribot/Peluso 2003). Yet, little to no research has interrogated the impacts of access on knowledge production and the broader links between research access and local resource access. In Myanmar, where this research is based, asking questions about researcher access is particularly salient. Due to a history of conflict and present conflicts, the forests along the Thanlwin River, particularly in Shan State, remains largely inaccessible to outsiders. This restricted access to researchers has made it difficult to investigate the impacts of decisions over natural resource access and use. In this chapter, I ask: in what ways does the changing political and governance conditions in Myanmar influence access to natural resources, not only concerning material objects like forest and river resources, but also forms of knowledge production?

To address this question, I carried out research between August 2015 to March 2016 in two villages in Shan State (Fig. 12.1: Map of Study Sites in Shan State, Myanmar) whose residents rely heavily on forest and river resources to provide for

K. B. Roberts, Ph.D. Candidate, York University; Email: arborpolitica@gmail.com.

C. Middleton and V. Lamb (eds.), *Knowing the Salween River: Resource Politics of a Contested Transboundary River*, The Anthropocene: Politik—Economics—Society—Science 27, https://doi.org/10.1007/978-3-319-77440-4_12

Fig. 12.1 Map of study sites in Shan State, Myanmar. *Source* Cartography by Chandra Jayasuriya, University of Melbourne, with permission

their livelihoods. Both villages are also located near to the Thanlwin River, and would be impacted by the proposed Mong Ton dam. Formerly known as the Tasang dam, the Mong Ton dam is set to be the largest dam in Southeast Asia, with a reservoir that would impact people beyond these two village, overall an estimated 12,000 to over 120,000 people (International Rivers 2012; Salween Watch 2016; Maung 2016).[1] Research in these two Shan State villages finds that both the resource user and researcher must navigate their access through state and

[1]Discrepancy between reservoir and impact lies in the difference between pro-dam and anti-dam information notices.

semi-autonomous institutions and a terrain of informal and formal governance arrangements, environmental degradation, and shared identity.

To draw out these multifaceted dimensions of access—including researchers' and resources users'—I present this work in the following sections. First, I provide an overview of the concepts of access I draw on, before that access also affects knowledge production in this case. I then discuss the collaborative methods this project used and the tools we implemented to increase transparency for the research process. Next, I delve into the two case studies of North and South Village to highlight how the informal and formal governance, environmental degradation, and shared identity affect access for resource users and researchers. The findings highlight the complicated and co-produced relationships and mechanisms that surround access influence access. Additionally, this research finds that as these state and semi-autonomous institutions exercise their authority, the informal and formal governance mechanisms and the increase in environmental degradation from extraction activities increases the vulnerability of local communities and reduces their ability to benefit from natural resources.

12.2 Concepts: Expanding a Theory of Access

In "A Theory of Access", Ribot and Peluso (2003: 153) "define access as the ability to benefit from things—including material objects, persons, institutions, and symbols."

Access is gained, they argue, not only through property rights, but also through influencing power relations that can also influence the practices of others. This chapter seizes on this examination power and relations, regarding not only material objects like forest and river resources, but also forms of knowledge production.

Aspects and effects of access discussed in this paper draw on theorizations of access to understand first, the roles of formal and informal governance (Sturgeon 2004; Ribot/Peluso 2003: 2), second, environmental degradation (Blaikie/Brookfield 1987; Dove 1993; Forsyth 1996; Peluso 1996; Rocheleau 1995), and third, shared identity (Elmhirst 2011; Ribot/Peluso 2003) in influencing the ability to derive benefits. Formal and informal governance influences access authority through claims to a territory, that is legally sanctioned, locally negotiated, or simply socially acquiesced (Sturgeon 2004; Ribot/Peluso 2003).

In these links with governance, this chapter analyzes how state and semi-autonomous institutions' access to resources results in the degradation of those resources, which inhibits local community members from deriving benefits from those resources. Myanmar has a history of discourse that blames forest degradation on local people (Bryant 1997), but this chapter, instead, looks at the day to day impacts on local communities and their discourse regarding environmental degradation (Dove 1993; Forsyth 1996). Yet, in questions of access and what determines access for local communities, shared identity is also paramount in Myanmar. Shared identity includes memberships to a community or group, by age, gender, ethnicity, or religion (Ribot/Peluso 2003) and can also help navigate access and increase cooperation (Drury/Stott 2001; Faria/Mollet 2014). Although people hold within themselves multiple and shifting intersections of identities that influence their interaction with others (Mohanty 2002; McDonald 2013), with the history of divisions along ethnic lines in Myanmar (Grundy-Warr/Wong Siew Yin 2002; Scott 2009) this chapter focuses on the shared identity of ethnicity.

Moreover, I augment this notion of access by also highlighting researcher access. These research sites are situated, and to ask questions about access involves asking questions not only about the resource users, but how we learn about them (Haraway 1988; Kobayashi/Peake 1994). This impact on researcher access requires a deviation from the self-determined forms of fieldwork geographers and anthropologists traditionally rely on (Caretta 2015; McKittrick/Peake 2005; Sundberg 2003). As such, this chapter takes political ecology's theorization of access and builds into it an inclusion of the researcher's role in the methods and analysis of these mechanisms. In doing so, I push research methods to be more transparent, particularly in conflict zones. This research seeks to avoid the practice that ignores both the role of the researcher in the research, but also the influence of politics, power, and authority on the research process (Faria/Mollett 2014; Hawkins/Ojeda 2011: 249).

Researchers, research assistants, interrupters and research participants are all part of the knowledge production process (Turner 2010: 216). Particularly in research that crosses cultures and languages, this knowledge is mediated through all actors and are influences by their own beliefs, assumptions, fears, and preferences (Caretta 2015; Temple/Edwards 2002; Temple/Young 2004; Turner 2010). A 'triple subjectivity' of inward, self-reflexivity and outward, reflecting on the relations of others is needed then to understand the position of and interactions between researcher, research assistant, and research participant (Caretta 2015; Temple/Edwards 2002; Turner 2010).

To accomplish this, the project involved a collaboration with two researchers from different Shan civil society organizations (CSOs), with the pseudonyms of Mai and Wah. I facilitated the research project, negotiated with Mai and Wah's organizations for their support and time, and paid Mai and Wah and their research assistants. Acknowledging the situated unequal power relationship between me as a white, educated female working in a country that was once a British colony, employing two Shan researchers, I sought to facilitate a co-created research project that would subvert my situated power and give more voice to Mai and Wah. The relationship between Mai and Wah, however, was also unequal, as Wah was educated in Bangkok and had a firmer command of English, which is the language that we operated in—baring a few specific terms in Shan and Thai. What is presented in the rest of this chapter is a result of how we negotiated access to research sites and how we negotiated the uneven power relations between researcher, research assistant, and the researched (Caretta 2015; Haraway 1988; Kobayashi/Peake 1994). I detail this further in my research methods in the next section.

12.3 Research Methods: Collaboration in a Restricted Research Setting

This research took place between August 2015 and March 2016 in two villages along the Thanlwin River in Shan State on the border between South and East Shan State. The two villages in this study are subsistence communities that are remotely located with limited access to health care, education, and markets. Instead they rely heavily on forest and river resources to provide for their livelihoods. As noted above, the proposed Mong Ton dam poses a threat to these livelihoods. North Village is located roughly 200 km north of the proposed Mong Ton dam site and would be impacted by the flood zone of the reservoir (Salween Watch 2013). South Village, while located downstream from the dam site, would also be impacted by the change in water flow. This proximity and connection to the site of the proposed dam also adds a layer of restriction to already restricted area for researchers. These villages were at the time of research listed as "brown zones" which require special permission to access (described further below).

Taking these concerns and restrictions into consideration, the design of the research was based on feminist collaborative methods (Sharp 2005) developed alongside local researchers. At the two village sites, research was conducted in Shan language (the primary language of the researchers and the villagers) and in each village the data collected included a key informant interview, a group interview, a community drawn resource-use map, participatory learning activities (PLA) about village concerns, 12 semi-structured survey interviews, and researcher diaries (see Table 12.1: Number of Survey Interviews by Gender and Age Divisions). Off-site data collection also included a literature review on the relevant topics of land and water policies and laws, pending and present, as well as reports and publications from state and non-state agencies and institutions, and reflexive interviews, in English, with research team members.

Table 12.1 Number of survey interviews by gender and age divisions

	South Village		North Village	
Total village population	199		116	
Total village households	60		27	
Total survey interviews	12		12	
Lead researcher	Mai		Wah	
	Male	Female	Male	Female
Total	3	9	6	6
No. of single young adults (a)	1	2	2	2
No. of married with children (b)	1	6	3	2
No. of elderly (c)	1	1	1	2

Source The author

12.3.1 The Negotiation and Collaborative Research Approach Developed

Mai, Wah, and I designed the research project and methods during a 3-day workshop, conducted predominately in English, in January 2016. During the workshop, we discussed our research backgrounds and to better understand our separate experiences and understandings of the research topic we conducted mock interviews. We also used participatory learning activities (PLA) to determine which research methods we had the most interest in using. Overall, the workshop allowed for the experiences, knowledges, and interests of Mai, Wah, and I to play a significant role in the design of the research.

Based on the workshop PLA activity we decided on using key informant interviews, group interviews that included PLA and resource mapping, survey-interviews, and a household inventory for our methods. Research questions centered on village histories, access to education and health care, agricultural practices, and use of forest and river resources. These questions were determined based on the sometimes separate and sometimes similar interests of Mai, Wah, and I. Selection criteria for the 24 survey interviewees were selected through a type of non-probability sampling via convenience/snowball where we ascertained which units should be observed based on our judgment about which ones will be the most useful or representative (Babbie 2007: 193). Mai, Wah, and I decided to try to interview an equal number of men and women, and as per their suggestion, we decided to break the age categories based on family status, i.e. single young adults, married with family, and elderly (see Table 12.1: Number of Survey Interviews by Gender and Age Divisions).

Mai and Wah selected sites based on their organizations' association or interest in certain communities. Mai selected South Village, because she is from the same district and her organization had previously worked in that same community and had an interest in continuing to build that relationship. Wah selected the North Village because her organization had an interest to begin working with the community for education outreach purposes and North Village, like her organization, fell under the control of the Shan State Army (SSA) (Wah, 6 January 2015).

12.3.2 Tools for Transparency

To 'standardize' and make transparent the research process across research sites, with different researchers, we relied on a range of tools, including daily journaling and regular follow up. After the 3-day workshop, Mai and Wah had the same research template to work from in their respective villages. Wah and Mai then went and trained their research assistants. Wah visited North Village with five other employees from her organization. Mai visited South Village with just one other employee from her organization. All village researchers (including Mai and Wah) wrote daily journals, noting what they did that day, what they learned, what challenges they encountered, and what they observed that they found interesting. These were then translated from Shan to English by Wah and Mai. Wah, Mai and I met again after the research was conducted, which gave me the opportunity to not only read all the translated interviews and journals, but to ask follow-up questions about the research and the research process, as well as the resource maps.

I also conducted reflexive "exit interviews" with Mai and Wah about their experiences in the villages and with the research process overall. The researcher journals and reflexive interviews allowed me the opportunity to have some insight into how the field research assistants and Mai and Wah positioned themselves within the research. Regarding data analysis, it allowed us to bring disparate data together, as a group spot trends and similarities across research sites, and it gave me an opportunity to clarify misunderstanding. I then coded the journals, reflexive interviews, key informant, group, and survey interviews using QSR NVivo11 software. Codes were development based on discussion topics and themes that emerged during the research process. The different sites and sources of data were then compared and contrasted. In the discussion below, all names are omitted from quotations.

12.4 Implications of Access

For North Village and South Village, we found that their access to natural resources are essential for their lives and livelihood. Both North Village and South Village have paddy fields along the riverbanks and practice *taungya* to grow food, including rice, for personal consumption. *Taungya* is the Burmese word for shifting cultivation, *taung* for hill and *ya* for cultivation. *Taungya* includes partial forest clearance, multiple cropping, shallow cultivation, and field rotation to produce food and sometimes cash crops (Bryant 1994: 226). It is a system, through use of a prolonged fallow phase that is longer than the cultivation phase, which allows woody vegetation to return to a site that had been cleared for annual crops (Brookfield 2015: 26). For swidden agriculturists, like both communities in this study, the 'forest' is a component of an integrated landscape that provides long-term and short-term benefits and products, including many non-timber forest products (NTFPs). As one interviewee stated, "*We use the forest resources every day. Because we have no market, we must go to the forest to find the food day by day*" (Interview 13, South Village, 3 February 2016). Yet, while both North and South village rely heavily on forest and river resources for their livelihoods, according to Myanmar State law,[2] all land in Myanmar remains state property and forest

[2]In January 2016, the National Land Law Policy was accepted by the previous government of Myanmar. For the scope of this chapter, it has yet to be seen how this will be implemented.

products may not be extracted without a permit (EIA 2015; Scurrah et al. 2015; SLORC 1992). According to these laws, the Myanmar government has de jure rights to North and South Village and the surrounding forest areas.

12.4.1 Informal and Formal Governance Impacts on Access

This official designation of land ownership by the Myanmar government held significance for us as researchers. In South and North Village, at the time of the research, both communities were in government listed "brown zones." As part of the previous military government's four cuts strategy, which sought to severe connections between resistance groups and local populations (Meehan 2015), Myanmar was divided into three zones. While these zones change and alter according to conflict, and have much broader ramifications, they do influence where foreigners can travel. White zones, which include most of the major cities and much of the central valley of Myanmar, are areas where foreigners can freely travel. However, much of the periphery of the country is brown or black zone. Black zones prohibit travel from foreigners and brown zones requires special permission (Grundy-Warr et al. 1997). While not all researchers respect these designations, and through relationships with semi-autonomous institutions can find a way into 'restricted' zones, for this research, we chose to obey this regulation. This meant that only the Shan researchers could travel to the research sites, which then had direct implications on research project design and knowledge production, as was discussed above.

Although as researchers we chose to respect the access designation as defined by the Myanmar government, for resource users in Shan State, the answer to the question of *who* is exercising authority over access for resource users is not that simple. In practice, authority of access is enforced by state and semi-autonomous institutions, who independently or jointly control their access through custom and at times force (Callahan 2009; Jones 2014; Sai Aung Tun 2009).

For North Village, while de jure rights also technically rest with the Myanmar government, the SSA both governs the village and spatially controls access to the area and its natural resources. For example, it is the SSA, not the Myanmar government, that provides social services. The SSA recently built a water tank in the village for cleaner drinking water and, although the village does not currently have a school, four or five years ago SSA provided two Shan teachers who taught for two years. Moreover, if villagers are extremely ill, they will travel roughly 50 km to reach the SSA's health clinic. In the past, the SSA also levied a 20% rice tax on the village; however, they now consider the village too small and no longer extract that tax (Group interview, South Village, 18 February 2016). In providing social services and extracting taxes, SSA enacts its spatial claim on North Village through custom and convention, albeit not law. According to the group interview, North Village acknowledges this territorial claim and 100% of interview respondents expressed that they did not feel that their access was restricted to forest resources.

Unlike North Village, South Village experiences more direct authority from a combination of the Myanmar government, Burmese military, and private-public and private-military partnerships. De jure land rights rest with the Myanmar government, and spatial access is maintained by force and coercion through the presence of the Burmese military and Lahu militia, which are financially backed through private-public partnerships (Kramer/Woods 2014: 65). The companies involved in the Mong Ton dam have hired the semi-autonomous Burmese military, who sub-contract with a local Lahu militia (Group Interview, South Village, 3 February 2016).[3] A 20-mile radius around the dam has been effectively cordoned off by both these armed groups (Samarkand 2015). The Lahu militia and the Burmese military both have an active presence around the dam site when workers are present, and their presence does influence South Village. As the headman described the situation: "*The Burmese military don't stay in the village, but they do control it*" (Interview 13, South Village, 3 February 2016). By law the Myanmar government has authority over South Village, but in practice villagers perceive that the Burmese military has territorial control.

According to the group interview, the Mong Ton dam companies linked to the project—described by local communities as the broad category 'Chinese'—did not directly control where villagers could travel. However, people did not feel like they could move about as freely as before and the Burmese military and Lahu militia will discourage them from traveling to certain places (South Village, 3 February 2016). While South Village did not report any conflict between themselves and the Burmese military or Lahu militia, 83% of interviewees in South Village discussed access limitations they experienced because of the Mong Ton dam. As one interviewee stated "*If we are near the dam on the river, we can't go further because it's limited by the Burmese military. Also, the Burmese military have limited where we can go into the forest near the Chinese's project*" (Interviewee 1, South Village, 4 February 2016). When the company staff were present, villagers only travel into the forest between 9:00am and 5:00pm, because when it is dark, they do not feel safe in the forest (Mai, 23 February 2016). This control, however, by the private-military partnerships, is a temporal one. During the time of the interviews, the company staff were not present around the dam. Therefore, individuals from South Village expressed that they felt freer to go into the forest and fish along the river (Mai, 23 February 2016). Yet, certain pressures remained. The Lahu militia, when not employed by the company, resided on the top of the mountain above South Village, but are still seen as having influence locally (South Village, Interview 13, 3 February 2016). These spatial and temporal restrictions limit villagers' ability to "derive benefits" from the forests and river that their livelihoods depend on. This

[3]For example, the Mong Ton dam is based on agreements signed by the Myanmar Parliament with the Chinese Three Gorge's Corporation, the Electricity Generating Authority of Thailand (EGAT) and the Burmese Ministry of Electric Power, with the Australian Snowy Mountain Engineering Company (SMEC) consulting on the construction of the dams (Salween Watch 2013).

spatial claim, laid by military-private partnerships, is enforced through law and force, yet, as the quotes suggest, it is socially acquiesced to the extent that villagers navigate their access to forested areas near the dam site based on military presence.[4]

12.4.2 Environmental Degradation Limiting Access

Yet, while the physical presence of the Burmese military and Lahu militia limit villagers' physical access to forest and river resources around the Mong Ton dam site, these military-private partnerships have also conducted activities that have resulted in environmental degradation that has further restricted access. Although South Village interviews respondents were not able to identify specific dates and company's names, the narrative they told indicated that the private-military and private-public partnerships that led to the selection of the proposed Mong Ton dam site (Kramer/Woods 2014) also enabled logging of the forests near the dam site and mining for gold within the river (Group interview, South Village, 3 February 2016).

Seventy-five percent of interviewees from South Village reported a negative change in the last ten years to forest and river ecosystem services. Of those who discussed a perceived change to the surrounding watershed, all mentioned the narrowing and drying of the river and 78% mentioned forest loss, both of which interviewees attributed to logging around the dam construction site. As one villager observed, "*The river is changing. In the rainy season, there is flooding, and people lose their land, and because of the logging, when people are doing agriculture, they cannot get enough water*" (Interviewee 5, South Village, 3 February 2016). While the majority of villagers in South Village practice *taungya* in the uplands, they also have rice paddy fields near the banks of a tributary of the Thanlwin River. However, as the quote suggests, villagers are reporting changes to the river and nearby land. Unlike *taungya*, rice paddy fields require irrigation, but the logging along the banks of the river has decreased the water retention of the soil and in turns increases the likelihoods of floods and droughts (Kandjj et al. 2006; Verchot et al. 2007). Additionally, 33% mentioned the difficulties they now experienced in hunting. "*The noise from cutting down the trees frightens the wildlife and causes them to run away from the village*" (Interviewee 7, South Village, 3 February 2016).

As villagers summarized during a group interview, South Village perceives negative changes to their surrounding environment:

> The Thanlwin River is narrower now and the water is not like before. The water level is lower and there is not enough for agriculture and it is harder to catch the big fish. The wildlife is also not like before and it is harder to hunt and there are fewer trees. This makes the weather hot. The forest and river are very important for our households and the next generation. The river is also important for our culture and tradition (Group Interview, South Village, 3 February 2016).

---

[4]This temporal forest access is also a result of the perceived threat women face as a result of military presence. See also, Hnin Wut Yee (2016).

As this quote suggests, South Village relies on the forest and river for their livelihoods, and yet, these changes have the ability to negatively affect the community's culture and traditions. Each of these environmental changes—lowered water levels, increased disturbances in hunting territories, and increased deforestation—are examples of South Village's inhibited ability to derive benefits from resources.

For North Village, while they did not feel a restriction to forest resources, they did express an inability to continue the practice of artisanal gold mining. Gold along the Thanlwin is collected by villagers to sell in a neighboring town's market when individuals need money to buy cooking oil and salt (Wah, 23 February 2016). According to interviews in North Village, the SSA allowed companies to dig for gold in the Thanlwin River. As the headman described in North Village "*A few years ago [3–4 years ago] there was a Chinese company digging gold from the river along the village, allowed by the military [SSA]. Recently there is no company nearby the village, because the gold is gone.*" (Interview 13, North Village, 18 February 2016). One-hundred percent of interviewees discussed the difficulty they now have in panning for gold along the river since that time. As one individual described:

> The gold left is less than usual. We could collect more before Chinese people went to dig gold at the floor of the river. After they used large machines and took most of the gold. Now, we can only collect one piece or two. However, we have to find gold for our lives to exist (Interviewee 2, North Village, 18 February 2016).

This degradation of gold in the banks of the river, because of a private-military partnership, reflects a shift in North Village's ability to derive benefits from river resources. As subsistence farmers, villagers did not grow cash crops and get very little earnings from selling NTFPs; therefore, gold was essential for them to have the money to purchase goods like salt and oil in town. With this increased challenge in finding gold, villagers expressed a difficulty in purchasing household items. This access, which these military, private, and semi-autonomous institutions have, and communities do not, resulted in a degradation of resources that limited local communities' ability to benefit from those same resources.

12.4.3 Access and Identity

As discussed above, access can be restricted through law and force and by the effects of commercial activities. Yet, these aspects and effects of access are also often navigated through social relations. This section discusses how relationships, such as shared-identity, can also impact access. Shared identity does not guarantee a socially acquiesced access, nor does it preclude it; yet, when a shared identity with authority exists it increases the likelihood of less resistance to the parameters established by authorities. This is particularly important in Myanmar, where the various states and semi-autonomous institutions are divided along ethnic lines.

Currently, South Village experiences control by the Burmese military and receives services, like primary school education, from the Myanmar government, but the villagers still have deep connections to Shan culture and tradition (Mai, 6 January 2015). During interviews with South Village, villagers shared an ethnic identity (Shan) with the interviewers. Yet, in their interviews they did not express a shared identify with the Burmese military or Myanmar government or private companies. For these military-private partnerships that exercise authority over access, this lack of shared social identity with South Village has resulted in resistance to their authority in the form of political elections and resistance to the hydropower project. During the 2015 elections, 100% of respondents who were able to vote, said that they voted for the Shan National League for Democracy (SNLD) party. As one interviewee stated: "*I got an opportunity to vote. I voted for SNLD partly because I believe this party will be do justice and we are Shan*" (Interview 4, South Village, 3 February 2016). This respondent's shared identity of Shan with the SNLD party caused them to vote for the SNLD instead of the military-backed Union Solidarity and Development Party (USDP) or the National League for Democracy (NLD) party that won elections in 2015.

Moreover, during the group interview in South Village, villagers discussed with Mai how they were not consulted for the Mong Ton dam. Some villagers did attend a meeting about the benefits and impact of the Mong Ton dam held by SMEC, but they did not feel like they were able to give their opinions (Group Interview, South Village, 3 February 2016). As a result of these mostly negative relations with the Burmese military and private companies, South Village does not support the Mong Ton dam. One-hundred percent of respondents opposed the dam, citing a belief that the dam will harm them and not benefit them. As one respondent stated: "*I don't want and need to build this project. If the big dam is built, we will get the impact more than the benefit*" (Interviewee 10, South Village, 3 February 2016). While these negative relations do not inhibit the ability of the Burmese military and Myanmar government to extract resources, they also do not improve or reproduce that ability. Instead South Village has voiced their opposition to the Burmese military and Myanmar government through their votes and resistance to the Mong Ton dam.

For North Village, however, access was further enabled through the SSAs shared social identity of Shan ethnicity with the villagers. During the 2015 elections, SSA supported the Shan National League for Democracy (SNLD) and the Shan National Democratic Party (SNDP) (Wah, 23 February 2016). One hundred percent of villagers who voted, reported voting for the SSA's associated political party of the SNLD, as one villager said, "*if we vote, we vote for Shan*" (Interviewee 2, North Village, 16 February 2016). This shared social identity is significant, yet villagers still must navigate the unequal power dynamics between themselves and the SSA. During the interviews, the most significant concern villagers raised was that of transportation. They are a very remote village and interviewees expressed a belief that improved roads would improve their lives. This is the sort of social service that the SSA can provide and villagers have requested help from the SSA; however, at the time of the interviews they had yet to receive approval (Wah, 23 February 2016).

This shared social identity of Shan also extended to the researchers who visited this village; although, sometimes this was not necessarily emerging freely from the researchers, but was ascribed to them. For instance, when Wah described her organizations relationship with North Village she said, "*They already knew about my organization and because they know we are from the SSA military too, they just take good care of us*" (23 February 2016). Wah's organization provides a boarding school for Shan students and much of the rice they use to feed the students comes from rice taxes extracted by SSA. Yet, Wah also acknowledged that villagers seemed "*afraid of the SSA…the military (SSA) seems to have so much influence on them*" (23 February 2016). From Wah's observations, interviewees were reluctant to share with her information that criticized the SSA. Here, although the shared identity of the researchers with SSA enabled their access to the community, it was complicated by uneven power relations between the SSA, Wah's organization, and North Village.

Bringing the research from these two sites together, it is evident that control over resource areas in Shan State is complicated. For South Village, spatial access and control over territory is maintained through law and the physical presence of the Burmese military and Lahu militia. Private-government and private-military partnerships that have enabled development efforts for the Mong Ton dam have also led to an increase in logging and mining for gold along the river, resulting in a degradation of these resources. Yet, these mechanisms of access face resistance from the lack shared social identity between South Village and the resource extractors. In North Village, spatial access and control over territory is maintained by the physical presence of the SSA. Private-military partnerships help finance efforts like gold mining in the river, inhibiting North Village's access to gold. Additionally, while a shared identity of Shan has resulted in an increased support of the authority of the SSA, the unequal power between SSA and North Village has left impressions of coercion and fear.

Throughout all of this, researchers must navigate access to sites and interviews with various overlapping authorities. As a foreigner, I adhered to Myanmar government laws and did not physically travel to South and North Village, while the two Shan researchers, Mai and Wah, had to navigate the nuanced aspect of their shared cultural identity with villagers. In this, what knowledge was produced was altered by these circumstances. Information shared with researchers was filtered by their alignment (or misalignment) with researchers. As with Wah's example, villagers' apparent reluctance to share information with her because of her organization's connections with the SSA suggests that how interviewees responded to questions was somewhat filtered with how they perceived they should respond in accordance with the SSA.

12.5 Conclusions

This chapter explored the dynamic and oft contradictory relationships that influence people's ability to derive benefits from resources, including knowledge production. The same private-public and military-private partnerships that influenced resource users' ability to benefit, also affected researchers' access. In countries like Myanmar, with brown and black zones that limit the presence of outsiders, the mechanisms that determine researcher access are especially important to dissect. The access that researchers must navigate is neither inherently good or bad; however, it must be navigated and acknowledged as essential to knowledge production and the research methods themselves.

What was presented above is just the start of the story of access in hard to reach places along the Thanlwin River in Shan State. As with most research projects, the knowledge produced about resource user access was contingent on researcher access. Further research could better tease out the nodes and points of relationships that determine the informal governance arrangements between private-military and private-public relationships. Additionally, the complexity of shared identity among ethnic populations and the corresponding armed groups needs to be further teased out.

From this research, however, what can be concluded is that the formal and informal governance arrangements, environmental degradation from resource extraction, and shared identity (or a lack of) that influence access do increase local communities' vulnerability. Be it decisions made by local and national authorities over a foreign company's right to extract gold from the river or log the forest, or the construction of a hydropower project, neither village has the power to determine who benefits. By asking questions of resource user and researcher access to river and forest resources and by increasing the transparency of the research process, it then becomes possible to better design research and analyze the impacts of resource extraction on the lives of local people.

References

Babbie, E. (2007). *The practice of social research* (12th ed.). Belmont, CA: Wadsworth.

Blaikie, P., & Brookfield, H.C. (1987). *Land degradation and society*. London: Methuen.

Bryant, R.L. (1994). Fighting over the forests: Political reform, peasant resistance and the transformation of forest management in late colonial Burma. *The Journal of Commonwealth & Comparative Politics*, *32*(2), 244–260.

Bryant, R.L. (1997). *The political ecology of forestry in Burma, 1824–1994*. Honolulu, HI: University of Hawaii Press.

Brookfield, H. (2015). Shifting cultivators and the landscape: An essay through time. In M. Cairns (Ed.), *Shifting cultivation and environmental change: Indigenous people, agriculture and forest conservation* (pp. 24–61). New York, NY: Earthscan.

Callahan, M. (2009). Myanmar's perpetual junta: Solving the riddle of the Tatmadaw's long reign. *New Left Review*, *60*, 27–63.

Caretta, M.A. (2015). Situated knowledge in cross-cultural, cross-language research: A collaborative reflexive analysis of researcher, assistant and participant subjectivities. *Qualitative Research*, *15*(4), 489–505.

Dove, M.R. (1993). A revisionist view of tropical deforestation and development. *Environmental Conservation*, *20*(1), 17.

Drury, J., & Stott, C. (2001). Bias as a research strategy in participant observation: The case of intergroup conflict. *Field Methods*, *13*(1), 47–67.

EIA (Environmental Investigation Agency). (2015). *Organized chaos: The illicit overland timber trade between Myanmar and China.* London: EIA.

Elmhirst, R. (2011). Migrant pathways to resource access in Lampung's political forest: Gender, citizenship and creative conjugality. *Geoforum*, *42*(2), 173–183.

Faria, C., & Mollett, S. (2014). Critical feminist reflexivity and the politics of whiteness in the "field." *Gender, Place & Culture*, *524*(July 2015), 1–15.

Forsyth, T. (1996). Science, myth and knowledge: Testing Himalayan environmental degradation in Thailand. *Geoforum*, *27*(3), 375–392.

Grundy-Warr, C., Rajah, A., Elaine, W.S.Y., & Ali, Z. (1997). Power, territoriality and cross-border insecurity: Regime security as an aspect of Burma's refugee crisis. *Geopolitics and International Boundaries*, *2*(2), 70–115.

Grundy-Warr, C., & Wong Siew Yin, E. (2002). Geographies of displacement: The Karenni and the Shan across the Myanmar-Thailand border. *Singapore Journal of Tropical Geography*, *23* (1), 93–122.

Haraway, D. (1988). Situated knowledges: The science question in feminism and the privilege of partial perspective. *Feminist Studies*, *14*(3), 575–599.

Hawkins, R., & Ojeda, D. (2011). Gender and environment: Critical tradition and new challenges. *Environment and Planning D: Society and Space*, *29*(2), 237–253.

Hengsuwan, P. (2013). Explosive border: Dwelling, fear and violence on the Thai-Burmese border along the Salween River. *Asia Pacific Viewpoint*, *54*(1), 109–122.

Hnin Wut Yee. (2016). Implications on women's lives and livelihoods: A case study of villages to be affected by the Mongton Dam Project in Shan State. In *International Conference on the Mekong, Salween and Red Rivers: Sharing Knowledge and Perspectives Across Borders* (pp. 303–333). Bangkok, Thailand.

Hunsberger, C., Corbera, E., Borras, S.M., Franco, J.C., Woods, K., Work, C., de la Rosa, R., Eang, V., Herre, R., Kham, S.S., Park, C., Sokheng, S., Spoor, M., Thein, S., Aung, K.T., Thuoun, R., & Vaddhanaphuti, C. (2017). Climate change mitigation, land grabbing and conflict: Towards a landscape-based and collaborative action research agenda. *Canadian Journal of Development Studies*, *38*(3), 305–324.

International Rivers. (May 24, 2012). *The Salween river basin fact sheet: Dam cascade threaten biological and cultural diversity.* Retrieved from: https://www.internationalrivers.org/resources/the-salween-river-basin-fact-sheet-7481.

Jones, L. (2014). The political economy of Myanmar's transition. *Journal of Contemporary Asia*, *44*(1), 144–170.

Koubi, V., Spilker, G., Bohmelt, T., & Bernauer, T. (2014). Do natural resources matter for interstate and intrastate armed conflict? *Journal of Peace Research*, *51*(2), 227–243.

Kandji, S.T., Verchot, L.V., Mackensen, J., Boye, A., van Noordwijk, M., Tomich, T.P., Ong, C., Albrecht, A., & Palm, C. (2006). Opportunities for linking climate change adaptation and mitigation through agroforestry systems. In D. Garrity, A. Okono, M. Grayson & S. Parrott (Eds.), *World agroforestry into the future*. Nairobi, Kenya: World Agroforestry Centre.

Kobayashi, A., & Peake, L. (1994). Unnatural discourse: "Race" and gender in geography. *Gender, Place & Culture*, *1*(2), 225–243.

Kramer, T., & Woods, K. (2014). *Financing dispossession: China's opium substitution programme in northern Burma.* Amsterdam, The Netherlands: Transnational Institute.

Le Billon, P. (2001). The political ecology of war: Natural resources and armed conflicts. *Political Geography*, *20*, 561–584.

Maung, L.M. (2016, March 15). NLD under pressure to scrap hydropower projects. *Myanmar Times*. Retrieved from: http://www.mmtimes.com/index.php/national-news/19461-nld-under-pressure-to-scrap-hydropower-projects.html.

McDonald, J. (2013). Coming out in the field: A queer reflexive account of shifting researcher identity. *Management Learning*, *42*(2), 127–143.

McKittrick, K., & Peake, L. (2005). What difference does difference make to geography. In N. Castree, A. Rogers, & D. Sherman (Eds.), *Questioning geography: Fundamental debates* (pp. 39–54). Malden, MA: Blackwell Publishing.

Meehan, P. (2015). Fortifying or fragmenting the state? The political economy of the opium/heroin trade in Shan State, Myanmar, 1988–2013. *Critical Asian Studies*, *47*(2), 253–282.

Mohanty, C.T. (2002). Under western eyes revisited: Feminist solidarity through anticapitalist struggles. *Journal of Women in Culture and Society*, *28*(2), 499–535.

Peluso, N.L. (1996). Fruit trees and family trees in an anthropogenic forest: Ethics of access, property zones, and environmental change in Indonesia. *Comparative Studies in Society and History*, *38*(3), 510.

Peluso, N.L., & Lund, C. (2011). New frontiers of land control: Introduction. *Journal of Peasant Studies*, *38*(4), 667–681.

Peluso, N.L., & Vandergeest, P. (2011). Political ecologies of war and forests: Counterinsurgencies and the making of national natures. *Annals of the Association of American Geographers*, *101*(3), 587–608.

Ribot, J.C., & Peluso, N.L. (2003). A theory of access. *Rural Sociology*, *68*(2), 153–181.

Ribot, J.C. (1998). Theorizing access: Forest profits along Senegal's charcoal commodity chain. *Development and Change*, *29*, 307–341.

Rocheleau, D. (1995). Maps, numbers, text, and context: Mixing methods in feminist political ecology. *The Professional Geographer*, *47*(4), 458–466.

Rocheleau, D., & Edmunds, D. (1997). Women, men and trees: Gender, power and property in forest and agrarian landscapes. *World Development*, *25*(8), 1351–1371.

Sai Aung Tun. (2009). *History of the Shan State: From its origins to 1962.* Chiang Mai, Thailand: Silkworm Books.

Salween Watch. (2013). *Current status of dam projects on Burma's Salween River.*

Salween Watch Coalition. (2016). *Current status of dam projects on the Salween River.*

Samarkand, A. (2015, June 11). Thailand's electricity utility may be complicit in human rights violations in Myanmar's Salween dams. *Mekong Commons.* Retrieved from: http://www.mekongcommons.org/thailands-electricity-utility-may-be-complicit-in-human-rights-violations-in-myanmars-salween-dams/.

Schroeder, R.A. (1999). *Shady practices: Agroforestry and gender politics in The Gambia.* Berkeley, CA: University of California Press.

Scott, J.C. (2009). *The art of not being governed: An anarchist history of upland Southeast Asia.* New Haven, CT: Yale University Press.

Scurrah, N., Hirsch, P., & Woods, K. (2015). *The political economy of land governance in Myanmar*. Mekong Region Land Governance. Retrieved from: http://www.mekonglandforum.org/sites/default/files/Political_Economy_of_Land_Governance_in_Myanmar_1.pdf.

Sharp, J. (2005). Geography and gender: Feminist methodologies in collaboration and in the field. *Progress in Human Geography*, *29*(3), 304–309.

SLORC (State Law and Order Restoration Council). (1992). The forest law (1992).

Sundberg, J. (2003). Masculinist epistemologies and the politics of fieldwork in Latin Americanist Geography. *Professional Geographer*, *55*(2), 180–190.

Sturgeon, J.C. (2004). Border practices, boundaries, and the control of resource access: A case from China, Thailand and Burma. *Development and Change*, *35*(3), 463–484.

Temple, B., & Edwards, R. (2002). Interpreters/translators and cross-language research: Reflexivity and border crossings. *International Journal of Qualitative Methods, 1*(2), 1–12.

Temple, B., & Young, A. (2004). Qualitative research and translation dilemmas. *Qualitative Research, 4*(2), 161- 178.

Turner, S. (2010). Research note: The silenced assistant. Reflections of invisible interpreters and research assistants. *Asia Pacific Viewpoint*, *51*(2), 206–219.

Verchot, L.V., Van Noordwijk, M., Kandji, S., Tomich, T., Ong, C., Albrecht, A., Mackensen, J., Bantilan, C., Anupama, K.V., & Palm, C. (2007). Climate change: Linking adaptation and mitigation through agroforestry. *Mitigation and Adaptation Strategies for Global Change*, *12* (5), 901–918.

Woods, K. (2011). Ceasefire capitalism: Military–private partnerships, resource concessions and military–state building in the Burma–China borderlands. *The Journal of Peasant Studies*, *38* (4), 747–770.

Chapter 13
Fisheries and Socio-economic Change in the Thanlwin River Estuary in Mon and Kayin State, Myanmar

Cherry Aung

13.1 Introduction

The Thanlwin or 'Salween' River Estuary is well-known in Myanmar for its rich fishery and wide range of fishes. The daily flood and ebb tides bring fresh and sea water together to support a diversity of fresh, brackish, and marine fishes, creating dynamic and fertile spawning and feeding grounds. Local communities in Mon and Kayin States have depended on the estuarine fishery as part of their local economies for generations, but at present a variety of factors, including geo-hydrological changes, overfishing, and unsustainable and illegal fishing methods, have depleted the fishery. While local workers seek opportunities elsewhere, particularly in neighboring Thailand, domestic migrants from other parts of Myanmar have moved to find work in this precarious estuarine fishery. This has meant that the estuary has changed dramatically: not only ecologically, but also in terms of labor and the socio-economic status of its resident communities.

This chapter focuses on community fishery livelihoods in this estuary, exploring the transformation of riparian communities in Mon and Kayin State. The chapter proceeds with the understanding that the Thanlwin is a significant source of food and livelihoods for the approximately 6 million people who live in the basin in Myanmar, particularly those who rely on the Myanmar stretch of the river (Johnston et al. 2017). It is well known that poverty in Myanmar is geographically positioned; specifically, poverty is higher in rural coastal and mountainous areas, particularly for agriculturalists who rely on the "monsoon crop" (Ministry of Planning and Finance and World Bank 2017). Much less is known about the links between poverty, recent socio-economic change in Myanmar, and non-agrarian livelihoods such as fishing. The Myanmar Department of Fisheries reports that export earnings

Cherry Aung, Professor and Head of the Department of Marine Science, Pathein University; Email: missaungmarine@gmail.com.

C. Middleton and V. Lamb (eds.), *Knowing the Salween River: Resource Politics of a Contested Transboundary River*, The Anthropocene: Politik—Economics—Society—Science 27, https://doi.org/10.1007/978-3-319-77440-4_13

for fish were in the range of more than 500 million dollars in 2013–2014 alone, and these earnings are rising with increased demand (World Fish 2016). These assessments, however, do not account for the local value of fish for subsistence or the trends seen locally. As such, this chapter provides insight into the changing livelihoods of fishers in this estuarine context, and adds nuance and detail to the broader overviews of Myanmar's socio-economic change in recent years.

The research presented in this chapter was conducted through interviews, focus groups, and household surveys across four villages located in the estuary over the period of 2015–2016, and investigates changes to fishers and livelihoods over a ten-year period beginning in 2005. The author is familiar with the livelihoods of these villages from her long-time work as a marine scientist, and from personal experience living along the Thanlwin River.

The chapter first provides background information on the current governance and fisheries situation along the Thanlwin before discussing the methods used to conduct this research. Through the study of four fishing villages, the author presents evidence of fisheries decline and the associated overfishing and environmental changes that have impacted fish habitat and species in the estuary. The following section examines the socio-economic differentiation linked to fishing and other events in the estuary. The chapter concludes with recommendations for better management and further research.

13.2 Background and Governance of the Lower Thanlwin

The Thanlwin estuary is a complex ecosystem, but little is known about it, particularly outside of Myanmar. The estuary sees the mainstem of the river connect to four tributaries and supports a wide diversity of fishes. For local people in the inshore fisheries zones of the Thanlwin estuary, catching fish for personal consumption and for market is a part of daily life. Men usually fish on boats, which require time away from home, while women and older people (above 60) are involved in other aspects of fish production. This production work ranges from repairing nets and preparing fish products, to selling fish in the markets. A range of fresh, brackish, and marine fishes comprise the local inshore fishing economy.

Offshore fishing is a bit different. The offshore fisheries require larger boats, and greater investments in equipment. While most offshore fishing is commercial, some local residents work in the fishery as laborers, which requires going offshore for weeks, and sometimes months, at a time. Nga pone na (Paradise threadfin) and Nga pyat (Coitor croaker) are the important commercial fishes around the estuary, but these fishes as well as others detailed below are in decline.[1]

[1]The fish species names presented in this paper include both Myanmar language names and scientific names in parenthesis.

Technically, the inshore fishery includes the regions between the low tide mark and ten nautical miles from the shoreline. The offshore fishery is beyond that point, and these distinctions have implications for management. The offshore fishery is managed by the national government, whereas the inshore fishery is the responsibility of the regional government (Soe 2008). At present, there are four laws related to fisheries in Myanmar that are relevant to both the inshore and offshore fisheries. This includes the inland fishery law, offshore fishery law, aquaculture law, and the foreign fishing investment law (Tsamenyi 2011).

While these four laws exist on the books, law enforcement is weak and some of the offshore fishing boats fish close to the shore (Tsamenyi 2011). This has created conflict between the small inshore fishing boats and the large offshore boats, with the smaller boats unable to compete with the larger boats. The small fishing boats do not have a way to address this problem, so many fishers involved in the inshore fishery either change their livelihoods or become workers on the larger boats.

In response to these issues and reported declines, the Ministry of Fisheries announced in February 2012 a seasonal fishing ban for three months from May 5 to end of July to prevent the declining of fish stocks in the river, and a corollary ban from June 1 to end of August for offshore fisheries. Moreover, protected areas for lobsters and fishes were marked. Using poison is not allowed. While these moves are significant, enforcement remains weak.

Related to these fishing practices and legal instruments, there is a need for research in addition to enforcement. As the director of the central Yangon Fisheries Department, U Tin Win Myint, told the *Myanmar Times*, "In our country, there is no research. We have poor conservation and recordkeeping. But now we're starting to take action" (Khin Wine Phyu Phyu, 7 July 2016).

While the author concurs that more research is needed, it is important to understand what kind of research needs to be conducted. For instance, how is the decline in fishing affecting riparian communities? Are they becoming increasingly connected to broader livelihood changes and external markets? Local livelihood strategies are not only changing in relation to fishing, but are increasingly driven by migration, remittances, and market opportunities. Social-economic change in the region is not entirely the result of decreased fish catch. Increasing mobility and strong external markets, both domestically and internationally, offer opportunities for higher wages than the local agricultural and fishing sectors, which has enticed many people to move. Many people in the estuary have moved or migrated to other places to work, especially to Thailand, but also Malaysia and Singapore. Poorer individuals who lack the means to travel or relocate for work remain in the region. They have not been able to invest in fishing equipment, which would allow them to compete with large scale fishers in more open water, and are thus facing increased livelihood vulnerability, inequality, and sometimes conflicts. This chapter highlights the interrelationships between these important changes in the Thanlwin estuary. The next section details how the author undertook the study.

13.3 Research Methods and Approach

The research data presented in this chapter on livelihoods and fishery status and trends were collected through surveys. Approximately 60 households were surveyed. The household survey consisted of four sections: amount of fish catch (and decline), impacts, resilience/response, and policy interventions. The author also brings expertise as a marine biologist who has studied estuarine environments for more than 10 years, which includes familiarity with fish species identification, habits, and habitat requirements.

The author also conducted focus group discussions and key informant interviews. Each focus group included seven to ten people in total, inclusive of fishers, and also included household heads from different social groups. Key informant interviews with local government officials, and focus group discussions with farmers associations, village leaders, and local government officials were also carried out. The majority of research was conducted in 2015–2016, with some follow up visits in 2017.

The analysis of the data included assessing the impacts of the fish catch decline and the related decline in income and food security. Livelihood activities and changes were assessed through focus group interviews and participatory rural appraisal surveys using different tools, including a household questionnaire, with the objective of acquiring baseline data on fisheries and the socio-economy of the targeted villages. Adaptive measures taken by communities were also identified.

To understand the fishery status across the estuary, the author selected four fishing villages in the estuary (See Figure 13.1: Map showing the four study areas in the Thanlwin River Estuary). Each of the four villages are known as productive fishing communities and are situated on the banks of the four tributaries of the Thanlwin River Estuary.

These four villages are Kayar village (along the Gyine River), Kyone-sein village (along the Attran River), Kau-mu-pon village (along the Mawlamyine River, on Bilukyun island), and Khindan village (along the Dayebauk River). Kayar village is situated within Kayin State and the rest are within Mon State. For the scope of this paper, the author will not go into great detail on each of the four villages, but these four villages are generally small, with populations ranging from 900 residents (Khindan village) to approximately 3,000 residents (Kayar village). The predominant occupation, over the 10-year period from 2005 to 2015, was in the fisheries or in fishing-related work.

The following diagram, Figure 13.2: Diagram of Fisheries and Livelihood Changes, represents the trend of fisheries and related socio-economic change. This research is positioned within an overall approach which seeks to understand "how livelihoods shift, evolve and adapt in villages where numerous actors are vying for access to the same natural resources" (Marschke 2015: 5). This diagram shows the links between various facets of the study, particularly in how the decline in fish is linked to both habitat loss and broader socio-economic changes.

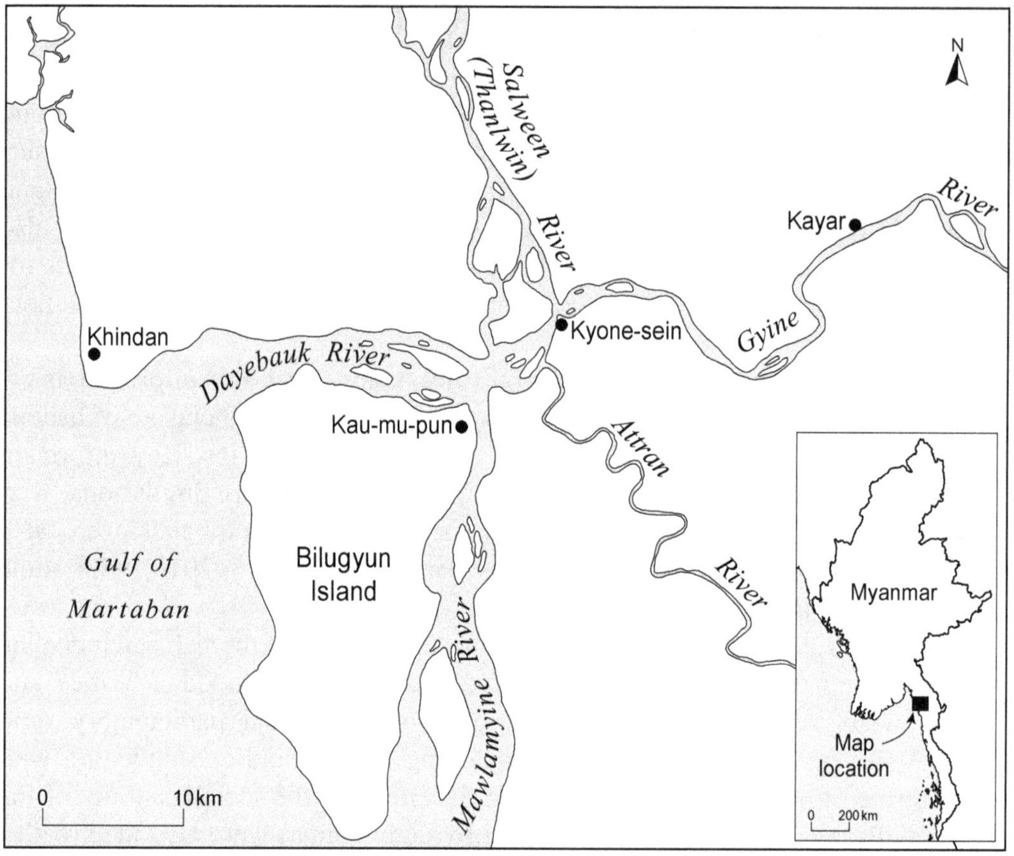

Fig. 13.1 Map showing the four study areas in the Thanlwin River Estuary. *Source* Cartography by Chandra Jayasuriya, University of Melbourne, with permission

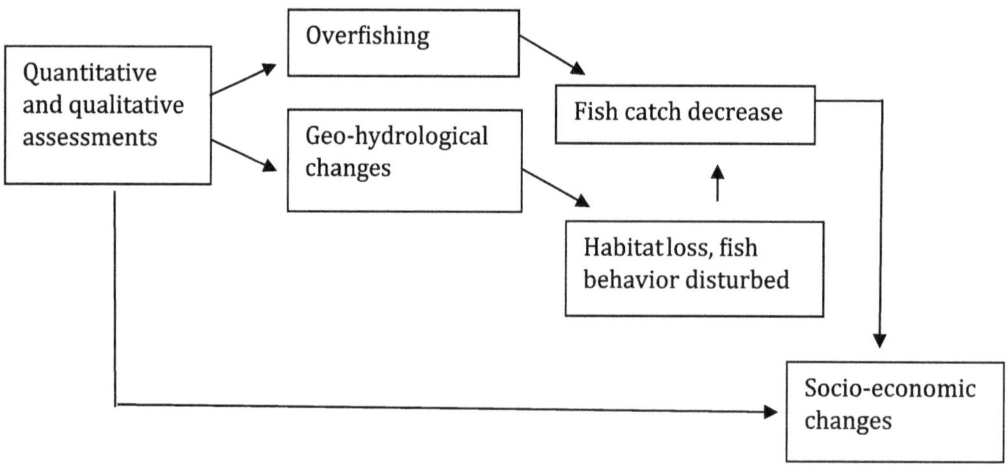

Fig. 13.2 Diagram of fisheries and livelihood changes. *Source* Created by Author

13.4 Estuary Region and Functions

Before moving onto the findings of this study, it is essential to outline the ecological functions and change in the estuary in relation to fisheries. All four tributaries of the estuarine region discharge into the Andaman Sea via the Gulf of Martaban. The Attran and Gyine rivers are at the upper region of the estuary and provide a variety of fishes and other resources. The Mawlamyine and Dayebauk rivers are at the lower region of the estuary, and function to discharge fresh water from the Thanlwin, Gyine, and Attran rivers to the Andaman Sea via the Gulf of Martaban.

All four tributaries experience a daily ebb and flood tidal influence, with a variable range of tidal inundation depending on the distance from the sea. The salinity level is high at the mouth of the Mawlamyine and Dayebayk rivers (where the more coastal villages of Kau-mu-pon and Khindan are situated), while the salinity is lower in the middle of the Thanlwin River and where the four tributaries meet. The more inland rivers Gyine and Attran have low salinity and are almost entirely fresh at some times of the year. This is where Kayar and Kyano-Sein villages are located. It is well known that the place at the border of the saline water intrusion zone in Gyine River, which is very near Kayar village, is one of the best fishing areas and is an important part of this investigation.

It is also important to understand the seasonal variation within the estuary. Local people report catching a variety of fishes according to the season and the tides. Rates of salinity and seawater intrusion vary according to season, which affects fisheries. Seawater intrusion can extend up to about 60 km in the Gyine River, particularly during the dry season.

Different types of fishes thrive within this range of habitats. Fresh water fishes and prawns are plentiful at the inner region of the Gyine and Attran rivers, and also in the upstream region of the Thanlwin River, while marine and estuarine fishes enter along with seawater into the inner region. This gives the estuarine region its unique environment, maintaining a variety of marine, brackish, and fresh water fish types that support the livelihoods of millions of people living along the estuary.

The next section presents the main findings from the study related to declining fish catch as evidenced by a range of factors.

13.5 Research Results: Fisheries Decline Evidenced by a Range of Indicators

At present, local communities are facing the problem of a declining fish catch in the river, the estuary, and even in the sea. The diversity, average size, and average catch of fish is reported to be declining. The reason for this decline is a variety of anthropogenic and environmental stresses. Local fishermen from the inner river and estuarine region are also facing many problems related to conflicts in the fisheries sector, which are linked to a variety of factors within the estuary.

Table 13.1 Fishery status of the estuarine region between the years of 2005–2016

Village	Main source of income (%)	
Kayar (Gyine River)	Fishery-6% Paddy field and farm-5% Working abroad-70% Other small scale jobs-19%	Fishery-25% Paddy field and farm-30% Working abroad-40% Other small scale jobs-5%
Kyone-sein (Attran River)	Fishery-25% Paddy field and farm-10% Working abroad-50% Other small scale jobs-14%	Fishery-60% Paddy field and farm-25% Working abroad-10% Other small scale jobs-5%
Kau-mu-pon (Mawlamyine River)	Fishery-25% Paddy field and farm-10% Working abroad-30% Poultry-10% Other small scale jobs-25%	Fishery-60% Paddy field and farm-20% Working abroad-10% Poultry-5% Other small scale jobs-5%
Khindan (Dayebauk River)	Fishery-60% Paddy field and farm-5% Working abroad-20% Other small scale jobs-15%	Fishery-80% Paddy field and farm-5% Working abroad-5% Other small scale jobs-10%

Source Reported in interviews conducted by the author

The fisheries status of the estuary over the past 10–15 years is represented across the four villages in Tables 13.1 and 13.2 (see Table 13.1: Fishery status of estuarine region between the years of 2005–2016 and Table 13.2. Fishing boats, gear, and overall fish catch between the years of 2005–2016). Examining these data, we can see the decrease in fish catch and efforts by local fishers to address this issue. For instance, responses from Kayar village show evidence of decline. Of the four villages in this study, Kayar is the most inland. In the past, Kayar was well-known for its fishery and the opportunity for a high catch along the Gyine River. Fishers have fished here and sold fish locally and to the city of Mawlamyine for decades. However, within the past 10–15 years, the fishery sector in Kayar has dramatically declined. Interviewees reported a steep decline, with fishing now representing only 6% of their main source of income, down from a more substantial 25% previously. This lost income is largely made up from working abroad, which is discussed below.

Kayar village was chosen as a case study because of its fishery status and its importance for the livelihoods of people living along the river. It is a place with a rich seasonal catch, and a variety of fresh, brackish and marine fishes. Freshwater prawns, Nga kyin (Carp), Nga phar mar (*Osteobrama feae*), Nga okephar (*Hemibogrus microphthalmus*, a fresh water fish which migrates from upstream of the river) and other commercial fresh water fishes migrate from upstream. Interviewees explained that Nga okephar has been locally extinct for the last 5 years. Additionally, according to interviews, one of the locally important commercial marine fishes, Nga Thalauk (Herring), is almost extinct in the river and Arius (Giant catfish) is totally extinct locally. Nowadays, fishers in Kayar village

Table 13.2 Fishing boats, gear, and overall fish catch between the years of 2005–2016

Village	Fishing boats and gear used		Catch/month	
	(2015–2016)	(Before 2005)	(2015–2016)	(Before 2005)
Kayar (Gyine River)	*Small boats* 20 (cotton, plastic net) 20 (purse seine net) 4 (large bag nets)	*Small boats* 40 (cotton net) 30 (purse seine net) 2 (large bag nets)	0–5 Viss (post monsoon) 0–2 Viss (monsoon)	50–200 Viss (post monsoon) 5–30 Viss (monsoon)
Kyone-sein (Attran River)	*Small boats* 100 (cotton, plastic net) 10 (purse seine net) 5 (large bag nets)	*Small boats* 200 (cotton net) 40 (purse seine net)	0–50 Viss (post monsoon) 0–20 Viss (monsoon)	200–500 Viss (post monsoon) 30–150 Viss (monsoon)
Kau-mu-pon (Mawlamyine River)	*Large boats* 15 (bag nets) *Small boats* 100 (bag nets)	*Large boats* 60 (bag nets) *Small boats* 160 (bag nets)	0–50 Viss (post monsoon) 0–20 Viss (monsoon)	200–500 Viss (post monsoon) 50–200 Viss (monsoon)

Source The dates and data were provided by local people, and may not be precise. Viss is a Myanmar measurement unit of weight equal to 1.63293 kg/3.6 pounds

also face the problem that fresh water fishes rarely travel down from the upper region of the river, and brackish and marine fishes rarely travel up from the sea. It is hard to survive as a fishing family in Kayar.

Closer to the sea, in Khindan village, we see a similar decline in fish catch from 80 to 60% nowadays. The Darebauk River, where Khindan is located, is also well known as a highly productive fishery resource. About 37 species of marine and brackish fishes were observed by the author. The fishers here catch fishes from the river and also from the sea. Interviewees reported that over the past ten years, not only the local catch but also the broader fishery industry has declined. The economy of local residents and migrantswho depend on fisheries has therefore changed, with harmful consequences for people in this area.

Another way that fisheries decline was communicated is related to changes in fishing practices. As the fish catch declines in the river, people must catch fish outside of the river at coastal inshore or offshore locations. Offshore fishing requires more powerful boats, as noted above, which represents a much higher investment for fishing gear and labor, and more time. The fishermen stay out at sea for one or more months on the larger boats, and they bring the fish catch in by using smaller carrier boats. The large fishing boat owners have to employ many workers. Poorer people cannot invest the equipment, time, and labor required to catch fish offshore. This is particularly true in Kau-mu-pon village, located on Bilukyun island along the Mawlamyine River. Here, the author was told that fishermen “leave their boats and become workers on large boats or work on carrier boats" (Kau-mu-pon village Interview, 21 July 2017).

In addition to a shift in practices, the overall amount of fishing boats is also decreasing, with multiple smaller fishing operations sometimes consolidated into larger ones. Interviewees explained that there were many more boats before. For instance, in Khindan village, there used to be more than 20 small boats, and now only about 5 remain. Moreover, the author was informed that in the past there were 600 large boats, but now (see Table 13.1: Fishery status of the estuarine region between the years of 2005–2016) only 40 small boats and 5 large boats remain in the Darebauk River near Khindan village. Most of the present-day fishermen are considered "rich people" who can invest in the fishery. Nowadays, some former fisherman now work under wealthy fishermen. This is because of the decreasing fish stock in the fishing grounds around the villages. Fishers must go further out, sometimes into the sea (beyond the estuary) to find fish, and in doing so, they must also compete with the high investment, modern large boats. Consequently, the socio-economy of the local population has also changed as fishers attempt to adapt to the situation, as discussed below.

Moreover, this shift in Khindan village is associated with the kind of fishing gear used. This is also the case in Kyone-sein village. Fishers in both villages recognize the fishing area where the Gyine and Attran rivers meet the mainstem of the Thanlwin as a key fishing area. Local people from these two villages explained in interviews that the fish catch has decreased since more than 10 years ago, after construction of the Thanlwin Bridge; they also note that illegal fishing has increased. Today, fishers report using almost entirely bag nets from boats (big and small). This is a more intensive form of fishing than traditional methods in response to fish declines. The large bag nets[2] are set up across the rivers or creeks; these nets are more like 'walls' that divert fish into traps, with much bycatch. Locals also report an increase in the illegal use of poison in the commercial fishery, causing high mortality in shrimps and creek fishes, which local residents rely on for subsistence. Illegal fishing by baby trawler and fishing in the spawning season at the inner region of the channels are also reported. Overall, according to interviews with the local community in these two areas, the major cause of depletion of fishes within the Gying and Attran rivers is overfishing due to these illegal practices. Consequently, most men have migrated outside the village for work as they cannot catch enough fish to feed their families.

In addition, key informant interviews highlighted that not only local fishers, but some of the private commercial fishing operators who own the large boats are also witnessing declines. They are still carrying out fishing operations, but are facing pressure to increase their catch, so they sometimes turn to illegal nets and net sizes (smaller mesh), the use of explosives and chemicals, fishing illegally in closed

[2]Bag nets are long (often several hundred meters), vertical 'nets' or 'walls' running at right angles to the shoreline; such nets are set up this way to interrupt fish as they swim and direct them into a series of traps.

seasons, and fishing beyond limited areas. These fishing operators are increasingly rare in the region as the fish catch has further declined.

Both commercial and local fishers sell their fishes in markets or to fish traders. Today, the traders are mostly ex-fishermen who stopped fishing because it did not provide an adequate income. These ex-fishermen also have the option to travel abroad to work and generally now have what they describe as a good income. As discussed further below, this has meant that only some people, including old people and women, are left to work in the fishery and to sell fishes to traders. However, even the traders report less income than in the past as the fish trading intensity is decreasing. The fish trade still continues, mainly during the high season.

13.5.1 But, Why? A Range of Reasons for Fisheries Decline

Interviewees explained that there is a long list of reasons for the fishing decline in the sea and the river, including: over-exploitation, high/increased sedimentation rates, the effects of the Tsunami, and changes in water direction. They also identify the move to modern fishing techniques as a reason for changes in fishing productivity, as this requires high investment and is more labor intensive, and requires fishers to change their fishing grounds. Indeed, local livelihoods are threatened by many factors, and not all are apparent locally. For example, destruction of the seasonally flooded forest to expend agricultural land use generates conflicts between fishing, agriculture, and environment conservation (Blake and Pitakthepsombut 2006; Hortle and Suntornratana 2008; Khumsri et al. 2006). A recent study on the Upper Salween, also known as the Nu Jiang, argues that environmental degradation has resulted predominantly from human activities. The research shows evidence of environmental impacts on the upper stretches of the River from land use and land cover change, land-resource utilization and regional economic development (Feng et al. 2010). Geo-hydrological changes to the river, such as increased sedimentation, also play a role (see more below). Yet, in the lower Salween (Thanlwin), perhaps the most pressing issue is overfishing. Overfishing has increased due to the need to meet the rising demand for sustaining local households, and has become a dominant threat to fisheries and fish stocks in the estuary.

13.5.2 Overfishing and a Changing Fishery

Gradually, the fishes and fish products trade has shifted to the wider market in the region. As noted, wealthier people collect or buy the fishes from the fishermen and sell the fishes to nearby cities or abroad. Later, the traders provide funding for the fishers to buy fishing nets, boats and engines. In return, the local fishers are obliged to sell the fishes only to the traders who invested in their practice, and often end up selling their catch to the traders at a low price. This arrangement allows for those who do not have funds for financial investment in equipment to continue work in fishing. We see the fishery sector become 'modernized' by more intense fishing practices, using high power engines, larger boats, and modern gear like large bag nets. These nets, as noted above, result in bycatch, and are generally seen as destructive when left in the water for long periods of time, rather than at particular moments of fish migration or seasonal flow.

Accordingly, fisheries operations have intensified, and the fish catch has declined over the past ten to fifteen years. This in turn has further intensified fishing activities and has increased irresponsible fishing practices, moving them further offshore and to spawning times when fishing is banned. Also of note is over-exploitation by using large stationary gill nets and modern techniques outside the estuarine basin and the increase of offshore fishing boats. This is one of the reasons that brackish fish are declining in the area, even though there is a seasonal ban and permits are limited. Fishers report that the numbers of marine and brackish fishes entering to the inner estuary, and the Dayebauk and Mawlamyine tributaries, have become fewer.

The major causes of depletion of fishes within the Gyine and Attran rivers are similar; interviewees noted overfishing by using small mesh sized nets, illegal fishing using poison which causes high mortality in shrimps and creeks fishes, and illegal fishing by baby trawl and by fishing in spawning season in inner regions of the channels. Local people from along these two rivers explained in interviews that fish catches have declined since over 10 years ago, after construction of the Thanlwin Bridge.

As noted above, the inland fishery law falls under the authority of state governments, so the Thanlwin River Estuary is under the authority of the Mon and Kayin State Governments. The closed season is May-June in inland waters and June-August in offshore areas. Fishing is prohibited in inland waters during these months due to the spawning times of most of the fishes. In fact, only licensed fishing boats limited in number and size are permitted, and fishers must pay for these licenses. Yet, because enforcement is weak and Mon and Kayin are mixed authority areas, local people say that some areas of the estuary are also governed by local ethnic armies, such as areas of Mon and Kayin State where ethnic armed groups play a role in governance, particularly in the villages. In some situations, it was reported that due to weak law enforcement of fishing bans, ethnic armed groups played a role in patrolling these areas for illegal fishing, adding a further dimension to governance.

13.5.3 Sedimentation and a Changing River

While overexploitation and governance challenges are key reasons for fisheries decline, there are other important factors. There is evidence of extensive mudflat formation by the gradual erosion of some parts of the estuary. While this is providing an opportunity for some fishers (particularly in Khindan) to access mudflat fisheries and other mudflat commercial resources, overall these sedimentation changes mean that some of the behavior patterns of estuarine life, especially fishes, has also changed. Some of the good fishing grounds within the estuary are degraded, and some fish species have not been found for the last 10–15 years. Due to these changes, some marine and brackish fishes cannot enter the inner region of the rivers and some of the fresh water fishes from upstream are now lost, according to fishers. This may also be because some algae or other detritus has been depleted from the river bottom due to increased rates of sedimentation. These algae and detritus are the main food sources for most marine fishes, so their loss leads to fish decline. Also, some brackish fishes (like Herring or Helsa) do not thrive in the lower water levels that have resulted from the increase in sedimentation in recent years.

Some local people directly identified links between fish declines and increased sedimentation. One fisherman from Kau-mu-pon village explained, "The water level has become lower over the past 10 years due to high sediment buildup on the bottom of the river so the algae, which is food for some fishes, has decreased which causes a decrease in fishes" (21 July 2017).

Other fishermen from Kau-mu-pon village explained in focus group discussions that in their decades of fishing experience, "Some of the fish habitat loss and fish migration behavior change [is because] some fishes don't like the shallower water with more sediment" (21 July 2017). Many interviewees also linked this increased sedimentation to the Thanlwin Bridge construction more than 10 years ago.

However, as noted above, this story of fisheries decline does not end with the changing river ecosystem or changes in fishing techniques. These declines are linked to broader social-economic shifts, such as in/out migration and shifting dimensions of gender and labor, as discussed in the final section.

13.5.4 Socio-economic Changes in the Village

While many activities have contributed to the decreased fish catch within the estuary and surrounding regions, the declining fish catch has in turn contributed to dramatic changes in the livelihoods of fishers and migrants. The declining fish catch and increased competition with larger fishing boats has caused some fishers to quit fishing, and some have sold their fishing boats and equipment. Some now work for wealthy fishers who can afford to invest in high power fishing boats and costly permits. At the time of the interviews, most of the fishers who had left the trade explained that they had made their decision based on the high investments required and decreased incomes they were experiencing. Depending on each individual's situation, they changed their livelihoods to focus on other activities, such as gardening, farming and other businesses. Some people decided to go abroad to work, mainly in Malaysia, Thailand, and Singapore. Some migrant workers who moved to the estuary area are now returning to their hometowns in other parts of Myanmar.

Some local people do continue fishing, mostly for local consumption, within the creeks and rivers by using traditional fishing methods like using baskets, as seen in Kayar village. However, the fish caught by local fishers are not enough for village consumption, and the village market now sells non-local fish from the Mawlamyine market.

In the villages nowadays, not only fishing families but also other families mainly depend on remittances from family members working abroad (see Table 13.1). It was reported that larger numbers of residents of each of the four villages are going to work outside the village, from 70% in Kayar village, which is closest to the Thailand border, to 50% in Kyone-sein, and less in Kau-mu-pon and Khindan villages, at 30% and 20% respectively. Thus, with only a portion of the total village population remaining in the village year-round, the situation in the village has changed. The population now consists of mostly young children, who stay for education, and elderly people, who still work in the small-scale fisheries or do gardening. Villagers who used to farm in the paddy and groundnut (peanut) fields, and workers cultivating palm sugar, have mostly left these livelihoods. A visit to the estuary reveals that most of the fields around the villages are no longer used for agriculture, except for vegetable gardens planted by elderly people for local consumption.

There is also a gender dimension to the changes in age composition of these villages. At least two shifts are observable. While both men and women migrate out of the village for work, it is more common for men to migrate. Also, household labor roles have shifted even for those families who remain in the village.

When the fish catch was more plentiful, the role of men in the villages included fishing, trading, working in fish processing like pounded fish and prawn, and making salty and dry fishes. The women largely prepared and mended fishing nets, made fish and prawn paste, salted fish, dried fish, and sold fish and fish products. However, as noted, at present many men have quit fishing. Those still involved are not doing it on a family scale, but are fishing for sale to traders or are employed as workers on larger fishing operations. As a result, many of the women in the village are now without jobs. When asked about the declining fish catch, one fisherman from Kau-mu-pon village explained, "Work in the offshore fishing industry is mostly for men. Women who used to work in the local fisheries have had to change jobs because of the declining fisheries" (Interview, 21 July 2017).

With the decline in fishing and the shift to fishing for sale and export, local communities do not have enough fishes for daily consumption. As a result, aquaculture fishes from other places are imported and women from fishing families have found work selling them instead of fishes from local rivers.

Moreover, this shift is not the same in each village. In Khindan, fishing families have focused on specific products for export at a high price to China. These high priced products include fresh water shrimps, the swim bladder of fishes, and eggs of particular fishes. In this case, the women work in preparing these special fish products for export. By contrast, in Kayar village there has been a switch to traditional fishing methods, focusing on smaller creeks.

13.6 Discussion and Conclusion: How to Make Sense of a Transforming Estuary?

In the author's assessment, overexploitation is a main cause of declining fisheries due to high demand and weak law enforcement. As fishing efforts increase and intensify, there are also growing problems concerning illegal fishing and, as a consequence, fishers are pushed out of the fishing profession. This decline in fishing as a livelihood has impacted the living conditions of both young and old, and men and women in communities along the Thanlwin Estuary.

The diagram introduced earlier (Figure 13.2: Diagram of Fisheries and Livelihood Changes) represents some of the relationships between fisheries trends and socio-economic changes, particularly how the decline in fish is linked to both river changes (like habitat loss) and broader socio-economic changes.

Results shown here indicate that some fishing communities are concerned about their economic situation due to the declining fish catch. Projected socio-economic trends are normally dependent on the income perspectives for individuals and households (e.g. Ministry of Planning and Finance and World Bank 2017). Also, in this case, the combined effect of declining fisheries and decreasing income has caused a significant proportion of the local people to move abroad for work.

Here, the combined effects of fish declining and increasing incomes have caused a significant proportion of the local population of workers to move abroad. At the same time, the inland and near shore fisheries are worth a lot to subsistence communities, even if not in terms of income. This is especially true for the rural riparian populace due to the low investment required in local fisheries, and the historically high yield, which plays an important role in local food security.

In this area, both the regional and national government have tried to manage the illegal inland and costal marine fisheries through various means such as imposing a seasonal ban on fishing. But there is much more to do. The inner estuary is under the fresh water fishery law, which is governed by the regional government. The law should focus on illegal fishing methods such as the use of poison, catching spawning fishes in the off season, illegal fishing by using large gear and nets along the rivers, and the use of bottom trawlers. The outer estuary is under the marine fishery law, overseen by the national government. It should be well managed with a focus on the very evident illegal fishing and illegal gear use.

Of course, there is work to do beyond management, patrolling, and policy efforts. The local government should improve research capacity on fish behavior and estuarine processes. This relates to the need to reduce high sedimentation rates and problems along the rivers and the channels to improve fish production and to predict any disturbances in the estuary. Cooperation between the local government and communities with the national research institutions is highly recommended, including collaboration with regional and international research partners. Further research is also needed to integrate knowledge on fisheries changes with broader livelihoods and socio-economic impacts in order to find solutions to these pressing issues (see also, Swuam Pyaye Aye Aung et al. 2016).

In sum, like many rivers in the world, the Thanlwin River Estuary is facing pressures from a number of ecological and anthropogenic stressors. In Myanmar, while poverty levels are high, projections show improvements, as seen by the assessment referred to in the introduction to this paper, prepared by the World Bank and the Ministry of Planning and Finance (as noted above, see also: World Bank 2017). Yet, this chapter shows that there remains much work to do to understand how these trends impact different people in Myanmar. It also shows how changes in fisheries affect the rural poor unequally, and how these changes are related to a range of ecological and socio-economic factors.

References

Blake, D., & Pitakthepsombut, R. (2006). *Situation analysis: Lower Songkhram River Basin, Thailand*. Bangkok: Mekong Wetlands Biodiversity Conservation and Sustainable Use Programme. Retrieved from: https://portals.iucn.org/library/node/8842.

Feng, Y., He, D., & Li, Y. (2010). Ecological changes and the drivers in the Nu River basin (upper Salween). *Water International*, *35*(6), 786–799.

Hortle, K.G., & Suntornratana, U. (2008). *Socio-economics of the fisheries of the lower Songkhram River Basin, northeast Thailand* (No. 17). Vientiane, Lao PDR: MRC.

IUCN (International Union for Conservation of Nature). (2013). *Ecological survey of the Mekong River between Louangphabang and Vientiane Cities, Lao PDR 2011–2012*. Vientiane, Lao PDR: IUCN.

Johnston, R., McCartney, M., Liu, S., Ketelsen, T., Taylker, L., Vinh, M.K., Ko Ko Gyi, M., & Aung Khin, T., Ma Ma Gyi, K. (2017). *State of knowledge: River health in the Salween*. Vientiane, Lao PDR: CGIAR Research Program on Water, Land and Ecosystems. Retrieved from: http://hdl.handle.net/10568/82969.

Khin Wine Phyu Phyu. (2016, July 7). Enforcement of fishing ban? It's complicated. *Myanmar Times*. Retrieved from: https://www.mmtimes.com/national-news/yangon/21261-enforcement-of-fishing-ban-it-s-complicated.html.

Khumsri, M., Sriputnibondh, N., & Thongpun, W. (2006). Fisheries co-management in Lower Songkhram River Basin: Problems and challenges. In *MRC Conference Series* (Vol. 6, pp. 121–126). Vientiane, Lao PDR: Mekong River Commission.

Marschke, M. (2015). *Life, fish and mangroves: Resource governance in coastal Cambodia*. Canada: University of Ottawa Press/Les Presses de l'Université d'Ottawa.

Ministry of Planning and Finance and World Bank. (2017). *An analysis of poverty in Myanmar (Part 2)*. Myanmar: World Bank. Retrieved from: http://documents.worldbank.org/curated/en/829581512375610375/An-analysis-of-poverty-in-Myanmar.

Mya Than Tun. (2001). *Marine fishes of Myanmar (Pelagic and Demersal)*. Yangon, Myanmar: Marine Fisheries Resources Survey Unit, Department of Fisheries.

Nelson, J.S. (1976). *Fishes of the world*. New York, NY: Wiley Inter Science.

Pratchett, M.S., Bay, L.K., Gehrke, P.C., Koehn, J.D., Osborne, K., Pressey, R.L., & Wachenfeld, D. (2011). Contribution of climate change to degradation and loss of critical fish habitats in Australian marine and freshwater environments. *Marine and Freshwater Research*, *62*(9), 1062–1081.

Soe, K.M. (2008). *Trends of development of Myanmar fisheries: With references to Japanese experiences*. No. 433. Institute of Developing Economies.

Swuam Pyaye Aye Aung, Hitke Htike, & Cauwenbergh, N. (2016, November 12). Evaluation of Mangrove Ecosystem Services in the Ayeyarwady Delta: A Step Towards Integrated Coastal Zone Management in Myanmar. Conference Paper for the International Conference on the

Mekong, Salween and Red Rivers: Sharing Knowledge and Perspectives Across Borders, Faculty of Political Science, Chulalongkorn University, Thailand. Retrieved from: https://static1.squarespace.com/static/575fb39762cd94c2d69dc556/t/5932f4b859cc6806cf1f5600/1496511675842/No.29_P700_712_Swuam+Pyaye+Aye+Aung.pdf.

Tint Hlaing. (1971). A classified list of fishes of Burma classification after Leo. S. Bery, 1949; Union of Burma. *Journal of Life Science*, *4*, 507–528.

Tsamenyi, M. (2011). *A review of Myanmar fisheries legislation, with particular reference to freshwater fisheries legislation.* Environmental sustainable food security programme. Rome, Italy: FAO.

World Fish. (2016). *Myanmar fisheries: Overview.* Retrieved from: http://pubs.iclarm.net/resource_centre/MFP-01-Overview.pdf.

Yin Yin Than. (2009). *Fish capture in relation to the types of fishing gear in some townships of Yangon Division.* (Zoology Department, Yangon University, Unpublished doctoral dissertation).

Chapter 14
The Impact of Land Cover Changes on Socio-economic Conditions in Bawlakhe District, Kayah State

Khin Sandar Aye and Khin Khin Htay

14.1 Introduction

In many regions of Myanmar, there is a close relationship between people and forests. A 2010 UN Food and Agricultural Organization (FAO) report found that 70% of Myanmar's total rural population, or approximately 30 million people, depend heavily on forests for their basic needs, and 500,000 people rely on forests for their employment (FAO 2010). However, according to the same study, between 1990 and 2010, a total of 7,445,000 ha, equivalent to 19.0% of Myanmar's total forest area, had been cleared. This gives Myanmar the seventh highest deforestation rate in the world (FAO 2010).

Forested land cover prevents soil depletion and erosion, sediment deposition in streams and rivers, and decline of biodiversity (Cunningham/Cunningham 2006). As such, changes in land cover have significance at global, regional and local levels (Lambin et al. 2001; Turner et al. 1990). Often a combination of economic, institutional, and political factors drive deforestation, including logging, agricultural expansion, infrastructure expansion, shifting cultivation and the extraction of non-timber forest products (NTFPs) and fuel wood (EIA 2015; Geist/Lambin 2001). In Myanmar, timber is an important export that occurs through both legal and illegal channels, especially hard woods such as teak. One recent study found that Myanmar exports 1.6 million tonnes of teak annually to neighboring countries such as India, China, Bangladesh, Thailand and Malaysia (EIA 2015).

Dr. Khin Sandar Aye, Professor, Geography Department, Loikaw University; Email: akhinsandar50@ gmail.com.

Dr. Khin Khin Htay, Researcher, Geography Department, Loikaw University; Email: Kkhinhtay.geog@gmail.com.

C. Middleton and V. Lamb (eds.), *Knowing the Salween River: Resource Politics of a Contested Transboundary River*, The Anthropocene: Politik—Economics—Society—Science 27, https://doi.org/10.1007/978-3-319-77440-4_14

This chapter investigates land cover change, forest depletion, environmental degradation and ecosystem damage as it affects the livelihoods of local residents in Bawlakhe district, Kayah state. It uses both quantitative and qualitative tools in field data collection, and remote sensing (RS) and global information system (GIS) mapping tools to assess the social, economic and land cover changes in the study area (Bryman 2001). Bawlakhe district is located within the Thanlwin River Basin in Kayah state in Eastern Myanmar. The Thanlwin River flows north to south through Bawlakhe district, and the proposed Ywathit Dam is also located in the district in Ywathit sub-township (Middleton et al., Chap. 3, this volume). Bawlakhe district is composed of three townships (Bawlakhe, Hpasaung and Mese), 20 village tracts, and a total of 86 villages.[1] This biodiverse area is home to a number of ethnic groups, including Kayah, Yintale and Shan, who depend on the watershed for food, water, security, fuel and income (Hla Tun Aung 2003). Within the study area, the majority of land is forested and the main source of income is shifting cultivation and employment in forest extraction and mining activities.

The research presented in this chapter finds that land cover conditions in this area have changed significantly since 1995, and especially after 2010. The dramatic decline of forest cover in this area has been caused by legal and illegal extraction of timber, in particular Teak (*Tectona grandis*) and Pyinkado (*Xylia xylocarpa*), together with over-cutting of fuel wood for domestic use and sale, expansion of shifting cultivation, and mining of lead. In this chapter, I identify three key time periods in which land cover, livelihoods, and governance changed significantly. The first period, prior to 2010, was characterized by armed conflict in Bawlakhe district. During this time, logging companies could not easily access the area due to risk of attack and land mines, and loss of forest cover was relatively limited. From 2010 to 2015, after a peace agreement was signed between local armed groups and the Union government, land cover changed dramatically as security improved and control of the territory was negotiated between the Myanmar military and armed groups. This allowed timber companies to log the area, rapidly depleting forest resources. Since 2016, the National League for Democracy (NLD) government has banned logging in the area.

While recent studies (EIA 2015; FAO 2010) have discussed some of the socio-economic factors that contribute to land use and land cover change, none have discussed Kayah state in detail, and to our knowledge this is the first research to report on the socio-economic impacts of land cover changes in Bawlakhe district. It is hoped that this research will provide local and regional decision-makers with accurate information to understand the implications of land use change, improve land use policy, and implement effective plans for regional sustainable development.

---

[1]Bawlakhe district is bounded to the east by Thailand, to the west by Kayin State, to the south by Thailand and Kayin State, and to the north by Loikaw district of Kayah state.

14.2 Methodology

Qualitative and quantitative interviews were conducted in four villages in Bawlakhe district. The villages were selected on the basis of their relative distance from the Thanlwin River and their current forest cover density (Fig. 14.1: Location of four villages studied in Bawlakhe district). Bhukhu and Hose villages are both located far from the Thanlwin River. Bhukhu village is less densely forested than Hose Village. Wanpla and Wanaung villages are near the Thanlwin River. Forest cover in Wanaung village is less dense than in Wanpla village.

Field surveys were conducted from September 2015 to November 2016. Our research group was composed of seven people: two lead researchers and five research assistants. A questionnaire was administered in the four villages with 180 households (Table 14.1: Summary of the number of interviews per village). Respondents were selected on the basis of their experience in the area (5 years, 10 years, and 15 years) (Table 14.2: Summary of the number of interviews per village, by period of experience). Furthermore, only respondents over the age of 18 were

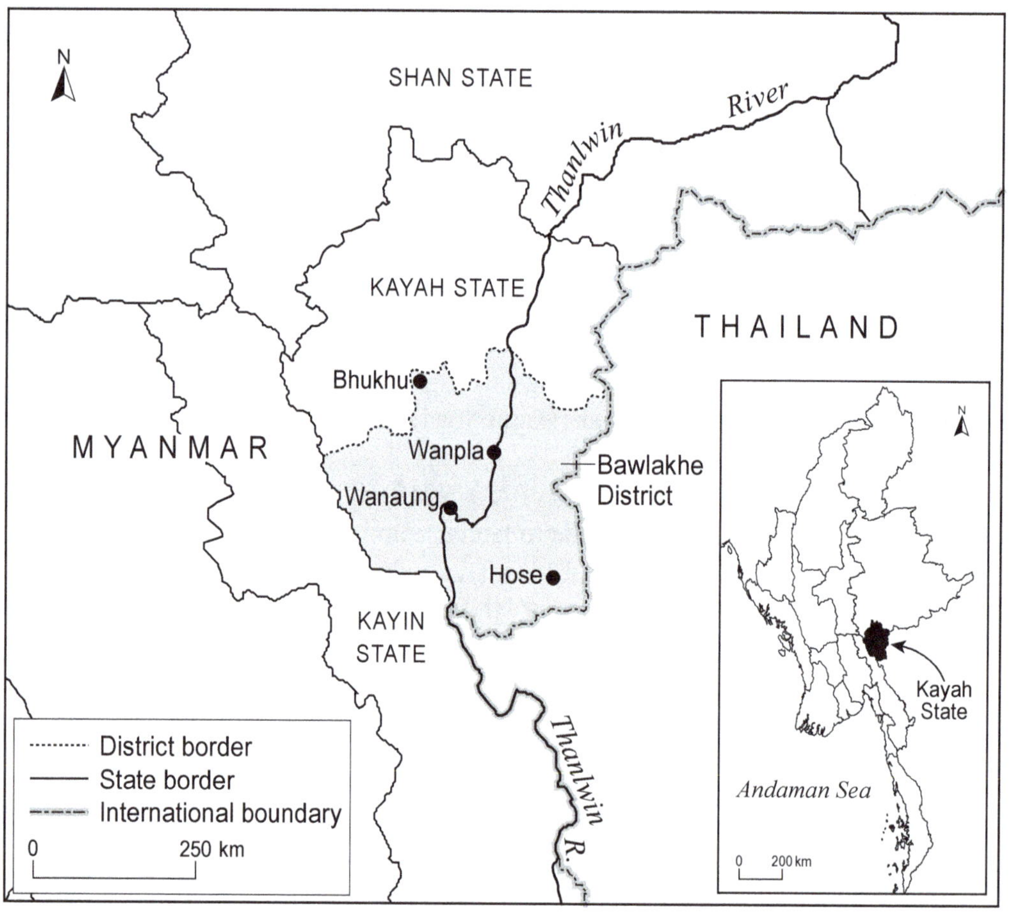

Fig. 14.1 Location of four villages studied in Bawlakhe district. *Source* Cartography by Chandra Jayasuriya, University of Melbourne, with permission

Table 14.1 Summary of the number of interviews per village

Village name	Questionnaires			Individual interviews (HH)		
	Male	Female	Total	Male	Female	Total
Wanaung	20	50	70	5	30	35
Wanpla	5	15	20	4	11	15
Bhukhu	3	12	15	2	8	10
Hose	33	42	75	18	12	30
Total	61	119	180	29	61	90

Source The authors' field survey

Table 14.2 Summary of the number of interviews per village, by period of experience

Village name	5 years experience		10 years experience		15 years experience		Total	
	M	F	M	F	M	F	M	F
Wanaung	18	56	4	10	3	14	25	80
Wanpla	3	18	3	7	3	1	9	26
Bhukhu	3	10	1	8	1	2	5	20
Hose	28	45	14	5	9	4	51	54
Total	52	129	22	30	16	21	90	180

Source The authors' field survey

interviewed, and gender balance was considered to ensure representation of both males and females. Sample selection by household ensured that 30% of all households in each village were represented (Table 14.3: Summary of village demographics and main livelihood activities). An additional 90 in-depth interviews were conducted with community members, government officials, and non-government organization staff (Table 14.1: Summary of the number of interviews per village). Finally, two or three focus group discussions were conducted per village. Each focus group discussion was composed of seven participants, and included village members who were aware of the history and patterns of settlement in the village. When the respondents could not speak Burmese, an assistant translated from Burmese to local ethnic languages.

Table 14.3 Summary of village demographics and main livelihood activities

Village name	Population	Households	Ethnicity	Main livelihoods
Wanaung	493	93	Shan/Yintale	Logging, shifting cultivation and livestock
Wanpla	268	87	Shan	Logging, shifting cultivation
Bhukhu	266	49	Kayah	Logging, shifting cultivation
Hose	1096	265	Kayah, Shan, Bamar, Kayin	Logging, shifting cultivation

Source The authors' field survey

The use of remote sensing and GIS technology can evaluate existing forest vegetation, land use and environmental conditions (Lillesand/Kiefer 2002). Many studies have been carried out using aerial photographs and Landsat TM remote sensing satellite (IRS-1A & 1B) data (Chaung 2006; Campbell 1996; Kushwaha 1990; Pant/Kharkwal 1995; Roy et al. 1993). Land cover changes in Bawlakhe district were assessed using Landsat TM images from 1995, 2005, 2010 and 2015. The approach used in this study to classify satellite images and change detection was based on satellite images from Landsat 7 ETM ± 1995, Landsat 7 ETM 2005 and 2010, and Landsat 8 ETM (2015) using a supervised classification method of the study area. Land cover classification was determined from satellite images using remote sensing software (Environmental Visualization Images (ENVI) version 5.0) and vector visualization using ArcGIS 10.2 software. Forest classification was defined according to Hla Tun Aung (2003). The classifications were triangulated with qualitative data gathered from interviews in the four villages and the researchers' observations. In this analysis, the use of images was the same, and the images were taken within two months of each other, which led to slight seasonal variation. Taking this into account, sub-tropical wet hill forest areas occur more frequently than other forest types in Bawlakhe district for 2015.

14.3 Land Cover Change and Livelihoods in Bawlakhe District

There are seven types of land cover in Bawlakhe district: (1) Tropical wet evergreen forest; (2) Sub-tropical wet hill forest; (3) Sub-tropical hill savanna forest; (4) Paddy land; (5) Water bodies; (6) Fallow land; and (7) Settlements (Hla Tun Aung 2003). According to Myanmar's legal classification of forest types, in Bawlakhe district 65% of the total area is considered forestland, and of this 7% is "reserved forest" and over 70% is "unreserved forest."

The majority of land cover in Bawlakhe district is forest land. The 3,000 feet contour is an important determinant of forest type in Myanmar. Hill forests are found above 3,000 feet, while evergreen and mixed forests occur below that altitude. However, due to human activities including logging, shifting cultivation, and extraction of NTFPs and fuel wood, the type of forest cover has changed in Bawlakhe district.[2] In general, forestland has become degraded and significant parts of the area have less vegetation cover than previously.

[2]Kayah State is also rich in minerals, especially tin, wolfram (tungsten) and antimony. The most important mine is the Mawchi mine, which produces tin and wolfram. This area lies in the Bawlakhe area and it is one of the causes of forest decline. British companies were the main extractors of minerals in the early days in the Mawchi area. In 1962, the government nationalized mining and took over the operation of the country's main mines on a joint-venture basis. The mining industry exists now as a joint venture. 150 tons of lead is extracted annually in the Bawlakhe area.

Before detailing the forest transitions that have occurred in Bawlakhe district, I briefly detail the types of forest. Tropical wet evergreen forest occupies the central part of Kayah State. Big, tall, isolated evergreen trees form the main canopy, which may be less continuous than in other areas. Single giant dipterocarps or other trees may stand above the level of the canopy. Middle and lower levels are characterized by dense green vegetation. Bamboos, dense masses of climbers, and canes occur abundantly. In 1995, this type of forest was found in the lower sub-tropical hill savanna forest areas. In 2015, these areas especially occupy the central part of the area.

Sub-tropical hill savanna forest is found in the western, northern, and eastern parts of Bawlakhe district, and in the north-central hills. These forest types are rich in biodiversity and valuable timber. Forest products found in this type of forest include teak and other hardwoods such as Pyinkado, Padauk and Ingyin, along with bat manure, lac, cutch, resin and honey. Pines are also scattered in sub-tropical hill savannah forest areas. In 1995, sub-tropical hill savanna forest covered about one third of Bawlakhe district, and this forest type area was found in the eastern and southeastern parts of the district in 2015. Big, tall, isolated evergreen green trees form the main canopy which may be less continuous than other areas. Single giant Dipterocarps or other trees may stand up above the level of the canopy. Middle and lower levels are green and dense. Bamboos, dense masses of climbers, and canes occur abundantly.

Sub-tropical wet hill forest is found in the central foothill areas. Several species of bamboo are found in sub-tropical wet hill forest areas, and woody climbers, root climbers and epiphytes are abundant. Forest products found in these areas include teak and other hardwoods such as Pyinkado, Padauk and Ingyin, as well as lac, cutch, resin and honey.

Within the study area, political conditions have had a direct impact on land cover change. The following sections detail land cover change during the period of armed conflict before 2010 and the five-year period following peace agreements (2010–2015), followed by a brief discussion of the most recent period after the establishment of the new government in Myanmar (2016–present). Each section details land cover conditions, then provides an analysis of the local political situation leading to the socio-economic conditions that have contributed to land use change.

14.3.1 Period of Armed Conflict (Before 2010)

From 1995 to 2005, sub-tropical hill savanna forest, tropical wet evergreen forest and sub-tropical wet hill forest were found in the western, northern and eastern parts of Bawlakhe district. Respondents who lived in these areas for a period of between 10 and 15 years described the previous land cover type as "forestland." According to data on land cover change and field surveys, there are both similarities and differences between the villages. Over the decade between 1995 and 2005, sub-tropical hill savanna forest and sub-tropical wet hill forest cover declined, with tropical

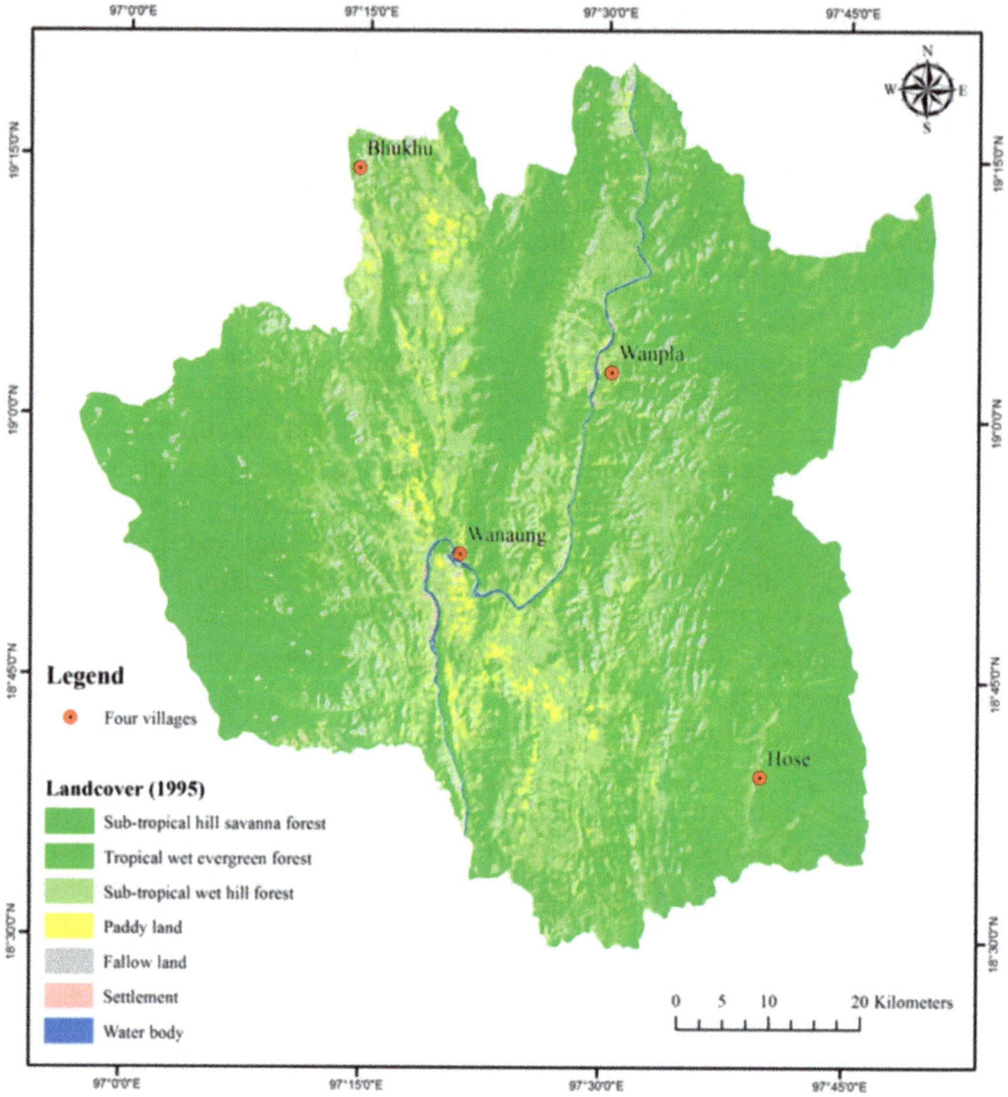

Fig. 14.2 Land cover conditions in Bawlakhe district, 1995. *Source* Landsat 7 ETM ± 1995

evergreen forest becoming more prevalent (Fig. 14.2: Land cover conditions in Bawlakhe district, 1995; Fig. 14.3: Land cover conditions in Bawlakhe district, 2005; Table 14.4: Land cover conditions in Bawlakhe district in 1995 and 2005).

Prior to 2010, Bawlakhe district had a wealth of forest products, and commercial forest production was moderate. Local people could provide for their livelihoods through the collection of fuel wood, pole wood, and NTFPs such as honey, orchids, and medicinal plants. Subsistence agriculture was also important, with most of the farming done by shifting cultivation. After clearing a forest area, the local people would use a plot of land for three to four years before clearing a new forest area and leaving the old field fallow (Bryant 1994). The main source of fuel for cooking was firewood, namely Pyinkado (*Xylia xylocarpa*). When a tree was felled, villagers

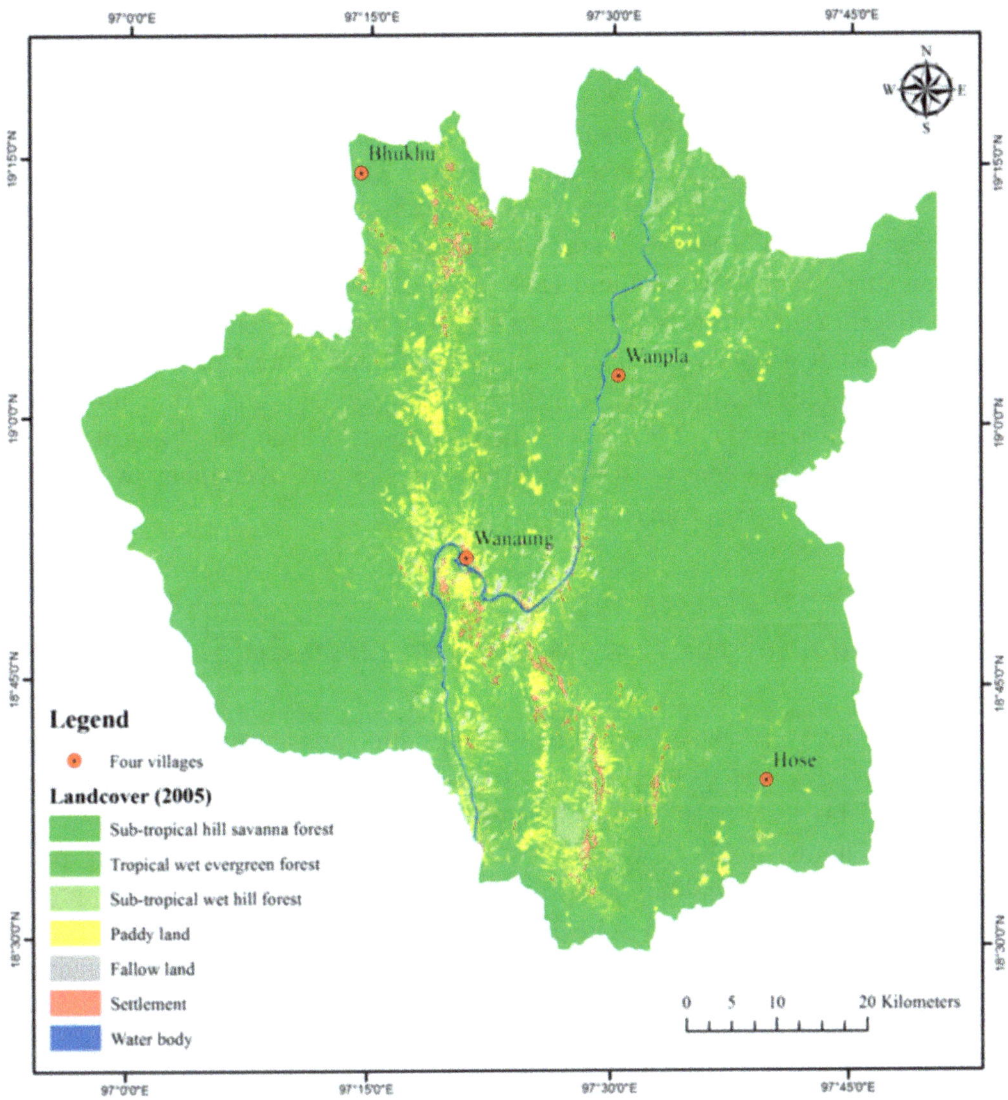

Fig. 14.3 Land cover conditions in Bawlakhe district. *Source* Landsat 7 ETM ± 2005

Table 14.4 Land cover conditions in Bawlakhe district in 1995 and 2005

Land cover types	1995 (km^2)	2005 (km^2)
Sub-tropical hill savanna forest	3066.9	639.3
Tropical wet evergreen forest	740.0	3847.8
Sub-tropical wet hill forest	940.8	251.2
Paddy land	212.8	199.7
Fallow land	162.4	149.7
Water bodies	22.3	20.8
Settlements	20.9	57.4

Source Based on Figs. 14.2 and 14.3

would remove the branches and cut the trunk, which could be up to two feet in diameter, into pieces to carry back to the village for firewood (Individual interviews, 2015 and 2016).

Prior to 2010, armed conflict was frequent in Kayah state. Armed groups would extract timber and trade it across the border to Thailand. However, conflict in Kayah state and the threat of attacks and landmines made it dangerous for larger timber companies to access the area. Kayah and Shan ethnic armed groups were dominant in Wanaung, Wanpla and Hose villages. During this period of conflict, many villagers relocated to safer areas within the region. Therefore, while armed groups and local people were able to access and extract forest products from the area, very few outside companies engaged in commercial wood extraction during this time. Starting in May 1997, five armed groups in Kayah state[3] began negotiating a cease-fire agreement with the Myanmar Union Government, which was signed in 2010 (Hla Tun Aung 2003).

14.3.2 Following the Peace Agreement (2010–2015)

The past ten years has seen an increase in deforestation and a decrease in the quality of timber in Bawlakhe district. This has changed the lives of villagers in each community. This section details the changes that have occurred in each village, and the socio-economic implications of those changes. After 2010, when the peace agreement between the military and armed groups was signed, security improved that allowed both private and state logging companies access to forests. Forest cover in Bawlakhe district dramatically changed as logging became more intensive. As more armed groups entered into ceasefire agreements, the area where there was no longer violent conflict extended. Reserved forest, nursery gardens and sawmills were established in government-held areas. There are now two sawmills: one in Bawlakhe township and one in Hpasaung township.

Between 2010 and 2015, the most significant land cover changes occurred in the study area, with an overall decline in forest cover (Fig. 14.4: Land cover conditions in Bawlakhe district, 2010; Fig. 14.5: Land cover conditions in Bawlakhe district, 2015; Table 14.5: Land cover conditions in Bawlakhe district in 2010 and 2015). By 2015, sub-tropical wet hill forest was more prevalent than other forest types in Bawlakhe district. According to our questionnaires, people who had lived in the study area less than five years had not witnessed much land cover change; however, people who had lived in the area longer than five years explained that the overall diversity and quantity of flora and fauna has decreased. Additionally, villagers who had lived in the study area longer than five years noted changes to the watershed, with some streams now dry, even during the rainy season. Some forestland has also

[3]Kayan National Defense Guards (KNG), Kayinni National Liberation Front (KNLF), Kayan Pyithit Party (KPP), and Kayinni National Progressive Party (KNPP).

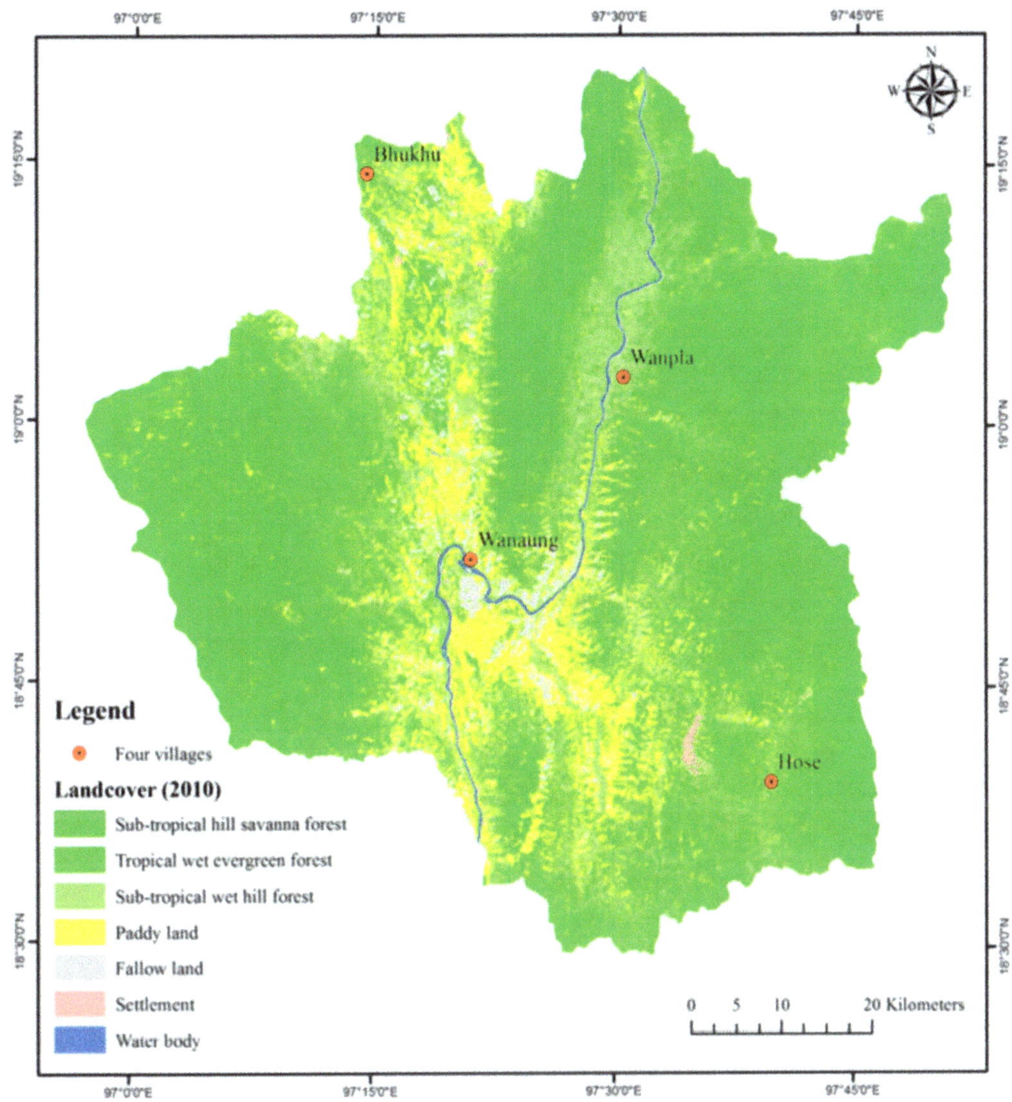

Fig. 14.4 Land cover conditions in Bawlakhe district, 2010. *Source* Landsat 7 ETM (2010)

changed to settlement land or cultivated land. The changes occurred gradually, with the forestland decreasing daily. Remotely sensed data shows that significant changes have taken place in the Bawlakhe district, with more degraded forest and settlement areas now apparent (Table 14.5: Land cover conditions in Bawlakhe district in 2010 and 2015).

There are both similarities and differences between the villages, including particular livelihoods and migration patterns. Before 2010, people who lived in this area depended on a modest amount of forest production and collecting NTFPs for their livelihoods. After 2010, when private and state logging companies began accessing the forests, many people from other places, especially central Myanmar, came to work in the area. Between 2010 and 2015, both local people and migrants worked in forest production for their main livelihood. A 2012 study shows that

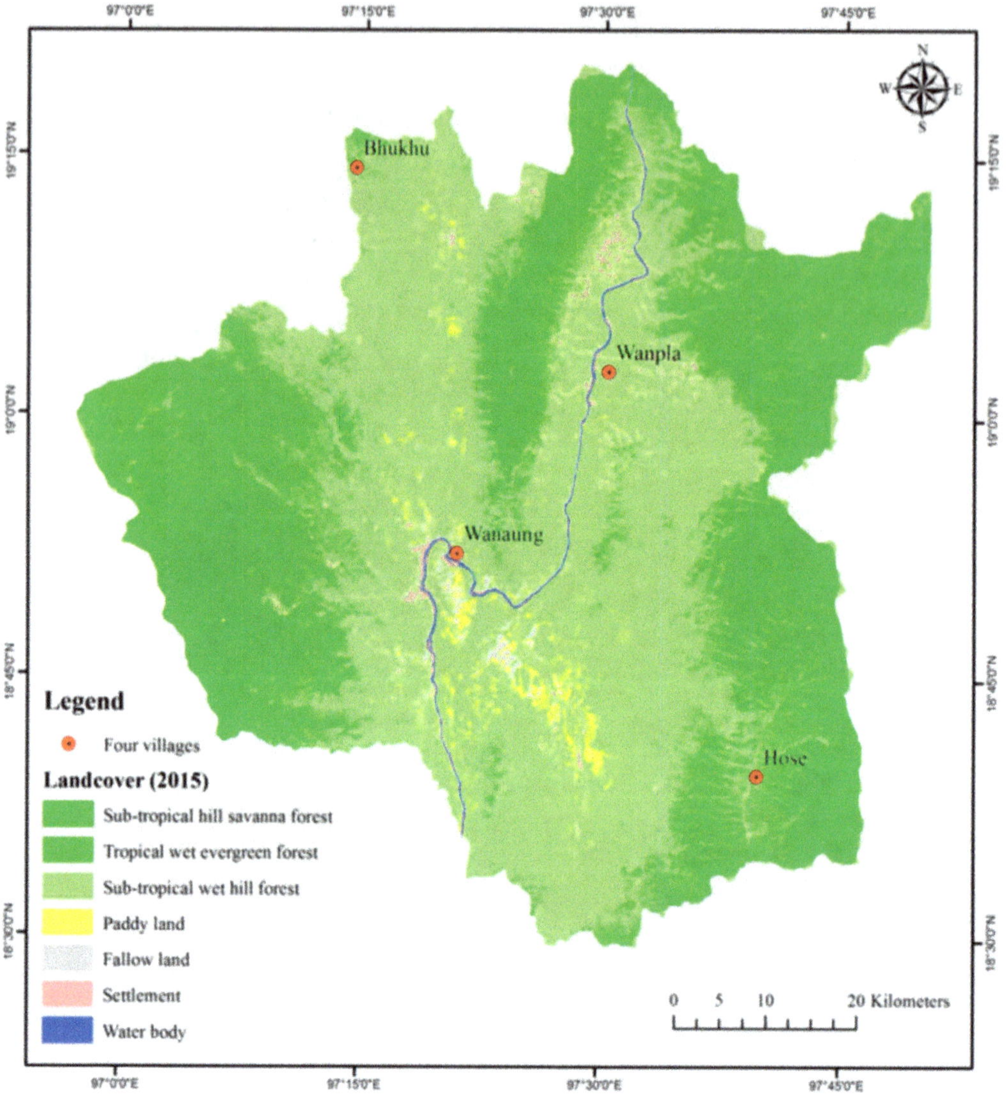

Fig. 14.5 Land cover conditions in Bawlakhe district, 2015. *Source* Landsat 8 ETM (2015)

Table 14.5 Land cover conditions in Bawlakhe district in 2010 and 2015

Land cover types	2010 (km^2)	2015 (km^2)
Sub-tropical hill savanna forest	501.9	444.5
Tropical wet evergreen forest	2312.6	1596.3
Sub-tropical wet hill forest	1706.0	2795.9
Paddy land	459.0	100.0
Fallow land	94.0	89.4
Water bodies	79.8	18.4
Settlements	13.2	121.5

Source Based on Figs. 14.4 and 14.5

during this time, 20,000 tons of rattan cane, 16,030 visses of bat manure,[4] 1,596 tons of charcoal, 310 tons of firewood, 345 visses of bamboo and 70 visses of honey were produced annually in Bawlakhe district (Politno 2012). According to our interviews, during this period all of the men worked in forest extraction, while women worked in housekeeping and cultivated vegetables, shifting agriculture, and paddy rice for household consumption.

The peace agreements had significant impacts on all aspects of peoples' livelihoods. Outsiders also migrated into the area in search of job opportunities, fueled by the border trade with Thailand, which increased the population. As markets in Thailand and other parts of Myanmar became accessible, reflecting evolving policies of the neighboring countries as well as improved infrastructure, occupations shifted to forest production, crop cultivation, the sale of charcoal, and animal husbandry (cows and goats) for export to Thailand. Most villagers interviewed had around two acres of cultivated land, and more families began to cultivate land as did new arrivals to the area. Residents in three of the four villages – Wanaung, Wanpla, and Bhukhu – also began growing sesame for sale to Hpasaung township and Thailand. In addition to increased forest extraction and changes in the agricultural sector, infrastructure development and road networks also contributed to land use and land cover change.

In 2012, the Union Solidarity and Development Party (USDP) government began to implement a "Greening Project" across the country. The Ministry of Agriculture, Livestock, and Irrigation and the Department of Livestock, Fisheries and Rural Development worked with the local government to implement the Greening Project. Through this project, the government provided loans to local people for agriculture and animal husbandry, and for building new houses. The aim of the project was to raise the standard of living in rural upland regions to a level more equitable with urban areas. The government cooperated with local organizations to raise funds for poverty alleviation. Through this project, the state and local government built housing, schools and hospitals to fulfill the needs of community members. The government's rural development plan also included "Rural Road Development," which entailed the construction of roads 30 feet in width directly connecting each and every village with nearby highways or motorways (Politno 2012). This government policy played a major role in the development of the town and villages in Bawlakhe district.

Changing political conditions after the 2010 peace agreements, along with the Greening Project, increased socio-economic opportunities for villagers in forest production and growing cash crops. However, these political conditions did not result in universal land use changes. For example, after 2010, villagers from Bhukhu village moved from the uplands to either the eastern lowland foothills or further away to Hpasaung township to facilitate easier access to markets. In the foothills, transportation and communication improved because there was no longer active conflict, and later the Greening Project brought some infrastructure

[4]One viss is approximately 1.63 kilograms (3.6 pounds).

development. However, some people were not able to support their livelihoods because their skills were mainly related to gathering forest products and shifting cultivation, which they could not practice in the lowland areas. According to one respondent from Bhukhu village, "some people returned to the old area because there was forest land and shifting cultivation land still available" (Bhukhu village interview, 23 October 2015).

In general, according to our interviews, additional work in timber extraction and forest production increased people's incomes in the four villages. Local people could now afford modern amenities like televisions, solar panels for electricity, water pumps, tractors, and motorbikes. Some people still remained jobless, however. Some young people, especially young adult men, went to Thailand to work.

Not all people welcomed the socio-economic changes brought about by the peace agreements and broader transformations in Myanmar, however. According to interviews, for some people the logging area was too far away and the work too difficult, that they did not want to move to another village. As one respondent stated, "We don't have modern facilities, but we don't want to move to another area" (Wanpla village interview, 2 December 2015). The enticement of money and modern amenities was not a sufficient reason for everyone to want to move. As one respondent explained "We are not educated and rich, but we are happy" (Wanaung village interview, 3 March 2015).

Overall, during the 2010–2015 period, the political changes that brought peace to the area also expanded border trade and increased opportunities for cultivating cash crops, and, most significantly, increased forest extraction by large private companies. Forest cover dramatically decreased and many forest products are now much harder to access or are gone. By 2015, forest cover had been degraded and depleted, and most of the logging companies moved away. Therefore, by the end of this period, some local people began moving to other areas of Myanmar and to neighboring countries to seek work.

14.3.3 The NLD Government Period (2016–Present)

After the 2015 election, the new NLD government came into power on April 1st, 2016 and implemented a one-year logging ban throughout the country (Trautwein 2016). According to 2016 records, the Forest Department established two state-owned forests and seven community forests in Bawlakhe district. Meanwhile, near the end of 2015, and as a part of the Greening Project, Hose village and local armed groups accepted the government policy to start replanting teak and hardwood.

As a result of the Greening Project, some people received new houses, while others moved to Hpasaung township in Bawlakhe district. In Wanaung village, the Greening Project has extended the settlement area to the western bank of the Thanlwin River. At the same time, Shan residents live on the eastern bank of the Thanlwin River, and an armed group is also based there. Before 2015, the economy of the study area was based on timber extraction, agriculture, animal husbandry, and mining. Timber extraction has moved outside of Bawlakhe district, limiting forest production in the area. However, charcoal production still provides local people with income. At present, limitations on forest production are leading to livelihood problems for local residents. According to questionnaires and individual interviews, over 80% of respondents described the forest area as far from the village and indicated that the collection of forest products has decreased. As the main economy of the area has long depended on the forest and forest products, many villagers are now jobless and some have moved to other parts of Myanmar and overseas to seek job opportunities. Some of the families who have lived in this district for more than five years have moved to other parts of Kayah and Shan states for their livelihoods because the local economy is dependent on forest production.

However, since 2015, some villagers in the district have started to raise more livestock and have changed their crops and farming methods. The agricultural land use changes that began in the 2010–2015 period have continued, with people extending their agricultural land to grow cash crops such as sesame and to raise livestock (goats and cows) for export to Thailand illegally. Hose village is located near the Thai-Myanmar border, and villagers began to grow sesame and garlic in 2014, and cardamom (*Elettaria cardamomum*) as a commercial cash crop in 2016. In 2017, the other three villages also began growing cardamom for sale mostly to Chinese markets.

14.4 Implications of Land Cover and Livelihoods Changes

Both our quantitative RS and GIS data, and observations gathered by interviews and group discussions, reveal that forestland has declined in the area (Table 14.6: Changes in land cover in Bawlakhe district from 1995 to 2015). Forcst products can no longer be collected near the villages because of the heavy logging that occurred, along with the expansion of settlement areas, grazing land, and agriculture. When timber companies arrived in 2010, many local people sought employment in logging, but the lack of a forest conservation and management plan led to significant forest depletion, which inevitably impacted the villagers' livelihoods and their broader relationship with the forest. Because the main source of income in this area

Table 14.6 Changes in land cover in Bawlakhe district from 1995 to 2015

Land cover types	1995 (km^2)	2015 (km^2)	Decrease (km^2)	Increased (km^2)
Sub-tropical hill savanna forest	3066.9	444.5	2622.4	–
Tropical wet evergreen forest	740	1596.3	–	856.3
Sub-tropical wet hill forest	940.8	2795.9	–	1855.1
Paddy land	212.8	100	112.8	–
Fallow land	162.4	89.4	73.0	–
Water bodies	22.3	18.4	3.9	–
Settlements	20.9	121.5	–	100.6

Source Data from Tables 14.4 and 14.5

depends on forest production, decreased forest cover has forced people to change their livelihoods. Some people have moved to other regions of Myanmar, while some people have migrated to neighboring countries.

Political changes in the country have been the greatest contributing factor to land cover change. The status of ceasefire agreements and location of conflicts have influenced when and where government agencies and non-state businesses are able to access forest resources for logging. Prior to 2010, most of the shifts in type of forest cover and land use changes occurred at a smaller scale because of conflicts. After 2010, however, most of the land use changes occurred from logging and resulted in the depletion of forest cover. Currently, depletion of forestland has reduced the availability of resources available for logging, and national policy discourages timber extraction. However, it is too soon to tell whether this will lead to a recovery of forest cover. These political changes, along with the sudden increase then decrease in forest production, has significantly impacted the livelihoods of local people, as they rely on forest products both for their own use and for economic opportunities through jobs in the forest sector.

In Myanmar, where populations are widely dispersed and infrastructural challenges limit the free flow of information, there is a great need to enhance awareness amongst national policy makers about local needs and aspirations. This was a key motivation for undertaking our field research. Before 2010, Kayah state lacked a clear plan for forest conservation and management. Thus, today, forest depletion is a key challenge, with strong impacts on community livelihoods. The remainder of this chapter will focus on the implications of our research for forest protection, land governance, and policy.

14.5 Implications for Forest Protection and Local Livelihoods

The economy of Bawlakhe district depends on forest production, and agriculture is mainly subsistence-based. Our RS data and surveys show that significant changes have taken place in the Bawlakhe district, in particular forest degradation and expanding settlements. The most remarkable changes in land use classes occurred from 2005 to 2015. Spatially, most land use change occurred in the central part of the study area due to easy access to commercial areas and increased settlement following road improvement and other infrastructure development. The decrease in forest cover has impacted water sources, and increased agricultural land cover and settlements have also contributed to ecosystem changes. Limited availability of wild animals and plants impacts local people who are dependent upon animals and plants for their livelihood. The lack of a plan for natural resource conservation and utilization has led to unsustainable land use changes, which gives rise to socio-economic and environment problems that are inadequately addressed by policy responses.

14.6 Changing Land Cover Governance

Deforestation is driven by regional economic, institutional, and policy factors which have led to agricultural expansion, logging, and infrastructure development (Angelson/Kaimowitz 1999; Lambin et al. 2001). Our research found that agricultural expansion and shifting cultivation contributed to forest depletion in all four villages. Local people rely on the forest for building materials, medicinal plants, food, orchids, fuel wood, and hunting. They also collect these products from forest areas and sell them for their livelihoods.

Timber extraction, particularly the production and export of teak wood, is the major economic activity in Kayah state, and the main cause of forest degradation is legal and illegal timber extraction by large private companies and the government. However, local people in this area also depend heavily on timber extraction for their incomes.

In January 2016, the USDP government passed a new National Land Use Policy that aims to systematically manage land use and tenure rights in the country, including both urban and rural areas. The objectives of the policy include harmonization and implementation of existing laws related to land. The policy also provides guidance to relevant departments and organizations on issues related to land use and tenure rights. While this policy has significant implications for the study area, at the time of the field research it had not yet been implemented and therefore its on-the-ground impact cannot be assessed. Furthermore, there is now a growing recognition amongst the government that a National Land Law is required to implement the National Land Use Policy (Pyae Thet Po 2018). This is particularly the case given a recent amendment to the Vacant, Fallow and Virgin Land Management Law in 2018 that civil society has flagged could exclude communities from customary land tenure rights (Gelbort 2018). There is also a recognition that historical and ongoing land grabbing in Myanmar is an urgent issue that must be resolved (Scurrah et al. 2015).

14.6.1 Future Policy Implications

Forest cover decline, land use changes, and the socio-economic conditions of local people in Bawlahke district are linked to political changes in the region and the country. Changes in forest cover have had a direct impact on people's livelihoods, and there is a great need to enhance policy makers' awareness of local needs.

The majority of deforestation and forest degradation has occurred from legal and illegal timber extraction by large private companies and the government, which employ both migrants and local people. A key conclusion of this chapter is that there is a large gap in knowledge about environmental conservation on the part of companies, local governments, and communities. Land use decisions need to fully consider the implications of deforestation on eco-system health, local livelihoods, and the sustainability of future timber extraction activities.

In addition to providing income through timber extraction, local people rely heavily on the forest for food, water, medicine, and fuel wood. Government conservation efforts should include the establishment of more community forests in the area. This would promote sustainable forestry management within local communities and allow continued access to the NTFPs that local people rely on for their livelihoods.

There is also limited access to formal education available to people in the study area. Nearly 95% of the people interviewed had no formal education, and the remainder had only primary school education. Government conservation policies should therefore include capacity building in communities, including access to formal education.

New approaches to regional development should also be considered. For example, in 2017, Mae Hong Song Province in Thailand signed a Memorandum of Understanding with Kayah state to increase tourism along the border area. Mae Hong Song and Loikaw have now been declared sister cities. The beautiful mountain scenery has high tourist potential, with good income prospects for local communities. Sparse forest settlement patterns along the Thai-Myanmar border create potential for eco-tourism, which could lead to new job opportunities.

Lastly, as the NLD party continues to implement the Land Use Policy (2016) passed by the previous government, the use of remote sensing techniques in combination with ground surveys is recommended for better planning and sustainable management of land cover area in Bawlakhe district. Remote sensing allows for monitoring of land cover changes, while ground surveys help explain the causes of land use cover change and understand the local context more thoroughly. Improved data collection methods could lead to greater clarity around socio-economic and environmental changes in the area. For future development, it is important to systematically evaluate previous patterns and dynamics of land use. Land use change based on time series analysis will provide insight into sustainable resource utilization and conservation.

14.7 Conclusion

Humankind must use the environment sustainably in order to survive. Without conservation practices, long-term environmental degradation sets in, threatening human survival (Wilson/Bryant 1997). For sustainable land use, access to reliable data and means to interpret and act upon it are vital (Anderson et al. 1996). Projections for future land-cover patterns are also needed to evaluate the implications of human action on the future of ecosystems (Turner et al. 1995).

This chapter is one of the first studies to assess in detail how land cover changes relate to the socio-economic conditions of the population in Bawlakhe district. The causes of forest degradation are many and varied. In using digitally processed

Landsat Imagery to compare each location over time, we could assess land cover changes in Bawlakhe district. What we found was that between 1995 and 2015, there was a significant decrease in forest cover. The field surveys and interviews added to our understanding of these changes by illustrating some of the socio-economic conditions that contributed to this land cover change.

A key finding of this study is that forest depletion and changes in land utilization have caused changes in the local economy. Sub-tropical hill savanna forest in particular, which is rich in species and valuable timber, has decreased in the area. As forest extraction increased to meet the demands for cash income, depleted forest resources have limited the availability of forest products. In the study area, legal and illegal extraction of timber, over cutting of fuel wood, and extension of agriculture land has already decreased forest cover area. Timber extraction has now shifted to other forested areas, and fuel wood must now be collected far from the village. Some people have moved to other areas, including neighboring countries, some have begun to cultivate different crops, and some are raising animals for export. Within the four villages studies, over 40% of the residents have changed their livelihoods from forest production to other economic activities.

A final key finding of this study is that forest depletion and changes in land use take place when political circumstances change and villagers seek to change their socio-economic status. There is a lack of planning for natural resource conservation and utilization on the part of all actors. The government must increase efforts to raise awareness of environmental conservation and land use development policy in the area, while supporting the development needs of the local people. For example, the government should promote community-based natural resource management in villages. Policies and guidelines for sustainable development that guide local and regional decision makers should be informed with accurate information to understand the impacts of land use change. Ultimately, policies that discourage deforestation and forest degradation and support local livelihoods are needed in Bawlahke district.

References

Anderson, L.E. (1996). The causes of deforestation in Brazilian Amazon. *Journal of Environment and Development, 5,* 309–328.

Angelsen, A., & Kaimowitz, D. (1999). Rethinking the causes of deforestation: Lessons from economic models. *The World Bank Research Observer, 14,* 73–98.

Bryant, R.L. (1994). Shifting the cultivator: The politics of teak regeneration in colonial Burma. *Modern Asian Studies, 28*(2), 225–250.

Bryman, A. (2001). *Social research methods*. Oxford, United Kingdom: Oxford University Press.

Campbell, J.B. (1996). *Introduction to remote sensing*. New York, NY: Guilford Publications.

Center for Diversity and National Harmony. (2015). *The State of harmony in Kayah State.* Yangon, Myanmar. Retrieved from: http://www.cdnh.org/publication/data/the-state-of-social-harmony-in-kayah-state-quantitative-data/.

Chaung, K.T. (2006). *Introduction to geographic information systems*. New York, NY: McGraw-Hill.

Cunningham, W., & Cunningham, M. (2006). *Principles of environmental science: Inquiry and application* (3rd ed.). New York, NY: McGraw-Hill Companies.

EIA (Environmental Investigation Agency). (2015). *Organized chaos: The illicit overland timber trade between Myanmar and China*. London, United Kingdom. Retrieved from: https://eia-international.org/report/organised-chaos-the-illicit-overland-timber-trade-between-myanmar-and-china/.

FAO (Food and Agriculture Organization). (2010). *Global forest resources assessment 2010: Main report*. FAO forestry paper (Vol. 163). Rome, Italy: FAO. Retrieved from: http://www.fao.org/docrep/013/i1757e/i1757e.pdf.

Geist, H.J., & Lambin, E.F. (2001). What drives tropical deforestation? A meta-analysis of proximate and underlying causes of deforestation based on subnational scale case study evidence. *LUCC Report Series No. 4*. Louvainla-Neuve, Belgium: University of Louvain.

Gelbort, J. (2018, December 10). Implementation of Burma's Vacant, Fallow and Virgin Land Management Law: At Odds with the Nationwide Ceasefire Agreement and Peace Negotiations. *Transnational Institute*. Retrieved from: https://www.tni.org/en/article/implementation-of-burmas-vacant-fallow-and-virgin-land-management-law.

Hla Tun Aung. (2003). *Myanmar: The process and patterns*. Yangon, Myanmar: National Center for Human Resources Development, Ministry of Education.

Kushwaha, S.P.S. (1990). Forest type mapping and change detection from satellite imagery. *ISPRS Journal of Photogrammetry and Remote Sensing, 45*, 175–181.

Lambin, E.F., Turner, B.L., Geist, H.J., Agbola, S.B., Angelsen, A., Bruce, J.W., & Xu, J. (2001). The causes of land-use and land-cover change: Moving beyond the myths. *Global Environmental Change, 11*(4), 261–269.

Lillesand, T.M., & Kiefer, R.W. (2002). *Remote sensing and image interpretation*. New York, NY: Willy.

Pant, D.N., & Kharkwal, S.C. (1995). Monitoring land use change and its impact on environment of central Himalaya using remote sensing and GIS techniques. *Journal of Hill Research, 8*(1), 1–8.

Politno. (2012). *Geographical study on the settlement patterns of Bawlakhe District* (Unpublished Master's thesis). Loikaw University, Loikaw, Myanmar.

Pyae Thet Phyo. (2018, October 8). Vice president says country needs National Land Law. *Myanmar Times*. Retrieved from: https://www.mmtimes.com/news/vice-president-says-country-needs-national-land-law.html.

Roy, P.S., Singh, I.J., Sarnam Singh, H.S.A., & Sharma, C.M. (1993). Monitoring growth conditions of monospecies forest plantations using satellite remote sensing and geographic information system. Paper presented at the International Symposium on Operationalisation of Remote Sensing. ITC, Enschede, The Netherlands, April 19–23, 1993.

Scurrah, N., Hirsch, P., & Woods, K. (2015). *The political economy of land governance in Myanmar*. Mekong Regional Land governance. Retrieved from: http://mrlg.org/wp-content/uploads/2015/12/Political_Economy_of_Land_Governance_in_Myanmar_FA_2.pdf.

Trautwein, C. (2016, May 5). One-year logging ban proposed. *Myanmar Times*.

Turner, B.L. II, Clark, W.C., Kates, R.W., Richards, J.F., Mathews, J.T., & Meyer, W.B. (Eds.). (1990). *The earth as transformed by human action: Global and regional changes in the biosphere over the past 300 years*. United Kingdom: Cambridge University Press.

Turner, B.L. II, Skole, D., Sanderson, S., Fischer, G., Fresco, L., & Leemans, R. (1995). *Land-use and land-cover change: Science/research plan*. Report no. 35/7. Stockholm, Sweden: Royal Swedish Academy of Sciences.

Wilson, G., & Bryant, R.L. (1997). *Environmental management: New directions for the twenty-first century*. Bristol, PA: UCL Press.

Chapter 15
Local Knowledge and Rangeland Protection on the Tibetan Plateau: Lessons for Conservation and Co-management of the Upper Nu-Salween and Yellow River Watersheds

Check for updates

Ka Ji Jia

15.1 Introduction

The Tibetan Plateau is the source of seven major rivers, including two of the longest rivers in Asia: the Salween and the Yellow River.[1] As a significant source of fresh water and wetland biodiversity, this area was established as the Three Rivers Source National Nature Reserve, the second largest nature reserve in the world and the world's highest and most extensive protected wetland area (Plateau Perspectives 2014). The Tibetan Plateau is not only a site for water and biodiversity; it is also the center of traditional Tibetan culture, identity, livelihoods, and belief systems. Here, Tibetan herders play an important role in environmental protection, and identify themselves as rangeland protectors and local experts. This chapter explores how herders' livelihoods are connected to the source of the Salween and Yellow Rivers, and how they have responded to environmental change on the Tibetan Plateau.

Land degradation has affected significant parts of China's rangeland including Qinghai Province, the Tibetan Autonomous Region, Inner Mongolia, and the Xinjiang Autonomous Region. Since the early 2000s, the government of Qinghai Province has sought to protect the rangelands from degradation due to 'overstocking,' 'overgrazing' and 'human intervention.' In some cases, this has meant resettlement of Tibetan communities. This chapter examines these terms and practices– i.e., 'overstocking' and 'human intervention' – critically. I also address

[1]In Tibetan language, the Salween is known as Gyalmo Ngulchu and the Yellow River is called Ma Chu.

Ka Ji Jia, Ph.D. Candidate, Department of Women's Studies, Chiang Mai University; Email: kathytibetan@gmail.com.

C. Middleton and V. Lamb (eds.), *Knowing the Salween River: Resource Politics of a Contested Transboundary River*, The Anthropocene: Politik—Economics—Society—Science 27, https://doi.org/10.1007/978-3-319-77440-4_15

how land degradation based on one form of knowledge, and neglectful of other ways of knowing, poses problems when applied as a standard for all rangeland contexts on the Tibetan Plateau regardless of differing social-economic and cultural contexts. Herders on the Tibetan Plateau have different interpretations of land degradation (Yeh 2003) based on their local knowledge, livelihoods, and geographic location. Land degradation from the perspective of herders is not only caused by 'overgrazing', but also by removing people and their knowledge from the land, as well as by introducing market interventions such as mining activities and land commercialization.

My research shows how herders use their local knowledge as a tool to negotiate with various stakeholders in an effort to maintain access to and authority over their natural resources in the face of a dominant discourse that places blame for land degradation on local resource users (Yeh 2003). I do this by examining local herders' daily life experiences and life stories alongside broader practices of governance on the plateau. I argue that Tibetan herders use their local knowledge as a form of cultural capital (Bourdieu 1986) to negotiate their relationship with the land, both in everyday life and in relation to outside actors, in order to protect their rangelands and cope with the impacts of development projects on their lives and livelihoods.

I include herders' perspectives on rangeland degradation and environmental conservation projects, and examine the impacts of these projects using qualitative methods. In the following sections, the chapter will address methodology, local herders' traditional knowledge, debates around land degradation, and examples from fieldwork to draw out herders' multiple forms of knowledge and the ways they use this knowledge to negotiate to protect the rangeland and sustain their livelihoods on the Tibetan Plateau.

15.2 Methods

This research was conducted over multiple field visits in 2014, 2015, and 2017. In 2014–2015, I focused on a herding community and an ecological resettlement community in Xinghai County near the source of the Yellow River in the Tibetan Autonomous Prefecture in Qinghai Province (Site 1). In 2017, I conducted further research at the source of the Salween River in the Tibetan Autonomous Region (Site 2). Both sites (see Fig. 15.1: Map of study sites 1 and 2 in the Tibetan Plateau) are located at the headwaters of different rivers, and the shared cultural and historical practices of the herders includes a shared cosmology of the surrounding environment. This cosmology has helped them to formulate their own cultural capital. Under the same cosmology, herders use their cultural capital in different ways according to their local contexts. While the scope and scale of the research conducted in both sites varied, the overall research project relied on qualitative methods, including informal and semi-structured interviews, personal observation, and group discussion.

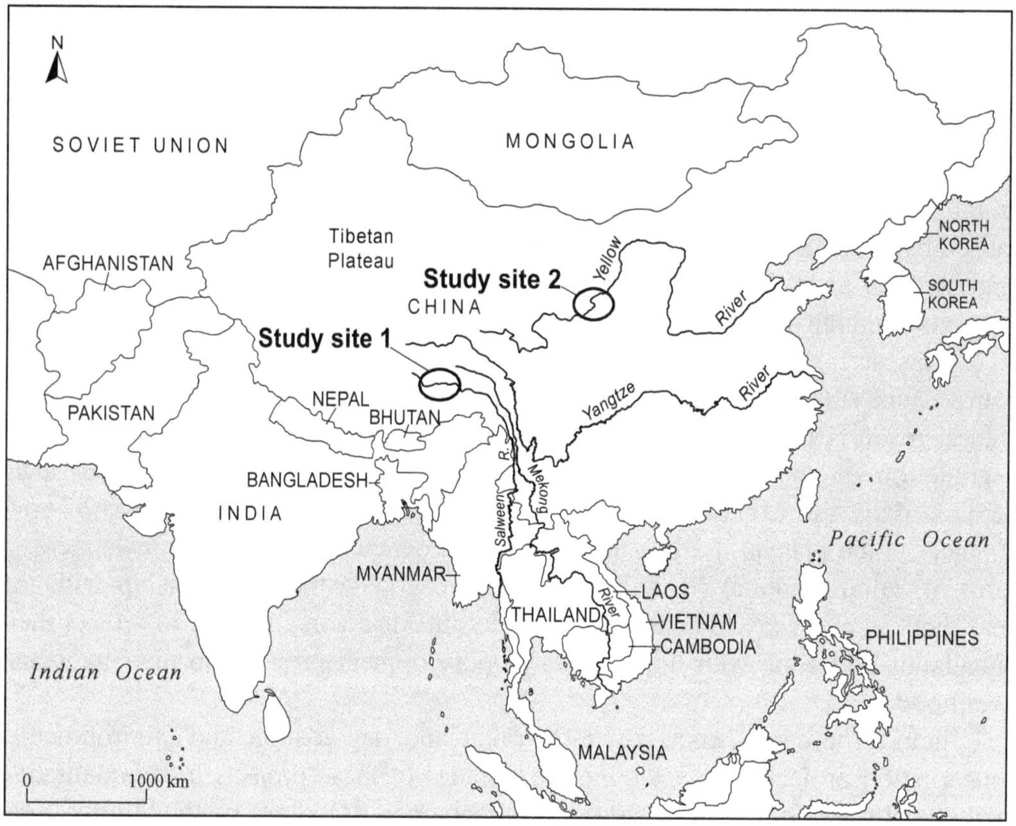

Fig. 15.1 Map of study sites 1 and 2 in the Tibetan Plateau. *Source* Cartography by Chandra Jayasuriya, University of Melbourne, used with permission

In total, I conducted 15 interviews, and I introduce these data relying on pseudonyms to protect the privacy of individual interviewees. At Site 1, over the course of 4 trips and 60 days, I conducted semi-structured interviews and open question interviews with 10 herders from a herding community relocated under an eco-resettlement project. Interviews were conducted in Tibetan language (Amdo dialect). I also held two group discussions with male and female herders on their rangeland separately. Research at the Yellow River Site focused on herders' local knowledge and how they use their local knowledge as 'cultural capital' to negotiate with various actors.

Building on my fieldwork at Site 1, I carried out a more targeted visit to the Salween River (Site 2). I spent 15 days in a town conducting five semi-structured interviews along with participant observation. At this site, the research focused specifically on herders' belief systems as a form of local knowledge that connects their worldview/cosmology with their environment.

Bringing the interviews and observations from these two sites together in this chapter, I present a range of approaches to understanding livelihoods and conservation on the rangeland.

15.3 Study Sites

The population estimates for the Tibetan population range from 6 to 7 million. There are three subgroups of Tibetan people who, according to their Tibetan dialects, reside in three regions: Amdo, Kham and U-Tsang. In these three regions, the people practice Tibetan Buddhism, and their religious beliefs are closely tied to nature. This has resulted in a strong environmental protection ethos that inspires people to take care of their natural resources including the water, land, mountains and animals. The two study sites are in Qinghai province and the Tibetan Autonomous Region.

At Site 1, most residents speak the Amdo dialect (see Fig. 15.1: Map of study sites 1 and 2 in the Tibetan Plateau). The community I visited is comprised of 305 households, in addition to a resettled community of 200 households. Here, local people practice Tibetan Buddhism, and the main economic activity is herding.

The second site, Sok Dzong (Su Xian County in Chinese), is a semi-herding county located at the source of Salween River, where local people speak Kham and U-Tsang dialects. According to local herders, this county is named after the Sok River, a tributary of the Salween. Here, the local people practice Tibetan Buddhism. Local livelihoods depend on wheat, potatoes, radish, beans and livestock such as yaks, sheep, goats, horses, and cows.

At present, communities across both sites also depend on caterpillar fungus, a native herb that is extracted for commercial purposes, and which I discuss below as a point of tension in the context of shifting livelihoods and environmental management strategies.

For these communities, as with the majority of Tibetan people, animal husbandry is a key livelihood activity. Livestock are generally animals that are resistant to extreme cold, and include sheep, yaks, horses, camels, and goats. Yaks are especially well adapted to extreme cold, and one third of the world's yak population live on the Tibetan Plateau (The Peoples Government of Qinghai 2007). Traditionally, Tibetan herders graze yaks, sheep, horses, goats and cows on the rangeland. These animals produce milk, cheese, butter, meat, and yogurt as food for herders. Herders also shear animals and use the wool to make Tibetan robes to wear in the winter. Even if a sheep or goat is killed by 'natural' causes, such as by wolves or in a snowstorm, the herders use the sheep and goatskins to make clothing, including robes, shoes, and hats. Tibetan herders give human names to their livestock and treat them as human beings. The children of herders play with young livestock and form intimate friendships with them. Herders even bring baby animals to eat and sleep with them if they feel that extra care is needed.

As I discuss further below, a number of changes have impacted herders' livelihoods, including a new property rights regime. Prior to the 1990s, rangeland on the Tibetan Plateau was managed through a common property rights regime. Local herders' livelihood activities were mainly for subsistence, and there was little to no direct involvement in the market. They raised animals for their own family consumption and exchanged excess products within their networks. After the

1990s, rangeland was placed under a private property regime, which helped herders to link their economy to the market. Today the market plays a vital role.

Tibetan herders maintain strong Buddhist and animist belief systems that involve worship of local deities including sacred mountains, animals, rivers, and trees. This belief system is linked to the Tibetan approach to livestock and animals in general. Tibetans show respect by not taking anything from sacred areas (Tsering 2006). For example, herders do not dig for gold, kill aquatic animals, cut trees, or hunt wild animals in sacred areas. Their environment and livestock not only have economic value, but also cultural and religious meaning. Tibetan herders free some of their livestock as part of religious activities and to show respect for local deities. Those freed livestock are considered holy animals and belong to local deities, and no herders will hurt or kill them. With this close connection to their animals, herders traditionally use a barter system for food rather than direct sale, and this includes exchanging their extra cheese, butter, meat, sheep wool, and yak fur for Tibetan farmers' barley, wheat and canola oil. These livelihood practices reveal a local knowledge significant for survival on the Tibetan Plateau.

Having introduced the study sites and the basics of Tibetan herder livelihoods and beliefs, in the next section I will address land degradation as defined by different actors and how these definitions tend to overlook Tibetan cultural and religious practices, including livestock rearing practices. I further explore how herders identify and define land degradation on their own terms.

15.4 Debating Land Degradation: Building on Political Ecology

The mainstream view of land degradation within the Chinese rangeland context is referred to in terms of desertification, overstocking, overgrazing and climate change (Yeh 2003; Nelson 2006). In particular, mainstream researchers perceive of 'overstocking' and 'overgrazing' as the main contributors to rangeland degradation in the highlands of China (Banks 2001: 718). Consequently, they place the blame on herders for severe pasture degradation (Yeh 2003: 505). While some researchers argue that rangeland degradation derives from rural reform dating back to the 1970s and that no 'tragedy of the commons' existed in commonly managed rangelands prior to this time (Banks 2001), policy makers claim that 'overstocking' results from unclear property rights on commonly managed rangelands. Various environmental conservation projects have been based on the assumption that herders' ignorance of proper pasture management has led to countless rodent infestations and overgrazing, resulting in infertile land and loss of productivity (Yeh 2003: 505).

As one approach to halting land degradation, some scholars (Liyu/Duoji 2008; Yu 2011; Niu 2011) have put forward strategies for conserving the rangeland. For example, they suggest that relevant authorities enhance the Grazing Ban Policy, which would prohibit herding in particular areas, reduce livestock numbers, establish protected areas, and eradicate the pika, a small rodent native to the rangeland (Liyu/Duoji 2008; Yu 2011; Niu 2011). Further suggestions (Liyu/Duoji 2008; Yu 2011; Niu 2011) include that fencing be introduced in order to allow for 'permanent rest' of the rangeland, which effectively means that the rangeland will be fenced in and thus not open for herders and livestock. This is linked to the assumption that to conserve the rangeland, 'human movement' and 'overgrazing' in the rangeland must be reduced. This has led to the resettlement of herders, as seen in the Eco-Resettlement Project in Site 1. Under the Eco-Resettlement Project, the state guaranteed long-term compensation subsidies to resettled people, with mixed results, as I describe below. Projects aimed at conservation have thus negatively impacted herders' livelihoods and property rights. This has stemmed from the application of one uniform theory of 'overstocking' in China, a theory which has failed to consider much of the research carried out to better understand 'land degradation' issues at different sites (Banks 2001; Harris 2010; Du 2012; Dkon mchog dge legs 2012; Dpal ldan chos dbyings 2012). As Wu et al. explain (2015: 11),

> It has been concluded in many studies that rangeland degradation associated with overgrazing and climate warming has increased the vulnerability of Tibetan livelihoods (Li and Huang 1995; Ma et al. 1999; Sheehy 2001; Zhou et al. 2003; Klein et al. 2004, 2007; Wang et al. 2006, 2007; Li et al. 2007; Shang and Long 2007; Fan et al. 2010; Harris 2010), and policies like rangeland privatization and ecological resettlement has caused uncertain consequences. Some of them also have accelerated the degradation of pastoral ecosystems on the [Qinghai-Tibetan Plateau].

As a result, they argue, "it is important to understand local people's views on how pastoral systems have been impacted by recent social, environmental and political changes and what actions and strategies they have undertaken to cope with these changes" (Wu et al. 2015: 11). Likewise, in examining sacred knowledge in Southwest China, Xu et al. (2005: np) "suggest that the long-term viability of the environment requires an interactive approach that involves local people as well as governments in the creation of environmental policy."

In this chapter, I aim to consider how herders' livelihoods are linked to the land and their access to land in particular pastoral contexts. Changes in land access not only impact herders' livelihoods, but also their local knowledge. Their knowledge is not static and unchanging; it is mixed with scientific knowledge within dynamic localized contexts wherein herders reinforce their local knowledge with outside information, and utilize their local knowledge as a form of cultural capital in negotiation with various stakeholders. Thus, to approach this, I draw on scholarship on local knowledge (Berkes 1999; Santasombat 2004; Xu et al. 2005) in addition work by Bourdieu (1986), who sees knowledge, culture, and various forms of negotiations and community networks as a kind of 'capital' built out of everyday life.

According to Bourdieu (1986), capital is accumulated in labor and through embodied forms. Capital takes time to accumulate and has a potential capacity to produce profits and to reproduce itself in identical or expanded forms. Capital in this sense means not only economic capital, but also other forms of capital including cultural capital, symbolic capital, and social capital. This could include exchange (such as trading) to expand networks, gain property rights, create space to negotiate with relevant actors, and exercise power. Social groups establish cultural capital from symbolic capital, which through social recognition legitimizes their knowledge (Bourdieu 1986). For example, a group of people with the same cultural background may inherit certain knowledge from their ancestors, but they also accumulate their own local knowledge through their everyday life practices. All of these knowledges and practices help to formulate their identities and help them to negotiate with others.

In some instances, Tibetan knowledge and practice is overlooked, while in others, Tibetan herders use their symbolic capital of mountain deities and other deities as cultural capital, which through social recognition legitimizes their local knowledge. They use cultural capital as a tool to negotiate with the different actors in their lives to benefit from and protect their natural resources.

In thinking about standard state approaches to land degradation, I examine the overlooked role of local knowledge in relation to Tibetan herders' livelihoods and their approaches to conservation of the plateau. In some instances, herders' local knowledge is indeed successfully used as a form of 'cultural capital' to push forward their own goals and insights into rangeland conservation, for instance, in their negotiations against mining on the rangeland.

In the following section, I describe the importance of local knowledge and local beliefs to Tibetan herders. This is a largely understudied topic, particularly as it relates to the Salween River on the Tibetan Plateau.

15.5 Herders' Local Knowledge and Practices

Local knowledge plays a vital role in determining how Tibetan herders promote cultural ecosystems on their rangelands and build harmony between nature and humans. As Miller (2008: 1) explains,

> [Tibetan nomads] have an intimate knowledge of their environment and an amazing ability to handle animals – a skill rare amongst most people today … Nomads' lives are finely tuned to the growth of grass, the births of animals and the seasonal movement of their herds … [they] have developed a close connection to the land and the livestock that nurture them.

Within traditional cosmology, culture and belief systems, Tibetans believe that mountains and rivers are deeply related to their lives. As explained above, mountains and rivers are sacred places where resident deities protect humans and other living beings. If nature is harmed, local deities are offended and this results in harm to human and animal populations. Therefore, when Tibetans celebrate religious festivals, harvest crops, or when someone needs to leave their community, Tibetans go to worship the deities and ask for their blessings (Tsering 2006). Consequently, Tibetan herders' local knowledge of resource management is directly connected to their belief system. At both the Salween and Yellow River sites, the Tibetan herders' cosmology is rooted in Tibetan Buddhism. Tibetan herders have different ways of practicing their religious activities, and they use this local knowledge as a form of cultural capital to support land protection. Tibetan herders worry that natural disasters, land degradation, and declines in vegetation/species are signs telling them that the local deities have been offended by their incorrect behavior and disrespect. In the next section, I provide examples of local knowledge along the Salween and Yellow Rivers on the Tibetan Plateau.

15.5.1 Salween Deities and the Spiritual Significance of the Land and River

The local informants from the Salween River Site, Sok Dzong, see land, mountains, animals, and rivers as all connected to their lives. According to interviewees, their belief system influences their cosmology and daily practices. As one interviewee, Mr. Karma, referred to the earth: "a female monster is sleeping here. The earth is like our mother. She is sleeping. If we offend her by polluting "Sok Chu" [the Salween River] and destroying the environment, she will wake up and become a devil and harm local people" (Personal conversation, June 2017). Furthermore, interviewees noted the ways they use religious activities to show their respect for local deities. For example, Mr. Karma explained that:

> When "Kar ma rak shi" [a special star] appears in the sky, we clean our horses in the waters of the Sok Chu [Salween], because during that time the river water becomes 'medicinal,' so our horses will not get illnesses. Also, on special days like the eighth of every month, we offer food to the fish in the Sok Chu [Salween River] (Interview in Sok Dzong, June 2017).

In this county entire communities, including girls, boys, men and women, also participate in religious activities like "Ma ni ser zam," which embody Tibetan cultural links with nature. When a community member passes away, the Ma ni ser zam ceremony helps the dead go to heaven. During the ceremony, small white stones and sand are used to write "Ma ni ser zam" on the ice once the river freezes over. Local people believe that when they do this on frozen rivers, the river becomes a bridge to help the deceased go to heaven. In this way, people pay respect to rivers, and regard them as special and sacred sites. This kind of activity shows how local beliefs and practices connected to death are linked to rivers. Such beliefs are part of the local knowledge system embodied within Tibetan culture, which denies the dualistic idea that nature is separate from human society. The fact that local Tibetan herders see themselves and their belief system as linked to nature invokes Berkes' assertion that belief plays an essential role in traditional ecological knowledge (Berkes 1999).

At the Yellow River Site, the Tibetan community has a similar belief system, which sees all living beings as equal to humans, and views some areas as sacred, including the highest mountains and largest rivers. In these sacred areas, local deities also exist. At the Salween River Site, the locals also believe in the existence of deities, and people at both sites commonly believe that offending deities may bring bad luck to individuals, families, or the community. The mining of minerals from a sacred mountain areas is considered a particularly offensive action for mountain deities. According to local beliefs, minerals belong to mountain deities, and the disrespect to deities shown by mining companies brings bad luck. Local communities thus see the establishment of sacred areas as a way to protect their natural resources.

15.6 Research Findings on Local Knowledge as Cultural Capital to Negotiate for Natural Resource Management

Building on their local knowledge and experience, Tibetan herders at both sites play important roles in environmental protection, and identify their local knowledge as a resource—in some instances, I argue that it becomes a kind of 'cultural capital'—to negotiate with various actors in order to maintain access and authority over the rangeland. With increased land degradation, these local practices are being challenged and misrepresented as contributing to degradation instead of protecting land and livelihoods. This tension—preservation of the plateau, traditional livelihoods, and land degradation—is at the center of this paper's interrogation of rangeland conservation protection debates. The three examples provided in the next section will show how Tibetan herders use local knowledge in natural resource management, and will explore some of the livelihood changes that herders are experiencing on the plateau as they struggle to maintain their land governance practices.

15.6.1 Tensions Between Experimental Zones, the "Holy Mountain," and Mining

Land use and land ownership on the plateau have changed for herders. There have been three shifts in land rights and management that are important to understand in the current context of natural resource management in the area. The rangeland has seen a transition from collective management of livestock on state-owned property to common property managed by villagers to the situation today where herders are granted individual, private property rights. As Nori et al. (2008: 11) argue "The institutional environment of Tibetan herders offers an eloquent example of how policy trends can adversely affect pastoral societies."

In short, in 1953, at the beginning of Chairman Mao's era, the local government in Xinghai County established herding sites for different livestock and hired herders to collectively herd livestock under a collective system. Herders were paid a monthly wage of around 60–70 RMB (10 USD), and were provided meals. The herders mainly raised sheep, yaks, cows, and horses on their own herding sites, and all of the livestock belonged to the state until 1956. This was a marked shift from Tibetan herder societies' customary management system, as although "land management had traditionally been communal, livestock were in fact household property" (Nori et al. 2008: 11). Private property as such ceased to exist during this time.

A second transformation came in 1956, when land reform was rolled out by the state at the national level. Property relations changed from a collective system of livestock management to a common property system with communal livestock (on communal land). Villagers on the plateau were able to manage the land until China's Open-Door Policy in 1978. From that time onwards, most of the communal land was divided into individual household plots and was privatized. In herding communities like the Yellow River Site, herders told me that they practiced this system, and that most communal land had been divided into households under privatization, which grants individual households land use rights.

Thirdly, by the 1990s, most communal land was divided into households under privatization, which meant that individuals had rights to manage or use their own land. Nori (2004) also explains how these property shifts led to a change in relations around the market and trade through state-controlled mechanisms (the quota system), and that market forces play a major role in reshaping local livelihood today.

Along with this shift to private property rights for individual and family land plots, resource extraction activities such as mining, and the rights to minerals, were also introduced under the private property system. The private property system allowed for the 'sale' of property, and allowed companies to access individual herders' land. Mining activities not only affected local livelihoods, but also weakened local collective power over land, land management and ecological sustainability.

These land governance and ownership changes, along with the introduction of mining, have impacted both herders' livelihoods and rangeland ecology. Interviewees told me that 40 years ago, grasses on the rangeland grew so tall that they could not see their sheep. This kind of phenomenon no longer exists. While herders attribute such changes to both climate change and mining activities, it was clear that when the mining companies began operating on the rangeland, it negatively impacted the grasses. For instance, after mining companies upgraded roads on the mountains, many small-scale deserts appeared. The quality of rangeland worsened and the environment became polluted by mining activities. With increasingly less rainfall and inadequate grasses on the rangeland (Wu et al. 2015), herders now have to buy grass to feed their livestock from people with few or no livestock on their land.

One interviewee explained,

> Mining has not only affected our area, but also people throughout our province, even the whole country. When companies take out minerals from the mountains like the holy mountain Amyi Rma Chin, these mountains are left with only their skins, but are empty inside, like a dead person who only leaves his corpse (Interview with XH3, Yellow River Site, March 2014).

Underlining the severity of the impacts, he explained further, "rivers have been contaminated, the fish are dead, and the mountain has collapsed."

Land privatization and the introduction of mining have not just led to environmental impacts; local culture has also been affected. Herding communities have responded to these impacts in a range of ways. Communities express various forms of open or hidden resistance. In these silent struggles, herders use their belief system, incorporating holy mountains, holy rivers and local deities as 'cultural capital' to access rights to natural resource management and to negotiate with various stakeholders to gain legal support for their claims.

A key example involves Amyi Rma Chin, a holy mountain that is well-known in Tibetan areas. This holy mountain is connected to herders' lives; local herders explain that the place where they live is Amyi Rma Chin's table, meaning that the holy mountain and their rangeland cannot be separated. However, local herders are suffering impacts from a large-scale copper mining project. With the project's establishment in 2001, the mining company told herders that they will help herders by building roads, creating jobs, and providing considerable payments. However, several years after the mining company began operations, local herders found that the company did not fulfil their promise. The herders also realized that the mining company had caused negative impacts on their rangeland, livestock, health, and belief system. For example, the mining company gradually expanded the area of their operations. This has caused a lot of sand to appear on the rangeland, the mountain has collapsed, and livestock have been negatively affected by tailings in the river and on the grasslands. Local herders report being affected by air pollution, and are concerned that their holy mountain has been damaged.

In response, herders have publicly reclaimed Amyi Rma Chin as a holy mountain, noting that the damage to Amyi Rma Chin is an offensive action against their religious beliefs.

Amyi Rma Chin is a place revered by the herders who believe that the mountain is master and protector of the area, and as such, will protect the lives of those who live in its presence. Amyi Rma Chin is also the herders' seasonal summer home and pastureland. Herders have not 'taken from' the holy mountain's rich store of mineral resources, as it is linked to their beliefs that it is a sacred place.

In addition, herders use their belief system to influence relevant laws in such a way that legitimizes their local knowledge. For instance, herders are making a case that their herding location belongs to a state experimental zone,[2] and in this experimental zone, mining activities are prohibited. One herder, Mr. Dorji, explained that herders can live in the experimental zone, and that even though the state encourages herders to limit their livestock numbers, and some land must be fenced in as protected land (where herding is not allowed), the establishment of the zone also means that mining operations are prohibited.

---

[2]There are a variety of "experimental zones" across China. They are not all the same, with different policies and different practices. I have explained here how local people see it, but it is not always entirely clear.

As a result of these community efforts, in 2017, local herders received legal support from the government, and mining in the experimental zone was permanently stopped. The herders used symbolic power to build collective solidarity and mobilize their community to protect the holy mountain.

These local efforts to engage with the mining company were not without effort and tension. There were different voices among the herders, with some herders supporting mining activities because, although they saw the mine's impacts on their livelihoods, they were able to access certain benefits. In order to build solidarity at the community level, the majority of herders not only focused on land rights and livestock, but also used their common belief in Amyi Rma Chin to mobilize the whole community to protect the holy mountain. This shows how herders built internal solidarity by using the symbolic power of the mountain as a sacred site, and this symbolic power became the cultural capital that enabled them to negotiate with the mining company.

In reflecting on these changes, some herders state that the common rangeland management system of the past was more efficient in protecting rangeland ecology than the current privatized rangeland management system. This is not only because of the impacts of mining. Some herders also experienced resettlement (discussed next), and some have resorted to allowing digging for previously sacred caterpillar fungus on their land in order to make money to support their families (discussed in the following section). They suggest that the common rangeland management system of the past enabled them to control their livestock size and make decisions based on what the land could support, and was not based on commercial or market forces. However, under the current private rangeland management system, there is an incentive for herders to make money by selling their livestock or allowing outsiders to use their rangeland.

15.6.2 Eco-resettlement as Alternative Livelihood

Through interviews with resettled herders and my own observations, I found that although there are some benefits to resettlement in terms of educational opportunities and limited subsidies, resettlement does not present a viable alternative livelihood. I found no evidence that resettlement furthers the aims of conservation, as it removes the herders and their care of the rangeland from the Tibetan Plateau.

Resettlement at the Yellow River Site began in 2005, when villages were moved under the state-sponsored "Eco-resettlement Project." The Eco-resettlement Project aimed to protect the rangeland and improve the livelihoods of local herders. However, it has created unintended consequences for herders' livelihoods, and most herders choose not to live in resettlement sites, as they are aware that the resettlement project cannot bring them sustainable livelihoods. Those living in resettled communities experience difficulties, such as water and electricity shortages, lack of jobs, discrimination, and lack of support for cultural practices.

Moreover, through interviews, resettled herders explained that resettlement is not an adequate solution to environmental protection. One herder explained, "*resettlement was supposed to protect our environment, right? But I'm not sure whether that's the right way to protect it.*" (Interview XH2, Yellow River Site September 2013). Across interviews, herders' comments highlighted the link between their roles as herders and environmental protectors. Herders felt that the outcomes of resettlement were contrary to the initial objectives of the project, which aimed to improve rangeland protection and herders' livelihoods in the face of what was perceived as severe rangeland degradation at the Three Rivers Source on the Tibetan Plateau.

Visiting the resettlement areas, I could meet only with the elderly and children. From my interviews with older herders, I learned that herders had resettled here because of a shortage of labor (lack of young people) to maintain their pastures and livestock. They required help from young people to manage their pastures, but most children had been sent to schools outside the community. In some cases, there were no young people in the family. According to herders experiencing labor shortages, resettlement was an alternative livelihood that would allow them to receive subsides from the local government. For example, government subsidies may include: support for children under 16 years old and for adults over 55 years old (hence the make-up of the resettlement community). They may also include: land subsidies depending upon the size of land, winter fuel subsidies, housing subsidies, and peasant household subsidies. However, these subsidies have not always improved living conditions. As one interviewee explained,

> When we resettled here, I thought there were no improvements to the environment and also our living conditions got worse. Our house is of really bad quality, and water often comes inside after it rains. I've had to rebuild my house three times and every time I've had to spend 4,000-5,000 RMB. In total I've spent 12,000 RMB from my savings (Interview XH6, Yellow River Site, September 2013).

Some resettled herders have been able to find alternative means of livelihood, such as working in construction, collecting caterpillar fungus (described below), and opening small shops. Many resettled herders, however, are not happy with their current life.

> I am not satisfied with my current life. I cannot find a job due to my physical condition, so I have no income- only expenses. For example, I took out loans from the bank for my child's education. In fact, I came here just for my child's education. Otherwise I would prefer to stay in my original house. (Interviewee XH2, Yellow River Site, September 2013)

Children's education was a main reason that herders provided for why they followed the eco-resettlement project. My informants claimed that they were voluntarily resettled, and that it was not permanent resettlement. Most of them explained that they led two ways of life. They would sometimes go back to their herding areas and sometimes come to the resettlement area. It was completely dependent on their own decision. However, in resettling, they also faced many social problems such as difficulty finding jobs, and lack of access to clean water, electricity and health care. One former herder explained, "*I've been doing nothing since I resettled here and my wife is gravely sick. Everything here I need to buy with money, such as water, electricity, meat and vegetables*" (Interview XH5, Yellow River Site, September 2013). And another former herder said, "*After I resettled here, my living conditions got worse. For example, I had cheese, butter, and milk before resettlement, but now I have none of these things to eat.*" (Interview XH2, Yellow River Site September 2013). Or as another interviewee explained, "*I feel that pastoral life is better than resettled life here. Before, I didn't need to buy meat, butter, potatoes or other vegetables*" (Interview XH6, Yellow River Site, September 2013). Their comments also underline the links between their pastoral livelihoods and the life satisfaction in their communities.

These aspects of "eco-resettlement" complicate standard assumptions that removing herders from the landscape will support environmental protection. In my analysis, this case also shows that herders rely on certain silent struggles and negotiations, as they may receive benefits from the eco-resettlement project, even as they may continue to try to practice pastoral livelihoods and do not fully support the project. In doing so, the majority of herders still keep their land and livestock as cultural capital in response to the resettlement project.

Moving forward, another alternative livelihood practice that has emerged is collection of the once-sacred caterpillar fungus. I consider this practice and the way it both relies on and impacts 'cultural capital' in the next section, drawing on work across the two research sites.

15.6.3 Caterpillar Fungus and Livelihood Change

The "Sok Chu" or Salween River Site is a main area for the collection of caterpillar fungus due to its high-altitude location. Local residents depend on a mixed set of livelihood activities. Here, a number of changes have led Tibetan herders to stop herding livestock, particularly sheep, due to frequent encroachments from feral dogs and wolves. Local people practice another form of livelihood based on their own cultural capital, that is access to caterpillar fungus. This practice represents a way for Tibetan herders to make a living in an increasingly difficult situation. It also represents a shift in local beliefs, for in the past, caterpillar fungus was not collected – it was considered sacred. According to local legend, Tibetans believe that caterpillar fungus is a female deity's braid. Because they saw caterpillar fungus as a holy plant, Tibetans did not eat or dig the caterpillar fungus for any purposes. Today, Tibetans have mixed feelings about collecting the product: "The opinions recorded varied between seeing it as an innocent activity not entailing any guilt and as a digpa [a sin or negative karma]" (Sulek 2016: 8).

Caterpillar fungus (*Ophiocordyceps sinensis*), known locally as 'Yar tsa gun bu', literally means 'summer grass winter worm.' It has been used widely in China since the 1990s (Shrestha/Bawa 2013). The caterpillar fungus boom is related to China's rapid economic growth; people believe that caterpillar fungus is a medicine that strengthens the lungs and kidneys and increases energy and vitality

(Holliday/Cleaver 2008: 225). Since the 1990s, the caterpillar fungus trade has been linked to national and global markets, and the fungus can be found in Nepal, Bhutan, and China (Tibet). In Tibetan areas, caterpillar fungus is mainly found at between 3,500 and 5,000 m altitude in Tibet (TAR), Qinghai, Sichuan, Gansu and Yunnan. It is mainly collected in Naqu and Changdu in Tibet, and Yushu and Guoluo in Qinghai (Xuan 2012). According to Tibet's Agriculture and Herding Office, in 2012, "150,600 people collecting the fungus in Naqu [District] harvested 16.3 tonnes – 3.7 tonnes less than the year before." Caterpillar fungus is declining, as it could be found above 3,500 m in the past, but can now only be found above 4,500 m (China Dialogue 2012). With the rise of the caterpillar fungus economy, Tibetan people's subsistence-based livelihoods have been altered in ways that pose further uncertainties.

Reasons for this shift towards collection, what some have termed the "caterpillar fungus boom" (Sulek 2016: 3; see also: Yeh/Lama 2013), include the rise in demand, better links to the market, and the shift in land ownership and control. While caterpillar fungus was not a commercial product in the past, with the introduction of a market-oriented economy on the plateau, caterpillar fungus has become a commercial product with high market value, and is collected on both private and common land. In some instances, herders abandon livestock production if they can earn enough from collecting caterpillar fungus, freeing all of their livestock except some for family consumption. Or as Sulek explains,

> The income from caterpillar fungus has made the pastoralists less dependent on pastoral production. They adjusted to the new situation by breeding fewer sheep and reducing the sale of dairy and other products. They also reduced the number of yaks sold for commercial slaughter (2016: 10).

The collection of caterpillar fungus is not only done by local herders, but also by outsiders. The outsiders are usually from other places in China. The local herders with land containing caterpillar fungus often hire diggers from other areas or rent their land to businessmen. The diggers might be Tibetans (famers and herders), Han Chinese, Muslim Chinese, and others. The diggers' backgrounds are very diverse and most of them may not have environmental protection awareness or may not want to take care of the land. In this case, local landowners and local people complain that the diggers do not protect the land as local people do.

Residents also link the history of state responses to livestock rearing to the rise of caterpillar fungus. At the Salween Site, for instance, the state introduced wire fences in many areas in order to limit the mobility of wild animals. The fence project affected wild animals' reproductive behavior, and the local food chain. This was all done with the goal of rangeland protection. However, this also had impacts on the herders' livelihoods. Herders reported that the feral dog population greatly increased, and that feral dogs often attacked herders' sheep, raising great concerns for their livelihoods. To respond to this, the herders have had to stop grazing sheep. This is another reason why they are increasingly turning to caterpillar fungus harvesting. The reduction of livestock numbers on the rangeland by both individual herders and conservation projects has resulted in feral dogs and wolves attacking other endangered animals like gazelle in the conservation areas.

In this case, I saw how local herders are drawing on their local knowledge to harvest caterpillar fungus in a way that is dynamic and responds to the local context. I also saw that many herders have had to stop grazing sheep due to wild animals' frequent encroachment onto their land, which is linked to state efforts to protect and 'fence in' the rangeland. While the ecosystem imbalance and increasingly market oriented economy pushes herders to stop herding, they draw upon their own cultural capital to access caterpillar fungus to maintain their livelihood.

15.7 Discussion and Conclusion

Tibetan herders from both the Salween and Yellow River sites share ethnic and geographical identities, livelihoods, beliefs and an overall cosmology that places them in a central position to protect nature and the rangeland. At these two study sites, local herders have mobilized these connections, using their cultural capital in the natural resource management process to respond to change and protect the rangeland. This can be witnessed through multiple forms of negotiation, strategies, and adaptation under different conservation regimes, property rights schemes, and market interventions.

At the Salween River Site, fences introduced to address ecosystem imbalances pushed herders away from raising sheep due to frequent encroachment by wild animals, however, they established other forms of livelihood based on their local knowledge of nature and landscape, what I see as cultural capital. The caterpillar fungus economy has had some impacts on the ecology, and in some places like Yun Ta village in Yushu Prefecture, Qinghai Province, local Tibetan people have developed their own environmental protection strategies and built strong mechanisms to prevent overharvesting of caterpillar fungus and reduce harms to the environment (Zhao 2015). These strategies are seen not only in the case of caterpillar fungus, but also in the ways herders strategically deployed 'experimental zones' for protection from mining, selectively participated in eco-resettlement projects, and made the conservation of the holy mountain explicit to outsiders. This is how local herders bring their knowledge to negotiate with state knowledge and to produce their own cultural capital.

The research presented here underlines the fact that herders' local knowledge in resource management not only lends a different perspective to the 'overgrazing' debate, but may also inform policy around such issues as mining, wire fencing, commercial harvesting, and climate change. My call to include herders' local knowledge in resource management recognizes the fact that the form this knowledge takes is constantly changing based on location, economic context, and livelihood strategies. Local knowledge is thus not based on one static idea. Herders develop different perspectives and livelihood strategies in response to various uncertainties and unintended consequences. It can be a struggle to work together with competing interests, but herders' actions support Yos Santasombat's findings that "knowledge production should be seen as a process of social negotiation involving multiple actors and complex power relations and must therefore be understood in terms of change, adaptation and dynamism" (2003: 43).

I also show the ways in which local knowledge in combination with symbolic culture can be mobilized to negotiate and care for the land, as seen when herders emphasize their holy mountains and local deities. Symbolic culture used as cultural capital can be legitimized to negotiate with relevant stakeholders to improve access to natural resources and seek better livelihood strategies based on individual and community interests.

In many ways, these findings are in line with other scholarship on the Asian Highlands, including a review by Xu/Grumbine (2014: 99) which showed that "In the Asian Highlands, a more integrated approach that blends strategies to secure more resource access for local people could pave pathways toward successful climate adaptation, more inclusive governance, and greater state stability."

However, what I add here are the personal accounts from both the Salween and Yellow River field sites which show how Tibetans' local knowledge is closely connected to their religious beliefs and their environment, reflecting that, "knowledge cannot be divorced from the natural and cultural context within which it has arisen, including their traditional lands and resources and their kinship and community relations" (Higgins-Zogib et al. 2010: 173). Tibetan cultural knowledge has long recognized that without nature, human beings cannot survive; people depend entirely on the earth to live. Similarly, without minerals, the grasslands and mountains will be empty and unable to survive. Projects and practices aimed at conserving the Tibetan Plateau could benefit by incorporating the beliefs and knowledge of herders.

Herders, stakeholders and relevant authorities need to reconsider 'co-management' and 'co-production of knowledge' to develop a shared analysis of rangeland environmental protection strategies. Through this collaborative process, the co-production of knowledge will help local people to meet their own educational, cultural and political needs while improving livelihood strategies and environmental conservation work on the Tibetan Plateau (Berkes 2002, 2009).

References

Banks, T. (2001). Property rights and the environment in pastoral China: Evidence from the field. *Development and Change*, 32(4), 717–740.

Berkes, F. (1999). *Sacred ecology: Traditional ecological knowledge and resource management.* Philadelphia, PA: Taylor & Francis.

Berkes, F. (2002). *The drama of the commons: Cross-scale institutional linkages, perspectives from the bottom up*. Washington, DC: National Academy Press.

Berkes, F. (2009). Indigenous ways of knowing and the study of environmental change. *Journal of the Royal Society of New Zealand, 39*(4), 151–156.

Bourdieu, P. (1986). The forms of capital. In J. Richardson (Ed.), *Handbook of research for the sociology of education* (pp. 241–258). Westport, CT: Greenwood.

China.org.cn. (2013). Retrieved from the Sanjiangyuan National Nature Reserve in China's Qinghai website: http://www.china.org.cn/travel/2013-03/06/content_28135757.htm.

Dkon mchog dge legs. (2012). China's pastoral development policies and Tibetan plateau nomad communities. *Asia Highlands Perspectives*, *18*(1), 37–72.

Dpal ldan chos dbyings. (2012). Plateau pika control in Santu alpine grassland community, Yushu Prefecture, Qinghai Province, China. *Asia Highlands Perspectives*, *18*(1), 108–124.

Du, F. (2012). Ecological resettlement of Tibetan herders in the Sanjiangyuan: A case study in Madoi county of Qinghai. *Nomadic peoples*. Retrieved from: http://www.case.edu/affil/tibet/tibetanNomads/documents/ecologicalresettlementfortibetanherderinthesanjiangyuan.pdf.

Harris, R.B. (2010). Rangeland degradation on the Qinghai-Tibetan plateau: A review of the evidence of its magnitude and causes. *Journal of Arid Environments, 74*, 1–12.

Higgins-Zogib, L., Dudley, N., & Kothari, A. (2010). Living traditions: Protected areas and cultural diversity. In N. Dudley & S. Stolton (Eds.), *Arguments for protected areas: Multiple benefits for conservation and use* (pp. 165–188). Abingdon, United Kingdom: Taylor and Francis.

Holliday, J., & Cleaver, M. (2008). Medicinal value of the caterpillar fungi species of the genus Cordyceps (Fr.) link (Ascomycetes), A review. *International Journal of Medicinal Mushrooms, 10*, 219–234.

Kumar, K.A. (2010). *Local knowledge and agriculture sustainability: A case study of Pradhan tribe in Adilabad district*. Center for Economic and Social Studies No. 81. Hyderabad, India: Center for Economic and Social Studies.

Liyu, H.Z., & Duoji, L. (2008). **青海省海南州三江源区草地退化成因及治理对策** [Qinghai Hainan Prefecture's rangeland degradation and conservation strategies]. **草业与畜牧** *Grassland and Animal Husbandry*, *2*, 49–51.

Miller, D. (2008). *Dropka: Nomads of the Tibetan Plateau and Himalaya*. Boonsboro, MD: Vajra Publications.

Nelson, R. (2006). Regulating grassland degradation in China: Shallow-rooted laws? *Asian-Pacific Law & Policy Journal*, *7*(2), 385–417.

Niu, L. (2011). **保护草原生态系统，促进牧区畜牧业可持续发展** [Protection of Grassland Ecosystem and Promotion of Sustainable Development for Animal Husbandry]. **中国畜牧兽医文摘** *China Animal Husbandry*, *5*, 14–15.

Nori, M. (2004). Hoofs on the roof: pastoral livelihoods on the Qinghai-Tibetan Plateau, the case of Chengdu county, Yushu prefecture. Rome, Italy: ASIA-ONLUS. Retrieved from: http://www.cwru.edu/affil/tibet/booksAndPapers/Hoofs_on_the_Roof.pdf.

Nori, M., Taylor, M., & Sensi, A. (2008). Browsing on fences: pastoral land rights, livelihoods and adaptation to climate change. *IIED Issue Paper,* No 148. Retrieved from: http://pubs.iied.org/12543IIED/.

Plateau Perspectives. (2014). *Sanjiangyuan (Three Rivers' Headwaters) National Nature Reserve (SNNR)*. Retrieved from the Plateau Perspectives website: www.plateauperspectives.org.

Prasert, T. (2008). Space of resistance and place of local knowledge in Karen ecological movement of northern Thailand: The case of Pgaz K' Nyau villages in Mae Lan Kham River Basin. *Southeast Asian Studies, 45*(4), 586–614.

Santasombat, Y. (2003). *Biodiversity: Local knowledge and sustainable development*. Chiang Mai, Thailand: RCSD Chiang Mai University.

Santasombat, Y. (2004) Karen cultural capital and the political economy of symbolic power. *Asian Ethnicity, 5*(1), 105–120.

Shrestha, U.B., & Bawa, K.S. (2013). Trade, harvest, and conservation of Caterpillar fungus (Ophiocordyceps sinensis) in the Himalayas. *Biology and Conservation*, *159*, 514–520.

Sulek, E.R. (2016). Caterpillar fungus and the economy of sinning. On entangled relations between religious and economic in a Tibetan pastoral region of Golog, Qinghai, China. *Études mongoles et sibériennes, centrasiatiques et tibétaines*, *47*, 1–17.

The People's Government of Qinghai. (2007). *Beautiful Qinghai.* Retrieved from People's Government of Qinghai website: http://www.qh.gov.cn/dmqh/glp/.

Tsering, B. (2006). The changing roles of Tibetan mountain deities in the context of emerging environmental issues: Dkar po lha bsham in Yul Shul. *Asian Highlands Perspectives, 40*, 1–33.

Wu, X., Zhang, X., Dong, S., Cai, H., Zhao, T., Yang, W., Jiang, R., Shi, Y., & Shao, J. (2015). Local perceptions of rangeland degradation and climate change in the pastoral society of Qinghai-Tibetan Plateau. *The Rangeland Journal*, *37*(1), 11–19.

Xu, J., & Grumbine, R.E. (2014). Integrating local hybrid knowledge and state support for climate change adaptation in the Asian Highlands. *Climatic change*, *124*(1–2), 93–104.

Xu, J., Ma, E.T., Tashi, D., Fu, Y., Lu, Z., & Melick, D. (2005). Integrating sacred knowledge for conservation: cultures and landscapes in southwest China. *Ecology and Society*, *10*(2), 7.

Xuan, C. (2012, March 9). How long can the caterpillar fungus craze last? *China Dialogue*. Retrieved from: https://www.chinadialogue.net/article/show/single/en/5143-How-long-can-the-caterpillar-fungus-craze-last.

Yeh, E.T. (2003). Tibetan range wars: Spatial politics and authority on the grasslands of Amdo. *Development and Change, 34*(3), 499–523.

Yeh, E.T., & Lama, K.T. (2013). Following the caterpillar fungus: nature, commodity chains, and the place of Tibet in China's uneven geographies. *Social & Cultural Geography, 14*(3), 318–340.

Yu, H. (2011). **青海省天然草地退化现状及可持续发展策略** [The situation of Qinghai's degraded natural grassland and sustainable development strategies]. **中国畜牧兽医文摘** *China Animal Husbandry*, *5*, 14–15.

Zhao, X. (2015). **云塔村的虫草——来自基层的管理逻辑** [Yun Ta village's caterpillar fungus - from local management logic]. Retrieved from Shan Shui Conservation Center website: http://www.shanshui.org/ArticleShow.aspx?id=264.

Chapter 16
Future Trajectories: Five Short Concluding Reflections

Chayan Vaddhanaphuti, Khin Maung Lwin, Nang Shining, Pianporn Deetes and R. Edward Grumbine

16.1 Concluding Commentary: State of Knowledge and Geographies of Ignorance of/in the Salween River Basin

Chayan Vaddhanaphuti

16.1.1 Introduction: Observations on the Status of "Salween Studies"

In the past five years, I have watched the work on the Salween take shape. The center I lead, the Regional Center for Social Science and Sustainable Development, and I have helped organize and support various events. The "First International Conference on Salween-Thanlwin-Nu Studies" held 14–15 November 2014 at Chiang Mai University, Thailand was a landmark meeting – the first of its kind to put "Salween Studies" on the map.

Chayan Vaddhanaphuti, Professor and founding Director of the Regional Center for Social Science and Sustainable Development (RCSD), Faculty of Social Sciences, Chiang Mai University, Thailand; Email: ethnet@loxinfo.co.th.

Khin Maung Lwin, Advisor, National Water Resources Committee, Government of the Republic of the Union of Myanmar; Email: drkmlwin98@gmail.com.

Nang Shining, founder and Director of Mong Pan Youth Association (MPYA) and co-founder of Weaving Bonds Across Borders; Email: shining.cu@gmail.com.

Pianporn Deetes, Thailand and Myanmar Campaigns Director, International Rivers; Email: pai@internationalrivers.org.

R. Edward Grumbine, Adjunct Faculty, Huxley College, Western Washington University; Email: ed.grumbine@gmail.com.

C. Middleton and V. Lamb (eds.), *Knowing the Salween River: Resource Politics of a Contested Transboundary River*, The Anthropocene: Politik—Economics—Society—Science 27, https://doi.org/10.1007/978-3-319-77440-4_16

Moving forward, I argue that what we need overall is to look at the whole Salween River Basin and identify certain areas for case study research so that we can learn more about the complex relationships between people and their environment, between water and culture. We need to use a multidisciplinary approach to these studies.

16.1.2 Knowledge 'Needs' and Gaps: What's Lacking in Our Salween Knowledge?

In terms of physical science, Myanmar has been ahead of us in Thailand. Professor Maung Maung Aye of Myanmar is leading a team to look at the geomorphology of this important river system along with others at the University of Yangon (e.g., Bird et al. 2008 foundational study on Irrawaddy and Salween geomorphology). In China, we are learning more about the river's biodiversity (Xiaogang et al. 2018). Whereas, in Thailand, we have seen a much greater focus on social science research. For instance, of late we have new case studies (e.g., Hengsuwan 2012, 2013; Lamb 2014a, b). Even as early as the late 1990s, though, we saw studies of the Salween start up in Thailand. These studies were not led by academics, but by a group of NGOs like SEARIN in collaboration with local residents, who led a study on the Thai side of the Salween River (e.g., TERRA 1999, or the Thai Baan 'Villager' research published by SEARIN in 2005). In Thailand, we have also seen emergent archaeological studies within the Salween River Basin (Shoocongdej 2006; Lampert et al. 2003), with ancient pottery uncovered near the Salween. This evidence of human and environmental history must also exist in Myanmar and China.

16.1.3 Social and Cultural Meanings of the Salween River

What of the people of the Salween? The diverse ethnic groups, who are they? Shan, Pa'O, Danu, Kayan, Karenni, Karen, Ta-ang, Inthar and more. What of the ethnic relations within or among these groups in the Salween River Basin? What of the conflict between the state and the ethnic groups? What is the likelihood their agricultural practices will continue?

The ethnic people in the Salween River Basin have different ways of adapting to the ecological systems. There are different types of shifting cultivation, different land uses, different types of land tenure systems, different types of community forest management, and the paths they are entering into, such as an agricultural transition. What has happened to these over and through time? We have limited knowledge on this, we should do more.

Another key issue is the cultural meaning of the river. Yu Xiaogang et al., Chap. 4, this volume, raise the issue of world cultural heritage sites and the way in which the Chinese government envisions the Nu Jiang as a 'green' watershed, an area focused on preservation of resources, not a dam area. But how do people fit into this vision? What is the perception or meaning given by local people? Do they recognize sacred places in the Salween River where people pay respect? In Thailand, local people ordain trees to protect the forests along the Salween. Tree ordination ceremonies link cultural practice to practices of forest conservation.

People also have special relationships with the river. They have local knowledge and certain ways of using the river. We need to know more about these complex relationships. People living along the Mekong River have an area where they call 'luang' or customary practice, or community rights to fish in the area. It is not private ownership of the river, but for common use. But there are also parts of the river that can be owned privately or owned by groups of people. What kinds of customary laws are in practice along the Salween? How do Peace Parks or World Heritage designations recognize those laws and customs?

There is need for intensive, qualitative or ethnographic studies on these communities building from local histories, and their livelihood practices.

16.1.4 Economy of the Salween River

I also have identified a need to study the Salween economy, based upon different resources such as fisheries, forestry, sand mining, rice-growing, handicrafts, riverbank gardening, local trade and cross-border trade. Can we estimate the income generated annually from the river basin? What are the main livelihoods of people in the Salween River Basin? The Salween fishing economy is a very interesting topic that I would like to take up personally. We should know how much income is generated by fisherman in the lower Salween river; even the economic values of the forest and so on, are just some of the things we do not know yet, but which demand further study.

16.1.5 Politics

When we talk about water governance, we must discuss politics. In Myanmar, for example, we cannot talk about only one state. There are states within a state. There are what Callahan (2003) called 'state-like agents', ethnic armed groups, BGF (border guard forces). What about the political economy of border land areas and cross border migration, dam construction in relation to militarization, and state and non-state actors in the contestations around dam construction? All of these issues need to be addressed. The question of refugees and internally displaced peoples (IDPs) has also been raised by many researchers, including some in this volume, but not as a topic of study on its own. This requires further work. It also requires recognition of the history of the Salween and resource development, that there is also a history of conflict, and that it is linked to dam building.

16.1.6 Gender

It is also worth highlighting that there has been movement to include gender perspectives in Salween Studies. At the 2018 Salween Research Workshop at least two papers brought in gender perspectives: one by Hnin Wut Yee (2018) about gender inclusion in social justice, and Nang Shining's (2018) ecofeminism evolution and the readiness of gender evolution in the case of Mongton Dam Project. We also have an important chapter presented in this book on intersectional feminism (El-Silimy, Chap. 8, this volume) and recent work on feminist political ecology (Lamb 2018). So, we have this important gender perspective emerging in the Salween River studies.

16.1.7 Pathways

I think it would be interesting to look at the whole Salween River Basin and identify certain areas, to take a closer look using case studies so that we can learn more. At the leading panel at the 2018 Salween Workshop, we heard several papers presented around the development model in the Salween River Basin. For instance, Middleton (2018) and colleagues introduced the distinct pathways of development for the Salween from local perspectives and from the perspective of states. This was a way to consider the multiple perspectives and modes of development for the river basin, and that they are ongoing simultaneously. It was the first time I heard about the pathways concept, but it helps to think about not only existing pathways, but the way to move forward.

16.1.8 Collaboration and Concepts

In bringing these diverse topics together – politics, knowledge, water, people, gender – I think that the challenging concept of Zomia is best positioned to help us in our thinking and collaborations.

van Schendel (2002), and later Scott (2009), have developed the term 'Zomia'. van Schendel argues, for instance, "In order to overcome the resulting geographies of ignorance, we need to study spatial configurations from other perspectives as well" (2002: 664).

If you look at this area, the Salween is representative of a 'border area'. There are people who do not want to be governed. You can look at the people along this border, between Thailand and Myanmar along the Salween River, who resist the control by the state. Not only in the Burmese/Myanmarese state but also the Thai state. What Saw John Bright, Chap. 5, this volume, shows us about the Peace-Park is that it can be understood as an attempt to set up self-autonomy or self-governance, which I also see as part of the art of *not* being governed. Of course, at the same time, people in the basin, and even in the Peace Park, may also want to have some state support. People in the basin should have the opportunity to do both.

Moving forward, what I see is that in the spirit of overcoming the 'geographies of ignorance' we need a kind of collaboration between scholars from different disciplines, to be not only spatially distinct but also distinct in approach. I want to emphasise this: scholars working together on certain issues in the Salween River Basin should be multidisciplinary. I am not speaking of a 'sum of studies' or a kind of systematic review, but there should be an attempt to cross-fertilize ideas between disciplines or to transcend disciplines. You cannot talk about fish species, river governance, and ethnic perspectives on ecology separately; these need to be in conversation with one another. Instead, more interesting would be to have, as a starting point, discussions about how ethnic groups make use of the river, and to highlight their local knowledge of fisheries. What is the meaning of fish, what kind of income (monetary or not) is derived? More than that, we can ask about the seasonal variation throughout the year, the different activities in dealing with the ecological change or river change and the different kinds of resources available at different times of the year. There is a need to work together from different disciplines.

In terms of multi-disciplinarity or transdisciplinarity, the emphasis is on people from different disciplines investigating on one issue, or more than that. This also includes working together with non-academic actors. Here is the call for collaboration between non-academics and academics. Academics from different disciplines working with non-academics, be they policy makers, local officials, or villagers.

I would like to conclude that this may be significant to find ways, resources, and further collaborations in learning more and I hope that this knowledge would be helpful for us to work to preserve or to protect the Salween River; to make it run free.

16.2 Positioning the Salween in Myanmar's River Politics

Khin Maung Lwin

16.2.1 *Myanmar's New River Politics*

With democratic change coming to Myanmar, its citizens are being pushed into "river politics." The first major event was the suspension of the Myitsone Dam in 2011, which is located at the confluence of May Kha and Mali Kha Rivers that form the headwater of the "Mother Ayeyarwaddy," as it is known by people from all walks of life dwelling in the river basin.

The suspension of the Myitsone Dam by order of President U Thein Sein encouraged citizens and activists. For the first time in Myanmar's history, people successfully raised their voices in protest against a mega-project. The government is lucky to hear the voices of the people, as these voices signal an early warning for the potential failures of development plans. The government should facilitate them to raise their voices and share evidence to ensure that decisions are properly informed.

Following the government's decision on the Myitsone Dam, waves of river politics followed. Project developers from India withdrew their plans to build the Htamanthi Dam on the Chindwin River, and on the Salween River the Hatgyi Dam was reportedly also suspended.

A growing body of research has been another important dimension of these new river politics. One important contribution of this book is to help us know the Salween River better, so as to keep our natural and social environment safe and livable, and to develop our economy with ethics. For example, Chap. 10, this volume, by Mar Mar Aye and Swe Swe Win shows how various medicinal plants are valuable to ethnic groups in the Lashio area of Shan State. Yet, unfortunately, this type of knowledge is commonly neglected by hydropower project developers. Hence, when the cost of constructing a hydropower dam is calculated, project developers often exclude valuable assets such as these. This insight reveals why hydropower seems to be the cheapest option on the menu when potential energy sources are considered.

16.2.2 *Conflict and Peace on the Salween*

The Salween River basin is one of fragmented sovereignties (Middleton et al., Chap. 3, this volume). Some actors talk of the long history of the basin and the people within it, while others want to push quickly for a modern vision of development. These different groups find it difficult to dialogue, and unfortunately some choose the way of armed conflicts. The victims are many ethnic people in the basin, who end up having to run for their lives from conflicts. Myanmar is attempting a process of national reconciliation to create a peaceful federation. However, the ceasefire arrangements to date are only a temporary and superficial solution. The Salween River basin must become a test area for more constructive engagement.

Here too, this book makes some useful contributions. For example, Saw John Bright, Chap. 5 this volume, offers the concept of 'Rights' and 'Rites' in his analysis of communities living near the proposed Hatgyi Dam in Kayin State. He shows how people there currently generate income without destroying nature, but are now facing a double threat from armed conflict and also the planned Hatgyi Dam.

Experts from the National Water Resources Committee (NWRC) have continued to work on a National Water Law since 2015, although it is still not yet finalized. This work is an effort to protect the rights of the people to receive timely information, be involved in project design processes and to receive a decent share of the benefits, and to ensure the right to claim for any losses to property and livelihoods.

As Saw John Bright rightly mentions in his chapter, Myanmar's National Water Policy, passed in 2014, did not address the subject of armed conflicts related to water. However, the draft law does seek to address this more directly. Legislating for the decentralization of water resources management to the state and regional governments, self-administered regions and local administration is a step that should enable various scales of dialogue. It should be based on the principles of federalism.

Another important step in Myanmar would be to review the current situation of each of the country's existing 581 dams of various sizes for their sustainability, for their revenue over investments, and for their outcome and impact. While awaiting the results of such an assessment, it would be advisable to keep the Salween River running freely without major dams built.

16.2.3 Five Salween River Scenarios

Reflecting on the future of the Salween River, there are at least five potential scenarios.

16.2.3.1 Scenario 1: Unregulated Development and Fragmented Sovereignty

This, unfortunately, is the present scenario in the Salween River basin. Though the Myanmar Union Government has legal sovereignty, there are many other actors also exercising political power all along the river. The past experience of ceasefires has resulted in a black economy including deforestation, narcotics, and the extreme exploitation of natural resources. If this continues, the Salween River will pass a point of no return in the not-too-distant-future, with elites benefiting at the expense of ethnic communities who will face poverty.

16.2.3.2 Scenario 2: Shared Ownership Symbolizing Sovereignty Under a Federal System

Recent developments in Myanmar's politics have opened a renewed process for peace negotiations with the goal of peaceful coexistence. It opens the possibility of working towards shared ownership of the Salween River among the Shan State, Kayah State, Kayin State and Mon State governments with the Union government in the form of a Federation. However, the task ahead should not be underestimated. While the cessation of fighting, when it happens, should be considered a success, nurturing a Federation will be even more painstaking.

16.2.3.3 Scenario 3: Diverse Actors' Influence over the Salween

As discussed above, Myanmar has entered a new phase of river politics, within which local people's voice and activism have gained more influence. Regarding more responsible business practices, initiatives such as the "Business for Peace" initiative is a game changer for domestic regulatory policies on Corporate Social Responsibility. Furthermore, the activities of university researchers who are collaborating between countries, as well as other regional research initiatives, are producing new insights that could shape present and future policy.

16.2.3.4 Scenario 4: A Nu Jiang-Salween River Commission

Recognizing that the Salween is an international river, there could be a regional engagement to integrate governance of the river that is shared between Myanmar, Thailand and China. There is, after all, already nowadays intergovernmental engagement, including under the Association of South East Asian Nations (ASEAN), and more recently the "One Belt One Road" initiative. Thinking positively about these geopolitical developments, the three governments sharing the Nu Jiang-Salween River should engage in a constructive engagement as soon as possible for defining meaningful sustainable solutions. One step to take would be to ratify the United Nations Watercourses Convention, which until now none of the countries have signed into national law. In the meantime, cooperation on the Salween River must be peacefully solved bilaterally in good faith as friendly neighbors.

A Nu Jiang-Salween River Commission could begin bilaterally and then expand to a tripartite arrangement among the Governments concerned. It should also partner with a wider range of other organizations including ethnic armed organizations (EAOs) in the Salween basin, local professionals and international experts, non-governmental organizations (NGOs), community-based organizations (CBOs), civil society organizations (CSOs), activists, businesses, including the tourism industry, and development institutions such as various United Nations agencies, the World Bank, and the Asian Development Bank (ADB).

While there is a logic for government cooperation, it is also important to recognize that the whole "Nu Jiang-Salween River" is animated by the people living along it, who often belong to each other as family members, relatives, and friends, including across borders. Their lived experience of the river, not always bounded by borders, should also count in terms of decision-making towards its future.

16.2.3.5 Scenario 5: The Nu Jiang-Salween River as World Heritage

Human beings depend on the environment for their survival. As many of the world's major rivers are under threat, the Salween River is a crucial environment that should be protected and nurtured for the present and future generations. To add to efforts at the local and national level, the global community should also advocate for the sustainable future of the Salween River and promote it as a "World Heritage" Site. Their support could be via conducting research to generate reliable data and information about natural resources, biodiversity and ecology, green development potential, and about ethnic diversity and livelihoods.

In Myanmar, in 2014, the National Water Policy developed by the NWRC set an objective to "Provide national policy and stand point on use of shared water resources and develop cooperation among riparian countries" (2.4 (vi)) (NWRC 2015). This policy objective should encourage Salween River stakeholders to coordinate towards a National Policy on the Salween River in Myanmar, and beyond that a Nu Jiang-Salween River basin policy. Developing such comprehensive policies seems to be almost impossible without strong support from the global community.

16.2.4 A Road Map for the Salween

From my perspective, it is time to look at the Salween River holistically. We need to quickly move beyond the first scenario described above of unregulated development and fragmented sovereignty, and work towards the status of "World Heritage" for the whole Nu Jiang-Salween River basin and protect it and develop it with sustainability and social justice. To conclude, I offer five suggestions for next steps.

16.2.4.1 Develop a National Policy for the Salween River in Myanmar

Building on the growing body of research about the Salween River, including the studies presented in this book, data related to the Salween River should be utilized as an initial input to formulate a National Policy on the Salween River in Myanmar. This should emerge from discussions with researchers, consultative meetings with key NGOs and CSOs, and individuals with different and diverse opinions. The draft should be submitted to policy makers and politicians for dialogue. Subsequently, a wider range of expert opinions with experience in river politics, for example from the Mekong River Commission (MRC), should also be invited to comment. This draft should also be widely deliberated with riverside communities to gather their contributions, and also consult with EAOs for their point of view and to identify entry points for collaboration.

16.2.4.2 Create a Plan for Collaboration

During the ceasefire period, regular engagements could take place between the EAOs and local governments representing the Union Government as well as themselves, with facilitation by representatives from the central government, to plan and implement activities according to the collectively identified priorities. The expected outcome would be a concerted effort to guarantee the ecosystems and livelihoods of the local population, while safeguarding their territories. Experiences learned from these actions can be utilized as tools for finalizing the National Policy on the Salween River in Myanmar that would be followed by feasible development programs.

16.2.4.3 Peaceful Coexistence on the Salween River

Involvement of third parties, such as researchers, civil society and government officials from Thailand and China could potentially be critical to play the role of arbitrator if there are conflicting ideas and actions in Myanmar. Since Myanmar's borders with Thailand and China are quite porous, and various actors have close relationships across the borders, this can serve as an opportunity for regional engagement in solving conflict and lead to sustainable development in the whole region.

16.2.4.4 A Salween Development Project

The Myanmar and Thai Governments could take the lead in forming a Bilateral Commission for Salween River Development. The shared part of the river along the border is 81 miles long, and the basin on both sides could be developed by communities there with the support of international agencies and development partners.

The Myanmar and Chinese Governments could also form a Bilateral Commission, and bring ethnic groups from both sides to strive for mutual development, for example using green technologies proposed by Chinese professionals. The idea of forming a national park in Yunnan Province (Yu et al., Chap. 4, this volume) could be replicated in Myanmar. Meanwhile, the Salween Peace Park was established recently in one ethnic area in Kayin State (Bright, Chap. 5, this volume). This initiative could be shared along the Salween River, while the Union Government builds closer ties towards forming a federal democratic state.

16.2.4.5 The Nu Jiang-Salween River as a "World Heritage"

Time is running out to protect the whole river as one. Beyond bilateral arrangements, all three governments could form a tripartite Nu Jiang-Salween River Basin Commission and develop partnership programs in collaboration with the global community and UN Agencies. This initiative could make the Nu Jiang-Salween River eligible for "World Heritage" status building on the "World Heritage Site" awarded by UNESCO for the Nu Jiang as part of the "Three Parallel Rivers of Yunnan Protected Area."

16.3 Concluding Commentary: What's Next for the River? Is the Thanlwin 'Under Threat' or 'on the Thread'

Nang Shining

16.3.1 Introduction

My name is Nang Shining. I am a Shan woman living in Mong Pan Township nearby the site of the proposed Mong Ton hydropower dam at the Thanlwin (Salween) River basin in Shan State, Myanmar. In this commentary for the book's conclusion, I want to assess the situation for the Thanlwin River basin from my perspective – as a woman who has carried out advocacy work and research across the basin, who has had the opportunities for graduate study, and who is now intentionally based in Taunggyi, my home-state in Myanmar. I also will put forward my thinking for future work and action on this significant river.

I am a Co-founder of Weaving Bonds Across Borders and Director of the Mong Pan Youth Association. Our mission is to collaborate with partner organizations and alumni of our programs to empower women and youth along the Thanlwin

River basin, and to support these groups in taking a more active role in peace building and sustainable development. We are also working to advocate for a healthy Thanlwin River through action-based research, awareness raising within the local community, engagement with government and policy makers, and networking with the people across the Thanlwin River basin.

We also deliberately chose the name of our group, Weaving Bonds Across Borders, to illustrate how we were thinking of developing networks as hopeful and meaningful, as work that spans across borders. The word 'thread', like in the title of this piece, is meant to link to the work of weaving, an activity known to many groups worldwide, particularly women. A weaver brings different threads together, to create something new, something strong, from disparate strands. Weaving, like networking and supporting young people, requires the work of transformation. Even if a small 'thread' alone may break, when combined together, in the process of 'weaving', it will be stronger.

We are strengthening women's participation in development decision-making as a way to provide capacity on social justice-related issues, and to support affected communities to advocate for sustainable development practices in their local areas. We really need young women who not just know the information and legal issues, but we must support them to find this information, connect with networks, and develop their leadership skills so that they can take action on behalf of their community. We strongly believe that if women have access and networks, and are supported, they will be able to build peace not only at home, but also in their community and society as a whole.

In order to do so, we also offer training programs to build leadership skills for marginalized and vulnerable young women in the Thanlwin River basin. At our office in Taunggyi, I work with youth volunteers from different ethnic groups such as Shan, Pa'O, Danu, Lahu, Tanaw, and Inthar across Shan, Mon, Kayah, and Kayin ethnic States along the Thanlwin River Basin. Many young people are interested in the environment, in development, and they want to document what's happening in their communities through film and other means, like action research or photography. The reason is that there are many big development projects, such as coal and mineral mining, coal power plants, and hydropower plants which will have social and environmental impacts. Yet, even though there are many development project activities, the majority of people have not been fully informed about the projects, nor have they been consulted.

This includes the Mong Ton dam project, near my hometown, a large hydropower dam proposed to be built on the mainstream Thanlwin River. It will be the biggest hydropower dam in Southeast Asia, but the potentially affected community like those from my hometown and villages next to the dam site, like Wan Sala, have not been consulted about the project. We produced research and video documentation to reveal overlooked gender issues of new development projects (Mong Pan Youth 2017), which can be shared to wider audiences, both international and domestic, at film festivals and network meetings.

In our own research, we found out that many people believe in the value of the river, but they have very little information about the big development plans for the river. Many do not have access to information related to the project as well as the impacts of the dam. One of our interviewees, a young woman from Mong Ton township, said, "Though Wan Sala village is the closest village and is only 19 miles away from the dam site, the consultative meetings have not been conducted and the people have not been targeted to participate in the meeting" (March 2016). Another interviewee, a middle-aged woman from Kun Heing township, explained, "Because of the Mong Ton Dam project, the natural beauty of one thousand islands will be flooded under the water forever. Not to mention other places, only in Kun Heng Township, the population around 50,000 will have to relocate to other places. It is only highlighting one small spot of the overall affected area" (March 2016).

The active participation of a diverse range of voices in local governance is critical to securing social justice and sustainable development. However, ethnic minority women in Myanmar, like those quoted above, face significant barriers to actively engaging in local governance. Even though women's daily life activities are strongly linked to the natural environment, they have little recognition in society and have little decision-making power within water and natural resources management. Women will be the one who suffer the most from development projects without meaningful participation in the decision processes. The rights of men and women appear to be equal when we look at them from the surface level, but based on the findings of our researchers, there are many facts that are not visible. Many women do not realize that their rights have been structurally violated within the existing system. Some regulations within the governance system also ignore the value of women. Thus, gender analysis should be considered in hydropower project planning because resettlement often means loss of access to resources like farmland, clean water, firewood and non-timber forest products which are regarded as the obligation and responsibility of women.

It is essential to understand the context of gender issues on the ground where the hydropower projects are proposed. The root cause of gender inequity derives from different aspects of society such as social customs and public policy. The social customs and traditional culture in Myanmar have a significant influence on attitudes towards women in leadership roles and in the public policies. Without addressing how social norms influence public policies, and without considering gender inclusion in hydropower projects, it will be a barrier in efforts to promote and protect the human rights of women.

16.3.2 Why Youth?

While there has been long-standing research on the role of women in the region, the roles of youth seem to be less visible. In South and Southeast Asia 15–24 year-olds make up around 20% of the population (Wagle 2018). The UN reports that in the countries of Indonesia, the Philippines, Vietnam, Thailand, and Myanmar "together have over 90 million youth or close to an average of 17 per cent of each country's population." But, young people deserve our attention for more reasons than just their numbers or that they are understudied. Youth[1] play a significant yet under-studied role in creating and shaping environmentalism, activism, and development across the region and we need to better include them in discussions about our future. In India, recent research reveals that youth are important change-makers (Dyson/Jeffries 2018: 573), and that young people "played important roles as mentors, political activists and social helpers. They served as exemplars, seeking to alter the behaviour and attitudes of those around them at the everyday level."

In Southeast Asia, youth in rural areas are leaving villages for a number of reasons, not limited to employment opportunities, education, and land scarcity (Rigg 2006; Barney 2012). That rural youth in Southeast Asia are more likely to travel for work and education is important to consider in decisions around policy relating to urban development, agricultural and land support, education, and broader governance issues.

Much work has focused on the challenges that young people face when they move outside of the rural village or as migrants to the city, and only few researchers are studying, instead, the new possibilities and dreams that youth are part of creating (Simone 2008; Guinness 2016). However, what my experience shows is that if we looked to the ways in which youth create possibilities—such as community organizing, developing cross-border networks, or the more informal work of mentoring and teaching children and siblings—we might better understand their potential contribution to civil society. We already have evidence of this work and its impacts in the region.

For instance, the Mekong Youth Assembly has emerged as a regional voice on the Mekong River. The Mekong Youth Assembly was developed by youth from the six countries along the Mekong River, namely: Cambodia, Myanmar, Laos, Thailand, Tibet/China, and Vietnam, with the aim to "encourage all youths in the Mekong countries to bring about change in environmental concerns by working with their communities." The work of Mekong Youth to include young people and support them as leaders is not dissimilar to the work of Mong Pan Youth and Weaving Bonds Across Borders.

In a group statement, they write of their dreams for the future, including freedom of expression, peace, equality, and participation in decision-making. In their own words, Mekong Youth (2017) explain that "We dream that today's adults, especially those in power, would bear in mind that "you do not inherit the Earth from ancestors, you borrow it from us, your children". Make sure our mother earth shall be returned to us with prosperous life elements. In this regard, always respectfully consider our lives." It is my opinion that we should be taking these dreams seriously.

[1]Youth are typically defined as part of a loose age range as those under age 30 or 35 (distinct from 'children' but not with a fixed age limit).

16.3.3 My Observations and Ways Forward: Crossing Borders Together

As I noted at the start of this piece, networking and developing women and young people, like weaving, requires the work of transforming disparate small pieces together, to be stronger when combined. In the examples above, of the Mekong Youth Assembly, Mong Pan Youth, and Weaving Bonds Across Borders, the networks being established are 'woven' across borders and basins. This is important because the Thanlwin River belongs to all ethnic groups, genders, and ages living along it and within its basin. Therefore, the development projects proposed on the Thanlwin River in Shan State are not simply a domestic issue for Myanmar, this goes beyond physical and political boundaries, beyond social and political demarcations.

A transboundary river requires cross-boundary agreements and treaties, a potentially higher level of cooperation among different stakeholders and collaboration than we have at present. From my perspective, the Thanlwin River is vulnerable and under threat. Despite its status as a transboundary river, it still lacks a treaty and cooperation among the water users along the river basin. The point I want to stress is that instead of competition, the basin countries would do better to collaborate in building a relationship or network. At the same time, it is essential to ensure that those who live in the basin and depend upon it – including women and youth – are not only included in those collaborations and networks, but are positioned to influence pathways and transformations going forward.

16.4 Concluding Commentary: The Salween as a Site for Transboundary Justice and Activism in the Face of Militarization

Pianporn Deetes

16.4.1 Introduction

Behind the enveloping morning mist lies the river that meanders around valleys and along the Thailand-Myanmar border. Against the backdrop of morning sunlight that glitters across the sandbars, a group of people who identify as "Karen-Thai" are wearing bright red, handwoven dress. They have positioned themselves along the water's edge to participate in a ceremony to 'extend the life' of the Salween River.

Nearby, a spirit house made of bamboo stands at the point where the Moei River meets the Salween, before it flows down into Myanmar's Karen State.

This ritual is held annually by residents in Sob Moei (or, 'Confluence') Village to mark the International Day of Action for Rivers. Today's rite is the 12th of its kind. Residents want to make clear that they stand for the preservation of river resources and that they seek to uphold their fundamental rights, including both UN fundamental rights and Thai constitutional rights, to make decisions that impact their lives, like on water, land, and matters of justice.

The only difference this year is they have been told by the Thai security forces to refrain from holding protest signs or to organize any acts of protest. The ritual proceeds even as these concerns of 'protest' are expressed, linked to the increasingly limited political space in Thailand under military rule; participants want to avoid their ritual being interpreted as an illegal act of political protest and so they focus on the ritual. After the river ceremony concludes, school children stand in a row silently. Each one holds a character, which when combined reads "Keep the Salween Flowing". While complying with the request of the Thai officials, the children are able to convey this message from this border community to the outside world.

In this concluding commentary, I am writing to highlight these small but long-standing acts of activism of and for the Salween River because I believe this work matters. More than this, I would also argue that these rituals present a way of seeing and relating to the Salween River, and to transboundary justice, that we—as civil society actors, international activists, development practitioners, researchers, and policy makers—should spend more time listening to.

After the ceremony, a member of the Sob Moei community explained to me that, "We live off the Salween. It helps us survive. Our family catch fish from the Salween and fish from the Salween swim into Moei and Yuam Rivers and other tributaries. During the dry season, we could grow food on the riverbanks. We can also grow tobacco there which helps to earn us tens of thousands of baht per year. It was enough for us to live without any problem" (March 2017).

Sob Moei's village leader, Mr. Somchai (pseudonym), is only in his early 30s. He explained that there are many of his sisters and brothers living in different villages inside Myanmar. Driven by armed conflicts, some decided to cross the Salween to seek refuge in Thailand. However, the community now in Thailand is concerned about proposed development projects along the river, sited both upstream and downstream of his village, as negatively impacting their community. One project, the Hatgyi dam, is proposed less than 50 km downstream. It is expected to impact Sob Moei village in terms of livelihoods, specifically fishing and riverbank agriculture.

"If this dam is built," he explained, "our lives would be completely destroyed. Our sisters and brothers who have fled elsewhere shall never return."

16.4.2 *Thailand's Role on "Both Sides" of the River*

As Mr. Somchai is aware, the Salween is one of a few remaining rivers that still run free.[2] Yet, as has already been described in detail in this volume, over the past two decades, efforts have been made by the governments of China, Thailand, and Myanmar (Burma) plus state enterprises and private companies to push for the construction not only of Hatgyi, but of more than 20 hydropower projects, both in China and Myanmar (Shan State, Karreni State, and Karen State) with seven projects on the Thailand-Myanmar border.

Upstream, China's stretch of the Salween is geographically significant and gorgeous, teemed with natural abundance and cultural richness. Efforts have successively been made by environmentalists and academics in China to oppose the 13 hydropower development projects in Yunnan.

On the lower reach of the Salween River, Thailand has been playing a major role. Thailand's electrical authority, the Electricity Generating Authority of Thailand (EGAT), has pushed forward two major hydropower development projects on the Thailand-Burma border in Mae Hong Son, the Weigyi and Dagwin Dams. The two projects would operate with a combined installed capacity of over 5,000 MW. This part of the Salween is an ecologically significant area, home to the Salween National Park and a Wildlife Sanctuary, as well as one of Thailand's richest teak forests.

It is not clear that EGAT is taking the significance of this area into account in its development plans, particularly in terms of ecological values or in terms of values to residents of Sob Moei. For instance, while EGAT has not been vocal recently, more than 15 years ago, EGAT's then Governor told the press that it is important to have these dams developed while democracy is yet to be restored in Myanmar. He was quoted as saying "I have promised myself to see the Salween Hydro project realized when I am the EGAT governor. The project is huge that I might not have another opportunity to be part of it [sic]" (Samabuddhi, 9 April 2003). He emphasized what a rare chance it was for EGAT. The message even back then was clear: Thailand's electricity utility put profits and energy development above the democratic rights of Myanmar citizens.

But, as a result of public campaigns by community-based organizations on both sides of the Salween River, we were able to shine a light on the impacts of these projects on the protected areas and on the constitutional right of communities in Thailand. To this day, these two projects proposed on the border have not been completed.

One of the hydropower development projects being publicly pursued at present is in nearby Shan State; it is the Mong Ton hydropower project (also, Mai Tong and previously Tasang). With an installed capacity of 7,000 MW, the dam may lead to inundation of a huge swathe of land—an area as large as the territory of Singapore.

---

[2]The mainstream of the Salween is not dammed, but there are hydropower projects on tributaries in Yunnan, China (see Middleton et al., Chap. 3, this volume).

The flood may even reach the 'thousand-islands' of the Pang River in central Shan State, an area of ecological and cultural significance. This is also the same area where, two decades ago, more than three hundred thousand people were forcibly displaced by the Burmese army. These individuals have become internally displaced people (IDP). Some remain living in IDP camps in Myanmar, while many have instead decided to cross the border to Thailand. This issue of displacement, not only of those directly displaced by a reservoir, but of a people displaced by conflict over the past several decades, is a pressing issue of justice that we need to strive to better understand and listen to.

In recent years, Shan State's human rights and environmental groups have taken opportunities to be increasingly in the spotlight to heap praise on the picturesque beauty of the thousand-island area and its geographical uniqueness marked by its large expanse of jade-colored water and cascades of waterfall covering tens of kilometers. Such unique beauty, possibly the only one in the world, would also be inundated upon completion of the Mong Ton Dam.

16.4.3 A Chance for Change?

These impacts have prompted EGAT to redesign the Mong Ton project, from one gigantic dam turned into two much smaller dams, with the plans at present still subject to negotiation with the Myanmar government. Human rights activists in Shan State, however, have expressed concern that any dam on the Salween, which proceeds without their input or decision, is not suitable for their hometown.

The Hatgyi project, noted above, is another of the projects with EGAT support. Located at the lower end of the basin, this hydropower development has had a contentious history. During the preliminary survey at the dam site, in 2007, EGAT lost two of its staffers as a result of land mines in Karen State. Such risk and loss of life should already send a signal that the area's conflicts need our attention and are not suitable for hydropower development.

As the project progressed, several years after this incident in 2009–2010, I was party to a series of meetings to get at the 'truth' of the Hatgyi project and its impacts. I was a member of the Thai Prime Ministers' Office (PMO) committee on Hatgyi dam. During one meeting at the PMO, a high-ranking officer from EGAT stated that a study was being conducted to ensure that no impacts could be felt in Thai territory. In the subsequent study report on the Hatgyi dam project, the report claimed that less than ten households in Mae Hong Son will bear the brunt from the project. As a member of the committee investigating the Hatgyi situation, I was skeptical. We could see in our own visits to the river-border, such as the village of Sob Moei, that any shift in the water levels and seasonal flows of the Salween would negatively impact the community who depended on the river for their livelihoods, particularly fishing and riverbank gardening.

This was reiterated in concerns expressed by community members. For instance, one female villager, Ms Salai, explained, "Even though they claim our homes will not be flooded, but fish from down there cannot swim [migrate] here. And there will be no land left for us to till. How could we survive? We have seen other examples where the villagers displaced by dams have been left to our own devices. They could not help themselves and have to work as hired labor. They have to even buy drinking and domestic water, not to mention rice."

I also wondered how the impacts of the project in this independent study could be so clearly 'known' before the outcome of the independent study. It made me argue that we needed, and still need, more fully independent academic and civil society research on these impacts.

For, it is not only the impacts, but also the idea that the project can proceed without proper assessment, that have implications for justice. As Ms Salai explained further, with her voice trembling, "[The experts] who came here to do the survey could not answer any of our questions. This has made us feel concerned. It has been worrying us for more than ten years. We wonder if they are going to build the dam or not."

Her comments attest to how little information about the dam's impacts have been made available, especially for affected peoples, who we know should be at the center of transparent and accountable—more just—development decision-making.

16.4.4 From Thailand to Myanmar: Concerns for Justice and for Peace

The overlapping and complex political powers in Myanmar's Salween River Basin have further complicated management of the river basin. The water as a resource falls under the domestic Myanmar jurisdiction, even though the river crosses international boundaries. Also, while the country has just undergone a transition from a military junta to a civilian-led government, many questions for justice and peace remain. For instance, there exist patches of land along the river under the influence of ethnic forces; these groups have been struggling to assert their territorial rights in the ethnic states of Shan, Karenni or Karen State. At the same time, while the ongoing peace process aims to bring these groups together, there exist efforts to rush through the proposed dam projects on the Salween River before agreements can be made. Such moves could increase militarization, which in turn would only jeopardize the path toward peace.

According to a military officer with the Shan State Army whom we interviewed by the Pang River in Shan State in June 2016, the Salween and Pang River Basins, and the area along their tributaries in the west which will potentially be impacted by the Mong Ton Dam, have already witnessed the catastrophic impacts from the Burmese Army's 'scorched earth' military campaign over the past two decades (see also, Shan Human Rights Foundation 1998). People have been displaced and have sought refuge in various places including in Thailand. Less than a half of those displaced have so far managed to return to their homeland.

He further explained, "On one hand, a narrative is being touted by the Burmese government to assure the international community that political transformation and a peace process are a work in progress there, but on the other, they are signing deals to build these dams claiming to give a chance for neighboring countries to invest there. I could only say that their actions will simply make life much more vulnerable among these peoples. This is yet to mention the extensive damage inflicted by the Burmese Army in central Shan State in the past 20 years, which has yet to be acknowledged or remedied. Now, they want to build dams that will bring us floods. No one could accept this" (June 2016).

These ongoing tensions he highlights reflect that this decades-long conflict is still very much a part of the present-day decision-making around resources in the Lower Salween.

In such a context, where agreements for peace need to be prioritized, it is quite challenging to develop a framework to strike the balance between the preservation and the development of the Salween River Basin as a transboundary basin.

16.4.5 Moving Forward

In closing this commentary, I suggest we return to the site of international action for rivers described at the start, where local residents have been demanding their rights to protect their natural resources. Here, the Moei River continues to slowly empty into the Salween, which in turn, meanders around the mountains in the distance. On the Burmese side, a small boat full of passengers is spotted navigating through the twilight toward a pier on the Thai side.

Mr. Somchai, the leader in Sob Moei, explains to us, as we wait alongside the river, "Our sisters and brothers are gradually returning to their hometown [in Karen State]". But as I noted above, this is happening while the terms and agreements for peace are ongoing, and Mr. Kyaw still has concerns regarding his brothers and sisters, and what Karen State will be like for their children. For me, such concerns underline that these plans for large scale development of the basin cannot proceed without first ensuring a peaceful existence for ethnic peoples in the basin, and clear pathways for justice.

16.5 Concluding Commentary: Salween as Source, Salween as Resource

R. Edward Grumbine

16.5.1 Introduction

River are alive. Of course, many people, especially those who live far removed from running waters, often see rivers only as flowing H_2O, or a nice place to take a cool bath on a hot day in the dry season, or an ecosystem that yields fish for the table and the marketplace. The nature of being a "natural resource" is that yielding benefits for humans, whatever those may be, comes first.

But, in general, the closer humans live to rivers, the more we tend to view them not as resources, but as sources—of sustenance, of foods and fishes, of the unique power to flow and flood and change the nature of land and livelihoods with silts and sediments. A 'resource' is a stock or supply of material assets that can be used by people. A 'source' is a place of origin, full of creative energy, mystery, and surprise.

The Salween River is no special watercourse in light of these basic characteristics. People did not create the Salween, make it up out of thin air. The river was following its watercourse way for at least several million years before people ever came to live on its banks and drink and bathe in its waters.

But we do construct stories about the Salween, and engage in multi-faceted relations with it. We do decide, knowingly or not, about the nature of these relations and whether they will be inclusive or exclusive, sustaining or degrading, broadly reciprocal or narrowly utilitarian. No matter how we make choices about these things, the Salween is always a source—of fish, food, and floodwaters for millions of people—before it is a resource. It pulses with the dry and wet seasons, and people learn how to live with it. And because rivers are alive and changeable, people living close to running water learn how to adjust and adapt to them over time.

Today, as the contributors to this book spotlight in great detail, we are doing poorly in our relations with the Salween. We are favoring the exclusive, the unjust, the opaque, and the narrow over inclusivity, justice, transparency, and the broad rights of people to healthy livelihoods on a free-flowing river. It appears that we humans in this modern age are less able to conceive of the Salween as an actor with its own mandate to flow. Upstream, until recently, China had plans to build mainstream dams with little regard for how these structures capture sediments that sustain fish, the riverside gardens and paddies of ethnic nationality peoples, and the biomass demands of thousands of wild animals and plants. Multilateral consultation across international boundaries is anathema to the Chinese government. Even current initiatives to create protected areas along the river in Yunnan Province to generate wealth for entrepreneurs and officials while local people scramble for service jobs in the new economy, have more than a whiff of industrial-scale tourism about them. And while the planned dams are off the table for now, they could reappear whenever the government wants them to; there is little in the way of participatory process in China.

Things are more complex downstream in Thailand and Myanmar. The Thai government believes that hydropower-generated electricity from five large dams on the mainstream Salween will provide the power that will drive the country into a new round of 21st century industrial growth. If all these dams were built, they would triple the megawatts now used in Myanmar. On the face of it, that would be a good thing; Myanmar has the lowest rate of per capita available electricity in Southeast Asia. But the Thai cut deals with the former Myanmar military government some years ago to buy the majority of this new power and send it to Bangkok once the dams are constructed. And Chinese companies would reap most of the economic benefits from constructing the dams. Given the growing impacts of climate change, maybe Thai leaders will finally notice that other forms of renewable energy offer more to bolster their country's future. Maybe not.

As for Myanmar, you have the Union government with The Lady as its nominal leader. But the military remains in control of important national decision making, and the *Tatmadaw,* the semi-autonomous military, wields power in many areas of the countryside, including parts of the Salween around proposed dam sites. Add in multiple Ethnic Armed Organizations and their various splinter groups, all fighting against central government control (and sometimes each other); state governments that are influenced by one or more of the above actors; national and transnational businesses looking for windows of profit; and, finally, local people who want to sustain their livelihoods in the face of the forces above. In fact, more and more river-based people cannot find work anymore in the Salween, and are migrating out of the watershed and Myanmar in search of cash income to send back home. And if people want to stay in their river homelands? The environmental and social costs of the dams would be great. The projects are expected to flood 1000 km^2 of land upstream of the new reservoirs and disrupt 690 km^2 of the main channel. No Environmental Impact Assessment for any of the five mainstream dams has ever been released for public review. How could one not side with most of the Salween's local ethnic minorities as the river loses its capacity to be a resource under so much ecological transformation in service of unsustainable development?

The odds are stacked against the Salween and local peoples' rights to participate in decisions about the future of their homelands.

But for more than a century, the political odds have always been against the river and local people. The British held colonial sway from 1886 to 1948 and residents of British Burma did not fare well. The next years of shaky independence represented a relative lull politically. But this was followed in 1962 by the infamous coup of a repressive military junta bent on creating a 'socialist state', ushering in the most repressive conditions in the country's history. Since the elections of 2015, Myanmar has struggled to open up and get out from under the legacy of the generals and the economic shambles they left the country in, even as the 2008 constitution continues to give the military ongoing power and control.

And yet, as this volumes details, there are alternative movements afoot that can bring a measure of environmental and social justice to life in the Salween. The dams are yet to be built and there are other pathways to develop than economic growth through flooding peoples' homelands in service of 'resource extraction' for outside interests. The first hint of this alternative future was Myanmar's decision to pull the plug on the massive Myitsone dam in 2011, alienating the Chinese government and its' dam building companies in the process. The second hint was

China's decision to hold off constructing the long-planned string of dams in Yunnan Province in the Salween's upper basin.[3] This was followed by the May 2018 release of the draft Strategic Environmental Assessment of hydropower development in Myanmar funded by the International Finance Corporation (World Bank Group) (available at www.ifc.org). This most detailed study of the benefits and costs of the nations' full suite of dam building proposals (including the Salween) cautions against constructing *any* large hydropower projects on the mainstems of Myanmar's major rivers. (Building dams on tributaries is another matter.)

If these several recent decisions and new advice hold, then the Salween and the people who depend on it have a chance to become engaged in the future of their riverine homes. The river has a chance to continue to flow relatively unimpeded, to continue to be a source of nourishment for all who live within the basin. It will take more than local, equitable, and just participation to secure the transboundary Salween; the river needs coordinated planning upstream and downstream, as well as strong links established between local, regional, and multilateral decision making. But that is what people must do when a source is a stake.

References

Barney, K. (2012). Land, livelihoods, and remittances: A political ecology of youth out-migration across the Lao–Thai Mekong border. *Critical Asian Studies*, *44*(1), 57–83.

Bird, M.I., Robinson, R.A.J., Oo, N.W., Aye, M.M., Lu, X.X., Higgitt, D.L., Swe, A., Tun, T., Win, S.L., Aye, K.S., & Win, K.M.M. (2008). A preliminary estimate of organic carbon transport by the Ayeyarwady (Irrawaddy) and Thanlwin (Salween) Rivers of Myanmar. *Quaternary International*, *186*(1), 113–122.

Callahan, M. (2003). *Making Enemies: War and State Building in Burma*. Cornell University.

Committee of Researchers of the Salween Sgaw Karen. (2005). *Thai Baan Research in the Salween River: Villager's research by the Thai-Karen Communities* (in Thai with English summary). [Documentary film]. Published and produced by SEARIN, Chiang Mai, Thailand. Retrieved from: http://www.livingriversiam.org/4river-tran/4sw/swd_vdo2-tb-research.html.

Dyson, J., & Jeffrey, C. (2018). Everyday prefiguration: Youth social action in North India. *Transactions of the Institute of British Geographers, 43*(4), 573–585.

Guinness, P. (2016). Dwindling space and expanding worlds for youth in rural and urban Yogyakarta. In K.M. Robinson (Ed.), *Youth identities and social transformations in modern Indonesia* (pp. 134–155). Leiden, Boston: Brill.

Hengsuwan, P. (2012). *In-between lives: Negotiating bordered terrains of development and resource management along the Salween River*. (Unpublished doctoral dissertation), Chiang Mai University, Thailand.

Hengsuwan, P. (2013). Explosive border: Dwelling, fear and violence on the Thai–Burmese border along the Salween River. *Asia Pacific Viewpoint*, *54*(1), 109–122.

Hnin Wut Lee. (2018). *Gender inclusion is key to social justice: A case study of two villages downstream of the proposed Mongton Dam, Shan State*. Paper presented at the 2018 Salween

[3]See China's National Energy Administration current 5-year plan at: http://www.nea.gov.cn/2016-11/29/c_135867663.htm.

Studies Research Workshop – The Role of Research for a Sustainable Salween River, Yangon University, Myanmar, February 26–27, 2018. Retrieved from: http://www.csds-chula.org/publications/2018/4/30/conference-proceeding-salween-studies-research-workshop-the-role-of-research-for-a-sustainable-salween-river.

Lamb, V. (2014a). "Where is the border?" Villagers, environmental consultants and the 'work' of the Thai–Burma border. *Political Geography, 40*, 1–12.

Lamb, V. (2014b). *Ecologies of rule and resistance: Making knowledge, borders and environmental governance at the Salween River, Thailand.* (Unpublished doctoral thesis), York University, Canada.

Lamb, V. (2018). Who knows the river? Gender, expertise, and the politics of local ecological knowledge production of the Salween River, Thai-Myanmar border. *Gender, Place & Culture*, 1–16. https://doi.org/10.1080/0966369X.2018.1481018.

Lampert, C.D., Glover, I.C., Hedges, R.E.M., Heron, C.P., Higham, T.F.G., Stern, B., Shoocongdej, R., & Thompson, G.B. (2003). Dating resin coating on pottery: The Spirit Cave early ceramic dates revised. *Antiquity*, *77*(295), 126–133.

Mekong Youth. (n.d.). *About us*. Retrieved from: http://www.mekongyouth.org/about/.

Mekong Youth Assembly. (2017, January 12). *Statement on the youth's dream for future Mekong*. Retrieved from: http://www.tdhgsea.org/web2017/2017/01/12/statement-on-the-youths-dream-for-future-mekong/.

Middleton, C. (2018). *Why think about Salween pathways approach overall?* Paper presented at the 2018 Salween Studies Research Workshop – The Role of Research for a Sustainable Salween River, Yangon University, Myanmar, February 26–27, 2018. Retrieved from: http://www.csds-chula.org/publications/2018/4/30/conference-proceeding-salween-studies-research-workshop-the-role-of-research-for-a-sustainable-salween-river.

Mong Pan Youth. (2017). *Water and Gender*. A Film by Sai Aung Nawe Oo. https://www.salweenstories.org/mongpan/.

Nang, S. (2018). *Ecofeminism evolution readiness of gender evolution in the case of Mong Ton Dam project.* Paper presented at the 2018 Salween Studies Research Workshop – The Role of Research for a Sustainable Salween River, Yangon University, Myanmar. February 26–27, 2018. Retrieved from: http://www.csds-chula.org/publications/2018/4/30/conference-proceeding-salween-studies-research-workshop-the-role-of-research-for-a-sustainable-salween-river.

National Water Resources Committee (NWRC). (2015). *Myanmar National Water Policy.*

Rigg, J. (2006). Land, farming, livelihoods and poverty: Rethinking the links in the rural South. *World Development, 34*(1), 180–202.

Samabuddhi, K. (2003, 9 April). Salween is Home to New Dam Row. *Bangkok Post.* Retrieved from: http://www.ibiblio.org/obl/docs/SW14.htm.

Scott, J.C. (2009). *The art of not being governed: An anarchist history of upland Southeast Asia*. New Haven, CT: Yale University Press.

Shan Human Rights Foundation. (1998). *Dispossessed: Forced Relocation and Extrajudicial killings in Shan State.* Thailand: Shan Human Rights Foundation.

Shoocongdej, R. (2006). Late pleistocene activities at the Tham Lod Rockshelter in highland Pang Mapha, Mae Hong Son Province, Northwestern Thailand. In E.A. Bacus, I.C. Glover, & V. Piggott (Eds.), *Uncovering Southeast Asia's past: Selected papers from the 10th annual conference of the European Association of Southeast Asian Archaeologists*. Singapore: NUS Press.

Simone, A.M. (2008). The politics of the possible: Making urban life in Phnom Penh. *Singapore Journal of Tropical Geography*, *29*, 186–204.

TERRA. (1999). The Salween – My river, my natural belonging. *Watershed*, *4*(2), 41–44. Retrieved from: http://www.terraper.org/web/en/key-issues/salween-river/.

van Schendel, W. (2002). Geographies of knowing, geographies of ignorance: Jumping scale in Southeast Asia. *Environment and Planning D: Society and Space*, *20*(6), 647–668.

Wagle, A. (2018, April 30). *Youth in South and Southeast Asia: A common denominator.* Retrieved from: https://www.internationalaffairs.org.au/australianoutlook/youth-in-south-and-southeast-asia-a-common-denominator/.

Weatherby, C. (2018). *Food, water, and energy in mainland Southeast Asia*. Presentation made at the Stimson Centre, Washington, DC, June 28.

Weaving Bonds Across Borders. (n.d.). *About us*. Retrieved from: http://www.weavingbonds.org/about-us/.

White, B. (2012). Changing childhoods: Javanese village children in three generations. *Journal of Agrarian Change*, *12*(1), 81–97.

White, B., & Margiyatin, C.U. (2016). Teenage experiences of school, work, and life in a Javanese village. In K.M. Robinson (Ed.), *Youth identities and social transformations in modern Indonesia* (pp. 50–68). Leiden, Boston: Brill.

Xiaogang, Y., Xiangxue, C., Middleton, C., & Lo, N. (2018). *Charting new pathways towards inclusive and sustainable development of The Nu River Valley*. Retrieved from: http://www.csds-chula.org/publications/2018/20/03/report-charting-new-pathways-towards-inclusive-and-sustainable-development-of-the-nu-river-valley.

Center for Social Development Studies, Faculty of Political Science, Chulalongkorn University, Bangkok, Thailand

The *Center for Social Development Studies* (CSDS) was established as a Research Unit within the Faculty of Political Science, Chulalongkorn University in 1985. It was established to undertake interdisciplinary research linking across the various fields of political science within the Faculty (government, public administration, international relations, and sociology and anthropology) and more broadly in the social sciences, and to provide support in education, research, and teaching. With the intention of linking the CSDS research to educational development, in 2006 the CSDS supported the launch of the MA in International Development Studies (MAIDS) program. Building on MAIDS, in August 2018, the Ph.D.-level Graduate Research in International Development (GRID) program was also launched.

The missions of the CSDS are: to produce interdisciplinary critical research on development policy and practice in Southeast Asia that aims to be innovative, inclusive and sustainable; to contribute to policy processes through undertaking research in collaboration with those involved in development policy and practice, participating in research, policy and academic networks, and by sharing our research publicly; to support young and mid-career researchers and public intellectuals via the MAIDS and GRID programs, other taught programs within the Faculty of Political Science of Chulalongkorn University, and through our fellowship programs and internships; and to organize public forums for debating critical issues on development by hosting seminars, conferences, and workshops.

The research of CSDS currently focuses on to five themes: resource politics; rethinking regionalization; human rights, human security and justice; transdisciplinary knowledge and innovation; and the public sphere, democratization and the commons. Our recent projects have been in collaboration with: universities in Southeast Asia and beyond; civil society organizations and community based

C. Middleton and V. Lamb (eds.), *Knowing the Salween River: Resource Politics of a Contested Transboundary River*, The Anthropocene: Politik—Economics—Society—Science 27, https://doi.org/10.1007/978-3-319-77440-4

organizations; research institutes; the media; and government organizations. To encourage collaborative research, knowledge exchange and capacity building, we host research associates, visiting researchers, and interns. Since 2018, the CSDS has also hosted the Chulalongkorn University Center of Excellence on Resource Politics for Social Development.

York Centre for Asian Research, York University, Toronto

York Centre for
Asian Research

The *York Centre for Asian Research* (YCAR) is a community of York University researchers who are committed to analyzing the changing historical and contemporary dynamics of societies in Asia, understanding Asia's place in the world, and studying the experiences of Asian communities in Canada and around the globe. Our inter-disciplinary membership includes faculty, students and other research associates from across the social sciences, humanities, health, education, creative/performing arts, law and business.

Some common themes characterize much of the research that YCAR fosters and supports. First, we adopt an explicitly transnational approach to research, meaning that we seek to understand connections within Asia, between Asia and the rest of the world, and between Asia and its diasporas. Second, we value research that is based on extended field and archival research, language study and the long-term development of expertise. Third, we emphasize a critical and engaged model of scholarship, attentive to social justice agendas that seek to address exclusions or inequalities based on class, gender, sexuality, 'race', caste, religion, region or environmental dispossession. Often, this involves collaboration with the communities being studied in the research process, and the mobilization of research findings to effect public education and social change.

The role of the Centre in the work of individual researchers is to create a space for interdisciplinary intellectual exchange, to provide administrative support for research projects, and to enrich student training through fieldwork and language awards and a graduate diploma programme. We also provide an access point for anyone interested in York expertise on Asia and Asian communities, and we actively seek to deliver research to the widest possible audience.

Founded in 2002, YCAR continues a strong tradition of internationally recognized research in Asian Studies at York, pioneered since 1974 by the Joint Centre on Modern East Asia, and the Joint Centre for Asia Pacific Studies (both in collaboration with the University of Toronto).

C. Middleton and V. Lamb (eds.), *Knowing the Salween River: Resource Politics of a Contested Transboundary River*, The Anthropocene: Politik—Economics—Society—Science 27, https://doi.org/10.1007/978-3-319-77440-4

About the Editors

Carl Middleton, Ph.D. is Director of the Center for Social Development Studies (CSDS), and Deputy Director for Research Affairs on the MA in International Development Studies (MAIDS) Program in the Faculty of Political Science, Chulalongkorn University, Bangkok, Thailand. Since 2018, the CSDS has also hosted the Center of Excellence in Resource Politics for Social Development of Chulalongkorn University, which he leads. His research interests orientate around the politics and policy of the environment in Southeast Asia, with a focus on nature-society relations, environmental justice and social movements, transdisciplinary research, and the political ecology of water and energy. His taught courses include on Development Theory and Practice; Politics of Public Policy; Environmental Policy and Politics; and Innovation for Inclusive Development.

Dr. Middleton has led or collaborated on various research projects in Southeast Asia. These include: a regional water governance study of the Salween River; a fellowship program in the Salween, Mekong and Red river basins; a study on the political ecology of flooding and migration in Southeast Asia; a project on knowledge co-production for the recovery of wetland agro-ecological systems in the Mekong region; and a curriculum development project on transdisciplinary research methods. He is currently undertaking research on the hydropolitics of knowledge production on the Lancang-Mekong River, the (tele-) connections between Thailand and Japan's water security (with Takeshi Ito), and on demarcating the public and private in resource governance in the Mekong Region (with Phillip Hirsch).

C. Middleton and V. Lamb (eds.), *Knowing the Salween River: Resource Politics of a Contested Transboundary River*, The Anthropocene: Politik—Economics—Society—Science 27, https://doi.org/10.1007/978-3-319-77440-4

His most recent book, co-authored with Jeremy Allouche and Dipak Gyawali, is titled *The Water–Food–Energy Nexus: Power, Politics and Justice* (Earthscan-Routledge, 2019). He also recently co-edited with Rebecca Elmhirst and Supang Chantavanich the book *Living with Floods in a Mobile Southeast Asia: A Political Ecology of Vulnerability, Migration and Environmental Change* (Earthscan, 2018).

His recent peer reviewed publications in the fields of political ecology and development studies include: "National Human Rights Institutions, Extraterritorial Obligations and Hydropower in Southeast Asia: Implications of the Region's Authoritarian Turn" in the *Austrian Journal of Southeast Asia Studies* (2018); "Branding Dams: Nam Theun 2 and its Role in Producing the Discourse of 'Sustainable Hydropower'" chapter in *Dead in the Water* edited by Bruce Shoemaker and William Robichaud (University of Wisconsin Press, 2018); "Water, Rivers and Dams" chapter in the *Handbook of the Environment in Southeast Asia* edited by Phillip Hirsch (Routledge, 2017); and "Watershed or Powershed?: A critical hydropolitics of the 'Lancang-Mekong Cooperation Framework'" co-authored with Jeremy Allouche in the journal *The International Spectator* (2016).

Address: Room 115, Faculty of Political Science, Building 2, Henri Dunant Road, Chulalongkorn University Pathumwan, Bangkok, 10330, Thailand.
Email: carl.chulalongkorn@gmail.com.
Website: www.csds-chula.org and www.maids-chula.org.

Vanessa Lamb, Ph.D. is a Geographer at the University of Melbourne. In research and teaching in the School of Geography, she focuses on human-environment geographies and political ecology of Southeast Asia, with a focus on water justice. Dr. Lamb completed her dissertation, Ecologies of Rule and Resistance, focused on the politics of ecological knowledge and development of the Salween River at York University's Department of Geography in 2014. Dr. Lamb was also the inaugural Urban Climate Resilience in Southeast Asia (UCRSEA) research fellow at the Munk School of Global Affairs at the University of Toronto, Canada and is an affiliated researcher with the York Centre for Asian Research, York University, Toronto, Canada. Across these institutional, Dr. Lamb has been working with collaborators in Southeast Asia on two major research projects focused on the Salween. This includes the CGIAR Research Program on Water, Land and Ecosystems (WLE) Greater Mekong research-for-development grant for "Matching policies, institutions and practices of water governance in the Salween River Basin: Towards inclusive, informed, and accountable water governance" which concluded in 2019.

Among her selected peer reviewed publications are multiple works in the field of political ecology, political geography, and international water governance, particularly of the Salween River, including: "Who knows the river? Gender, expertise, and the politics of local ecological knowledge production of the Salween River, Thai-Myanmar border" in *Gender, Place & Culture* (2018), "Whose border? Border talk and discursive governance of the Salween River-border" chapter in *Placing the Border in Everyday Life* edited by Corey Johnson and Reece Jones (Ashgate, 2016), "Making governance 'good': The production of scale in environmental impact assessment and governance of the Salween River" in *Conservation & Society* (Online April 2015), and "Where is the border? Villagers, environmental consultants and the work of the Thai-Burma border" in *Political Geography* (2014).

Dr. Lamb has co-authored works on Chinese investment in hydropower, with Dr. Nga Dao, "Perceptions and Practices of Investment: China's hydropower investments in mainland Southeast Asia" in the *Canadian Journal of Development Studies* (2017). Most recently, Dr. Lamb collaborated with Melissa Marschke and Jonathan Rigg to publish the first study on the sand trade affecting rivers in Southeast Asia in Geography's flagship journal, the *Annals of the American Association of Geographers* (2019). She also edited a special issue, with Professor Amrita Daniere, on "Water and Asia's Changing Cities: Marginality, Climate, and Power in the Transformation of Urban Space" in the *International Development Planning Review*.

Address: Dr. Vanessa Lamb, Lecturer, School of Geography. Level 2 (Rm 2.35), 221 Bouverie Street. The University of Melbourne, Carlton, Victoria 3053 Australia.
Email: vanessa.lamb@unimelb.edu.au.
Website: http://vanessalamb.org.

About the Contributors

Alec Scott

Alec Scott is the Regional Water Governance Technical Officer for the Karen Environmental Social Action Network (KESAN). He began researching armed conflict and natural resource governance while working as a teacher along the Thai-Burma border in 2008, and focused on the links between large dams, armed conflict and forced displacement while attending the School of Oriental and African Studies in 2011. Following a period of fieldwork in the Salween, Sittaung and Upper Irrawaddy river basin areas, Alec began working with ethnic civil society organizations on issues related to large dams, land and natural resource governance, armed conflict and human rights.

Email: scottsangpo@gmail.com.

Carl Middleton

Dr. Carl Middleton is an Assistant Professor and Deputy Director for Research on the Master of Arts in International Development Studies (MAIDS) Programme, and Director of the Center for Social Development Studies (CSDS) in the Faculty of Political Science of Chulalongkorn University, Thailand. Since 2018, the CSDS has also hosted the Center of Excellence in Resource Politics for Social Development of Chulalongkorn University. Dr. Middleton's research interests orientate around the politics and policy of the environment in Southeast Asia, with a particular focus on nature-society relations, environmental justice and social movements, transdisciplinary research, and the political ecology of water and energy.

Email: carl.chulalongkorn@gmail.com.
Website: www.csds-chula.org.

Chayan Vaddhanaphuti

Professor Chayan Vaddhanaphuti is the founding Director of the Regional Center for Social Science and Sustainable Development (RCSD) in the Faculty of Social Sciences at Chiang Mai University, Thailand. He has degrees from Chulalongkorn University and Stanford University in Anthropology and in International

C. Middleton and V. Lamb (eds.), *Knowing the Salween River: Resource Politics of a Contested Transboundary River*, The Anthropocene: Politik—Economics—Society—Science 27, https://doi.org/10.1007/978-3-319-77440-4

Development and has been teaching and advising students in the Department of Social Sciences and Development at Chiang Mai University for several decades. He is a renowned advocate for the rights of marginalized communities and ethnic or indigenous people in Thailand and across Southeast Asia, working hard to ensure that their voices and knowledges are part of broader academic and policy conversations which affect their lives and livelihoods.
Email: ethnet@loxinfo.co.th.

Chen Xiangxue

Chen Xiangxue has seven years work experience with Chinese environmental NGOs. Since 2015, she worked for Green Watershed based in Kunming. She focuses on policy analysis of hydropower development on the Nu River, and community research in the Nu River basin. She also works on safeguard policies advocacy towards multilateral development banks such as the Asian Infrastructure Investment Bank.
Email: chenxx417@163.com.

Cherry Aung

Professor Cherry Aung is Head of the Department of Marine Science at Pathein University, Myanmar. Prior to moving to Pathein University's Marine Science Department (one of only three in Myanmar), she also worked at the Department of Marine Science at Mawlamyaine University, at the mouth of the Salween River. Professor Cherry Aung's research interests include socio-economy of fisheries, sedimentation in the Irrawaddy Delta, and coastal resources like mangroves and corals.
Email: missaungmarine@gmail.com.

Hannah El-Silimy

Hannah El-Silimy is a trainer and consultant for non-profits and social movements, and one of the co-founders of Weaving Bonds Across Borders, an international women's network for peace and justice. She has a Bachelor's degree in Politics from Oberlin College in Ohio, and a Master's degree in Global Politics from the London School of Economics. She is currently in her first year of the Ph.D. program in Political Science at the University of Hawai'i at Mānoa, where her research focuses on Indigenous Politics, women and transnational social movements.
Email: hannahel@hawaii.edu.

Johanna M. Götz

Johanna M. Götz is a Ph.D. candidate in Development Studies at the University of Helsinki. Educated as a critical geographer in Leipzig, Yogyakarta, Bonn and beyond, she aims to critically scrutinize hegemonic assumptions and paradigms by asking how multiple waters can give new insights into ongoing processes of state-formation and the peace negotiations in Burma. Her involvement and interest in Myanmar and the Salween build upon her work at the Heinrich-Böll-Foundation in Yangon, her Master thesis entitled "The politics of water governance in

Myanmar – a hydrosocial approach", and her further academic engagement at the Center for Social Development Studies, Chulalongkorn University.
Email: johanna.goetz@helsinki.fi.

K.B. Roberts

K.B. Roberts is a geography doctoral candidate, with research interests that address issues within and between the political ecology of resource frontiers, rural livelihood change, deforestation and forest degradation, and feminist research methods. Roberts holds a B.Sc. in Environmental Studies and a M.Sc. in Resource Conservation.
Email: arborpolitica@gmail.com.

Ka Ji Jia

Ka Ji Jia is a Ph.D. candidate in the Department of Women's Studies, Faculty of Social Sciences, Chiang Mai University, Thailand. She was born in a Tibetan community in China and she received the bilingual education (Tibetan and Chinese) at Tibetan schools in Qinghai Province. She holds degrees from Qinghai Normal University (Undergraduate) and Chiang Mai University (Masters). Currently, she is pursuing her doctoral degree with research focus on Tibetan herders' local knowledge and gender in natural resource management, environmental conservation and development contexts. She writes on the Tibetan herders' traditional culture, local knowledge, and ecological conservation in Tibetan areas in China.
Email: kathytibetan@gmail.com.

Khin Khin Htay

Dr. Khin Khin Htay is Associate Professor in the Geography Department of Loikaw University, Kayah State, Myanmar. She specializes in the use of Global Information System (GIS) and Remote Sensing (RS) methods.
Email: Kkhinhtay.geog@gmail.com.

Khin Maung Lwin

Dr. Khin Maung Lwin is a water expert, medical doctor, and current advisor to Myanmar's National Water Resources Committee (NWRC). He is also a former Myanmar Ministry of Health Director and has worked to secure access to sanitation and safe drinking water for Myanmar's people for over 20 years. He started his career as writer and journalist in 1968, and has published eleven books and many other materials related to health and the environment. In 2008, he received the Tun Foundation Literary Award. He was elected as a Global Steering Committee Member representing Southeast Asia and the East Asia region of the Water Supply and Sanitation Collaborative Council (WSSCC) based in Geneva, Switzerland, affiliated to UNOPS, between 2011 until 2018.
Email: drkmlwin98@gmail.com.

Khin Sandar Aye

Dr. Khin Sandar Aye is a Professor in the Geography Department of Loikaw University, Kayah State, Myanmar. She was also a Salween Research Fellow with the Center for Social Development Studies, Chulalongkorn University from 2015 to 2018.

Email: akhinsandar50@ gmail.com.

Laofang Bundidterdsakul

Laofang Bundidterdsakul is a human rights lawyer, and co-founder and Director of the Legal Advocacy Center for Indigenous Communities (LACIC) based in Mae Hong Son Province, Northern Thailand. He is an ethnic Hmong who has completed a Bachelor's of Law at Chiang Mai University. He has also studied at the EarthRights International Mekong School, and joined the Indigenous Fellowship Program of OHCHR in 2016. In his role, he encourages ethnic minority communities to use legal mechanisms to protect their rights, and has played an important role defending against criminal cases in which indigenous people or ethnic minorities have been sued for alleged violations of Thailand's Forest Law.

Email: phaajxyaaj@yahoo.co.th.

Mar Mar Aye

Professor Mar Mar Aye (retired) was Head of the Department of Botany at Lashio University, Northern Shan State, Myanmar, until May 2018. Prior to that, Prof. Mar Mar Aye contributed to and was affiliated with the Departments of Botany at Yangon University, Taungoo University and Lashio University. She was awarded her Ph.D. from the University of Yangon (2012) with a thesis focusing on pharmacognostic studies (studies of medicines derived from natural sources).

Email: pro.marmaraye@gmail.com.

Maung Maung Aye

Professor Maung Maung Aye is an expert in fluvial geomorphology and public intellectual in Myanmar. He has served various University positions, including as the Rector-in-Charge (Retd.) at the Yangon University of Distance Education as well as a former professor and head of Geography Department for the Dagon University. He is also a member of the Myanmar Academy of Arts and Science, and a member of the Revitalization (Software) Committee of the University of Yangon. He has also served as a Patron of the Geographical Association of Myanmar. He holds positions as a member of the Ph.D. Steering Committee in Geography at the University of Yangon and a member of the National Education Policy Commission (NEPC) in Myanmar. Outside academia, he is also a Patron and Chief Advisor to the Myanmar Environment Institute (MEI), a member of the National Commission of Enquiry on the Hydropower Projects in the Myitsone Area and the Upstream Ayeyarwady River Basin, a full-time member of the Advisory Group for the National Water Resources

Committee (NWRC), and a delegate of the Union Peace Conference-21st Century Pang Long (Land, Resources and the Natural Environment Sector).
Email: maungaye3246@gmail.com.

Nang Aye Tin
Nang Aye Tin is a member of and researcher with Mong Pan Youth Association located in Shan State, Myanmar. They were part of a team producing a documentary film and research on "women and water access" as part of the Salween Water Governance project, "Matching policies, institutions and practices of water governance in the Salween River Basin: Towards inclusive, informed, and accountable water governance." Nang Aye Tin also works in research and data collection along the Thanlwin River Basin.

Nang Hom Kham
Nang Hom Kham (Hom) is from Southern Shan State, Myanmar. She is working with EarthRights International. Before ERI, she worked in Shan State, Myanmar with Mong Pan Youth Association. Hom Graduated with a BA (Taunggyi University) and holds a diploma on Human Rights and Environment (EarthRights International School). Her professional experience includes producing a short documentary on the development of Salween, writing, and research on the Salween River. In research, her main focus is on the forestry and shifting cultivation, culture, gender, and water management affected by Mega-development projects along the Salween River in Shan State. A key interest for her is supporting the women's participation in water management.
Email: aeliterann@gmail.com.

Nang Sam Paung Hom
Nang Sam Paung Hom is a member of Mong Pan Youth Association located in Shan State, Myanmar. She was part of a team producing a documentary film and research on "women and water access" as part of the Salween Water Governance project, "Matching policies, institutions and practices of water governance in the Salween River Basin: Towards inclusive, informed, and accountable water governance." Nang Sam Paung Hom also has interests in financial management for NGOs. Currently, she is serving as Financial official at the Mong Pan Youth Association.

Nang Shining
Nang Shining is a Shan ethnic from Southern Shan State in Myanmar. She holds advanced degrees Chulalongkorn University, Thailand; University for Peace, Costa Rica; and Ateneo De Manila University, Philippines. She worked for four years as EarthRights International's School Training Coordinator and Alumni Program Coordinator. She is a founder and Director of Mong Pan Youth Association (Shan State, Myanmar) and co-founder of Weaving Bonds Across Borders (Chiang Mai, Thailand). She is interested in the intersection of peace and conflict transformation, human rights, gender equality, and environment. She has conducted research on

Gender and Hydropower, and Water Governance in the Salween Basin as a Salween Fellow.
Email: shining.cu@gmail.com.

Naw Aye Aye Myaing
Naw Aye Aye Myaing (Ku Ku) works as a researcher for the Karen Environmental and Social Action Network (KESAN), a community-based, non-governmental, non-profit organization that works to improve livelihood security and to gain respect for indigenous people's knowledge and rights in Karen State, Myanmar. She has been a key part of KESAN's local research team, producing research on water access and presenting these findings as part of the Salween Water Governance project, "Matching policies, institutions and practices of water governance in the Salween River Basin: Towards inclusive, informed, and accountable water governance."
Email: nawayeayemyaing@gmail.com.

Paiboon Hengsuwan
Dr. Paiboon Hengsuwan is a Lecturer at the Department of Women's Studies, Chiang Mai University. He has a Ph.D. in Social Science from Chiang Mai University (2013). His recent publications include "Living in Fear of Violence at the Thai-Burma Border" published in "Uncertain Lives: Changing Borders and Mobility in the Borderland of the Upper Mekong" (2015), and "Living with Threats and Silent Violence on the Salween Borderlands: An Interpretive and Critical Feminist Perspective" (Regional Journal of Southeast Asian Studies vol. 2 no. 1).
Email: paiboon1998@yahoo.com.

Pianporn Deetes
Pianporn (Pai) Deetes works as a part of the International Rivers global team, based in Thailand, to protect rivers by focusing on Thailand's role abroad as a dam developer, and Thailand's position as the main intended market for hydroelectricity in Mainland Southeast Asia. A former Ashoka Fellow, since 2002 Pai is a known as a pioneer in her work with Thai and Burmese ethnic grassroots organizations, to organize river communities to slow down dam projects along the Salween River that threaten the local ecosystem and the livelihoods of those who depend on the river and its resources.
Email: pai@internationalrivers.org.

R. Edward Grumbine
Ed Grumbine has worked on bringing conservation science into natural resources policy and land management since the 1990s. From 2010 to 2014, he lived in China as a senior international scholar for the Chinese Academy of Sciences, focusing on environmental security in river basins (with an emphasis on the Salween and Mekong), water governance in the Asian Highlands, and protected areas in China. Ed is the author of multiple scientific papers and three books including *Where the*

Dragon Meets the Angry River: Nature and Power in the People's Republic of China (Island Press, 2010).
Email: ed.grumbine@gmail.com.

Sai Aum Khay
Sai Aum Khay is a member of Mong Pan Youth Association located in Shan State, Myanmar. He was part of a team producing a documentary film and research on "women and water access" as part of the Salween Water Governance project, "Matching policies, institutions and practices of water governance in the Salween River Basin: Towards inclusive, informed, and accountable water governance." He is interested in filmmaking and documentation. Currently, he is in charge at Video Documentation Program as Program Associate at Mong Pan Youth Association.

Saw John Bright
Mr. Saw John Bright has been working with the Karen Environmental and Social Action Network (KESAN) as Water Governance Coordinator. KESAN is an ethnic Karen organization in Myanmar working with communities in Karen State to achieve environmental sustainability, gender equity, local participation and local ownership in the development process. Saw John has been actively engaging in Community Based Water Governance at both the local and national level, and has also been working on trans-boundary water issues at the regional scale. His work mainly focuses on research and policy advocacy towards inclusive, informed, and accountable water governance in the Salween River.
Email: hserdohtoo@gmail.com.

Saw Tha Phoe
Saw Tha Phoe is the Campaign Coordinator for Karen Environmental and Social Action Network, a community-based, non-governmental, non-profit organization that works to improve livelihood security and to gain respect for indigenous people's knowledge and rights in Karen State, Myanmar. Prior to working with Kesan, Saw Tha Phoe worked with a civil society networks, such as Karen Rivers Watch. Since 2013, he has been a key part of KESAN's local research team, coordinating research and campaigns activities on grassroots-led water governance and presenting these findings as part of the Salween Water Governance project, "Matching policies, institutions and practices of water governance in the Salween River Basin: Towards inclusive, informed, and accountable water governance."
Email: phoehein@gmail.com.

Saw Win
Professor Saw Win is retired Rector of Maubin University, an Executive Director of the Renewable Energy Association Myanmar (REAM), a Director of Myanmar Environment Institute (MEI), and as a Senior Research Associate with Chulalongkorn

University's CSDS program, he has helped coordinate the Salween fellowship program.
Email: prfswwin@gmail.com.

Swe Swe Win

Dr. Swe Swe Win is currently a Lecturer in the Department of Botany at Lashio University, Myanmar. She completed her doctoral degree (Ph.D., Botany) in 2008 at the University of Mandalay. Since that time, Dr. Swe Swe Win has gained experience working in Botany departments around the country, including a recent position at Loikaw University. Her research interests include medicinal plants and traditional medicine in Myanmar.
Email: Drsweswewin.91@gmail.com.

Vanessa Lamb

Dr. Vanessa Lamb is a Lecturer in the School of Geography at the University of Melbourne. In research and teaching in the School of Geography, she focuses on human-environment geographies and political ecology of Southeast Asia, with an emphasis on water justice. Dr. Lamb is also an affiliated researcher with the York Centre for Asian Research, York University, Toronto, Canada. She led (along with Dr. Carl Middleton and range of contributors in this volume) the CGIAR Research Program on Water, Land and Ecosystems (WLE) Greater Mekong's research-for-development funded project, "Matching policies, institutions and practices of water governance in the Salween River Basin: Towards inclusive, informed, and accountable water governance."
Email: Vanessa.lamb@unimelb.edu.au.
Website: http://vanessalamb.org.

Yu Xiaogang

Yu Xiaogang established the NGO Green Watershed in 2002. In the same year, his research report 'Social Impact Assessment on the Manwan Dam' pushed China's government and the dam owner to compensate 80 million RMB to affected people. Since 2003, he has been advocating to protect the World Natural Heritage area in the Nu River Basin. In 2006, he initiated an NGO network pushing Chinese banks to take the environmental and social responsibility. His work with minority communities in support of sustainable development and poverty reduction was awarded the Equator Prize by the UNDP in 2015. He was also awarded the Goldman Environmental Prize in 2006 and the Ramon Magsaysay Award in 2009.
Email: greenwatershed@163.com.

Index

C. Middleton and V. Lamb (eds.), *Knowing the Salween River: Resource Politics of a Contested Transboundary River*, The Anthropocene: Politik—Economics—Society—Science 27, https://doi.org/10.1007/978-3-319-77440-4

www.ingramcontent.com/pod-product-compliance
Lightning Source LLC
LaVergne TN
LVHW082116200926
848484LV00008B/1962
* 9 7 8 1 0 1 3 2 7 6 4 3 9 *